Schlutter

Injection Mold Coating Technology

Ruben Schlutter

Injection Mold Coating Technology

HANSER

MIX
Papier aus verantwortungsvollen Quellen
FSC
www.fsc.org
FSC® C083411

Print-ISBN: 978-1-56990-266-0
E-Book-ISBN: 978-1-56990-271-4

Bibliographic information of the German National Library:
The German National Library lists this publication in the German National Bibliography; detailed bibliographic data are available on the Internet at http://dnb.d-nb.de.

www.hanserpublications.com
www.hanser-fachbuch.de
Editor: Dr. Mark Smith
Production management: le-tex publishing services GmbH, Leipzig
Cover concept: Marc Müller-Bremer, *www.rebranding.de*, Munich
Cover design: Max Kostopoulos
Cover picture: © KIMW-Forschungs GmbH
Typesetting: Eberl & Koesel Studio, Kempten
Printed and bound by: CPI Books GmbH, Leck
Printed in Germany

Contents

Preface

Plastic molded parts must meet increasingly more demanding requirements. It is irrelevant whether these requirements relate to the molded part itself (such as a reduced wall thickness, an increased flow path length, or the masking of weld lines), to the material used (such as the use of reinforcing materials or aggressive additives), or to the injection molding process (such as a reduction in cycle time or an increase in the lot size). In order to meet these requirements, the injection molding tools must also be continuously optimized and pushed to their limits. In addition to the use of high-quality materials, coatings are one way of achieving the necessary functionalization.

The coating of metallic materials can be carried out in different ways and represents a key competence of the non-profit KIMW Forschungs-GmbH. The presentation of the current state of science and technology is always problematic. In-depth knowledge of the process technology for the deposition of coatings, as well as the validation of the coating properties, is necessary in order to produce high-quality coatings. Over several years, experience had to be gathered as to which properties can be achieved by coatings, and how suitable coatings can be applied (and also examined). The necessary information often had to be obtained from a large number of technical articles, many of which are not publicly accessible, and which usually deal with specific applications that are not directly transferable to molding tool technology. Moreover, the authors are usually chemists, physicists, or experts in the field of coating technology, meaning that the technical articles are usually difficult for engineers to understand and comprehend. Up to now, there has been no reference work that adequately summarizes the relevant coating technologies for injection molds, and test methods for characterizing the coatings.

This motivation gave rise to the idea of combining the knowledge available at the non-profit KIMW Forschungs-GmbH and making it accessible in a basic resource. The result is this book – a guide that summarizes the relevant information on coating technology in a way that engineers can understand.

In several chapters, this book deals with the relevant aspects of coating technologies. Following a general introduction in Chapter 1, and an overview of tool steels in Chapter 2, in Chapter 3 we discuss the various possibilities for depositing coatings with regard to their application in injection molding tools. This is followed in Chapter 4 by a discussion of the measurement methods used to characterize the comparatively thin coatings encountered in the field of tool and mold making. The book concludes in Chapter 5 with an overview of the various applications of functional coatings in injection molding tools, aiming to give the reader an impression of the possible applications. This enables the reader to select the most suitable coating and the resulting deposition process for their application.

I would like to thank all those who have contributed to the success of this book, in particular Rebecca Wehrmann and Dr. Mark Smith at Hanser Publications for their editing and patience during the production of this book.

I would also like to thank all the authors for providing the respective chapters, which bring the content of this book to life and have made even these highly complex issues easy to understand. Finally, I would like to thank you for your patience with me.

Ruben Schlutter

Lüdenscheid, 2024

The Authors

The Editor

Dr. Ruben Schlutter

Dr. Ruben Schlutter is a freelance lecturer in the field of plastics technology and simulation. He primarily teaches the subjects of injection molding simulation, structural mechanical simulation, and design with plastics.

After studying mechanical engineering with a focus on product development and design at Schmalkalden University of Applied Sciences, he completed a cooperative doctorate between Chemnitz University of Technology and Schmalkalden University of Applied Sciences under Prof. Dr. Michael Gehde and Prof. Dr. Thomas Seul in the field of pressure loss analysis in injection molding simulation. After completing his doctorate, he moved to the Kunststoff-Institut für die mittelständische Wirtschaft in Lüdenscheid. There he supervised various research and development projects, such as the internationalization of the existing network for injection molding in the "Industry 4.0" environment (MONSOON).

In 2018, Dr. Schlutter moved to the non-profit KIMW Forschungs-GmbH, and supervised projects on the production and use of biocidal nanoparticles and the development of a non-destructive testing method for the qualitative assessment of the foam structure of molded plastic parts. At the same time, he was involved in the training courses offered by the Plastics Institute for SMEs, focusing on shape and position tolerances, plastic-compatible molded part design, and plastic tolerancing in accordance with ISO 20457. He has been a freelance lecturer and member of the Association of German Tool and Mold Makers (VDWF) since 2022.

The Co-Authors

Dr. Andreas Balster

Dr. Andreas Balster received his doctorate in organic chemistry from the Ruhr University Bochum in 2001 and has worked at the Kunststoff-Institut (KIMW) in Lüdenscheid since 2002. Among other things, he headed the department for material and failure analysis, the Polymer Training Center (PTC) and the German Institute for Interlaboratory Tests (DIR) before moving to pro**3d**ure medical GmbH in Iserlohn in 2023 as head of the analytics laboratory.

Dr. Balster has been a freelance lecturer at FH Südwestfalen since 2008 and taught online at Frankfurt University of Applied Sciences in 2020 and 2021. From 2015–2022 he was an active member of the DIN Standards Committee Plastics (FNK), Division NA 054-01-03 AA (Physical, rheological and analytical tests).

Dr. Anatoliy Batmanov

Dr. Anatoliy Batmanov studied semiconductor technology at the Lviv Polytechnic National University in Ukraine. He then completed his doctorate at the University of Magdeburg on the subject of "Design, modeling, optimization and manufacture of high-frequency microelectromechanical switches and coplanar filters". Dr. Batmanov has been working in the field of deposition technologies of thin films of various types using CVD and PVD for 20 years, and has been a project manager in the Coating Technology department of the non-profit organization KIMW Forschungs-GmbH since 2019.

Non-profit KIMW Forschungs-GmbH
Lutherstrasse 7
58507 Lüdenscheid

KUNSTSTOFF INSTITUT LÜDENSCHEID
GEMEINNÜTZIGE KIMW FORSCHUNGS-GMBH

Dr. Veronika Brune

Dr. Veronika Brune is a post-doctoral researcher with a teaching assignment at the Institute of Inorganic Chemistry at the University of Cologne.

After studying chemistry at the University of Cologne, Dr. Brune completed her doctorate under the supervision of Prof. Dr. Sanjay Mathur at the University of Cologne. She currently works in Prof. Dr. Mathur's laboratory, where her research interests lie in the design and synthesis of metal chalcogenide molecular precursors for gas-phase deposition.

Institute of Inorganic Chemistry
University of Cologne
Greinstrasse 6
50939 Cologne

Dr. Martin Ciaston

Dr. Martin Ciaston has a background in analytical chemistry, and is currently working as a research associate in project management at the non-profit KIMW Forschungs-GmbH. Here, Dr. Ciaston is responsible for the development of process and equipment technology in MOCVD and has developed and characterized coatings for injection molds, as well as developing and validating various test methods.

Dr. Ciaston carried out his Ph. D. studies at the Thünen Institute in Braunschweig, Germany, where he worked on the catalytic derivatization of itaconic acid for polyester synthesis. During his time as a research assistant at the Thünen Institute, he developed chemical-catalytic processes for the production and modification of bio-based monomers from fermentation products. Dr. Ciaston also holds a degree in chemistry from the Technical University of Braunschweig, and his thesis at Volkswagen AG focused on the suitability of polymers and amorphous carbons for implementation in lithium-ion cells as a coating material for temperature and pressure sensors.

Patrick Engemann

Patrick Engemann first completed an apprenticeship as a tool mechanic and then studied production engineering at the South Westphalia University of Applied Sciences. After completing his Bachelor of Engineering at the University of Applied Sciences, he continued his studies at the University of Wuppertal in the field of mechanical engineering. Since 2017, he has been employed by the non-profit KIMW Forschungs-GmbH, where he focuses on the further development of tool temperature control for injection molding tools and CVD system technology.

Non-profit KIMW Forschungs-GmbH
Lutherstrasse 7
58507 Lüdenscheid

KUNSTSTOFF INSTITUT LÜDENSCHEID
GEMEINNÜTZIGE KIMW FORSCHUNGS-GMBH

Dr. Thomas Fischer

Dr. Thomas Fischer is an academic advisor at the Institute of Inorganic Chemistry at the University of Cologne and is responsible for gas-phase coating technologies in the laboratory of Inorganic and Materials Chemistry. After studying chemistry at the Julius-Maximilians-Universität Würzburg, he completed his doctorate under Prof. Dr. Sanjay Mathur at the University of Cologne.

His current research interests include the development of new gas-phase coating methods and in-operando investigations of the resulting gas phase and surface reactions. Together with Prof. Dr. Mathur, he founded the Steinbeis technology transfer company "Materials Alliance Cologne". As a member of the Sustainability Council of the University of Cologne, Dr. Fischer is responsible for the area of sustainable research and is an active member of the American Ceramic Society.

Institute of Inorganic Chemistry
University of Cologne
Greinstrasse 6
50939 Cologne

Vanessa Frettlöh

Vanessa Frettlöh studied chemistry in Siegen and has been working in the field of coating technology at the non-profit KIMW Forschungs-GmbH since 2013. In addition to supervising research projects in tool and coating technology, at Lüdenscheid she has also spent time in the further development of CVD technology. Since 2020, she has been Head of Surface Technology Tools and thus responsible for coating technology with a focus on chemical vapor deposition.

Non-profit KIMW Forschungs-GmbH
Lutherstrasse 7
58507 Lüdenscheid

KUNSTSTOFF INSTITUT LÜDENSCHEID
GEMEINNÜTZIGE KIMW FORSCHUNGS-GMBH

Marko Gehlen

Marko Gehlen studied plastics engineering at RWTH Aachen University from 1990 to 1996. He then worked for several years in the mechanical development of DECT telephones at Siemens AG in Bocholt, before taking over the management of mold design/NC programming at Gigaset Communications GmbH (formerly Siemens). Finally, he was in charge of plastics processing for four years.

Gehlen joined Kunststoff-Institut Lüdenscheid in 2015, where he was initially responsible for the cross-divisional development of joint projects. Since 2019, he has been working in the research department of the Kunststoff-Institut Lüdenscheid, focusing on the development of functional tool coatings. Among other things, he is responsible for strategy and innovation.

Non-profit KIMW Forschungs-GmbH
Lutherstrasse 7
58507 Lüdenscheid

KUNSTSTOFF INSTITUT LÜDENSCHEID
GEMEINNÜTZIGE KIMW FORSCHUNGS-GMBH

Matthias Korres

Matthias Korres completed his training as a process mechanic for plastics and rubber technology in injection molding in 2010. From 2012 onwards, he studied mechanical engineering, specializing in plastics technology, at the South Westphalia University of Applied Sciences in Iserlohn. As his final thesis in 2015, at the Kunststoff-Institut Lüdenscheid, he developed and optimized a test bench for measuring the adhesion and sliding properties of plastics at the time of demolding.

After successfully completing his studies, Korres initially worked in the research department of the Kunststoff-Institut Lüdenscheid on projects for the substitution of metal components in the field of compressed gas generators and hybrid tool material concepts. He then moved to the area of application and tool technology at the institute, where he developed a recording unit for recording and analyzing emissions in the injection molding process. Among other things, he deals with the evaluation of coatings with regard to their effectiveness in reducing adhesion and coating problems in the plastics processing industry. Furthermore, he has headed and coordinated the additive manufacturing department at the Kunststoff-Institut Lüdenscheid since 2021.

Ameya Kulkarni

Ameya Kulkarni studied Computational Mechanics at the University of Duisburg-Essen. He then worked in research and development at the non-profit KIMW Forschungs-GmbH, undertaking flow simulations of CVD reactors and multiphysics FEM calculations in various research projects. He has been with the Winora Group since 2022, and works in the development department as an FEM engineer.

Winora Group
Max-Planck-Strasse 6
97526 Sennfeld

Dr. Angelo Librizzi

Dr. Angelo Librizzi initially completed vocational training as a tool mechanic, studying mold technology. He then studied mechanical engineering, with a focus on plastics technology, at the South Westphalia University of Applied Sciences in Iserlohn and at the University of Paderborn. In addition to his professional activities, he completed his doctorate in 2015 in the field of thin-film sensor technology for temperature measurement in injection molds at Kunststofftechnik Paderborn, within the University of Paderborn. From 2008 to 2015, he worked as a project engineer and head of surface technology at the non-profit Kunststoff-Institut Lüdenscheid GmbH. In 2015, he moved to the research center at the Kunststoff-Institut (non-profit KIMW Forschungs-GmbH). There, he is currently responsible for tool and process technology, and is a member of the management board as an authorized signatory.

Non-profit KIMW Forschungs-GmbH
Lutherstrasse 7
58507 Lüdenscheid

Dr. Orlaw Massler

Dr. Orlaw Massler initially studied materials science at the Friedrich-Alexander University in Erlangen. He then worked as a project and department manager at the Diamond Research Laboratory of the De Beers Industrial Diamonds Division in Johannesburg, South Africa. From 1999, he then worked at Balzers AG in Balzers, Liechtenstein (now Oerlikon Surface Solutions) in various positions, most recently as Head of Development. From 2008 to 2015, he worked as a tribology expert in Group Research at Hilti AG in Liechtenstein. Since 2015, Dr. Massler has been Head of Research and Innovation at the De Martin Group, headquartered in Switzerland. His focus is on innovative and hybrid functional coatings with electrolyte, PVD, and PACVD technologies, as well as disruptive technologies in surface science.

De Martin AG Surface Technology
Froheggstrasse 34
CH 9545 Wängi

Prof. Dr. Sanjay Mathur

Prof. Dr. Dr. (h.c.) Sanjay Mathur is Chair and Director of the Institute of Inorganic Chemistry at the University of Cologne in Germany. He is also Director of the Institute of Renewable Energy Sources at Xi'an Jiaotong University in China and a Distinguished Professor at Chonbuk University in Korea. He is also a visiting professor at the Institute of Global Innovation Research at Tokyo University of Agriculture and Technology (TUAT) in Japan and co-director of the "Stimulating Peripheral Activity to Relieve Conditions (SPARC)" project at IIT-Madras in India.

Prof. Dr. Mathur's research work focuses on the targeted application of nanomaterials and advanced ceramic materials for energy technologies. He holds eleven patents and is (co-)author of more than 500 scientific publications (h-index 71), and has also edited a number of books. He is editor of the scientific journals *Journal of Electroceramics* and *NanoEnergy*. He is an academic member of the World Academy of Ceramics as well as the American Ceramic Society (AcerS) and ASM International.

In 2016, Prof. Dr. Sanjay Mathur was awarded an honorary doctorate from Vilnius University. He is currently on the board of the European Materials Research Society (E-MRS). In 2020, he was awarded the R.C. Mehrotra Lifetime Achievement award by the Indian Science Congress Association. He was honored as a member of the European Academy of Science in 2020 and as a member of the National Academy of Science, India, in 2021. He was also honored with the Woody White Prize of the Material Research Society (MRS) in 2021 and received the Indian Chemical Research Society Medal in 2022. In the same year, he received the Materials Frontiers Award from the International Union of Materials Research Society (IUMRS). He is currently the President of the American Ceramic Society (ACerS 2022–2023), USA, and has been considered for the Orton Jr. Lecture (2022/23) of the American Ceramic Society. Prof. Dr. Sanjay Mathur is also a board member of the German Ceramic Society (DKG) and an elected member of the review board ("Fachkollegiat") of the German Research Foundation (DFG).

Institute of Inorganic Chemistry
University of Cologne
Greinstrasse 6
50939 Cologne

Markus Pothmann

Markus Pothmann is a graduate of Münster University of Applied Sciences and has extensive experience in the field of materials technology. During his time as a research assistant at the Laboratory for Materials Engineering, he focused in particular on the development of tribological test benches. Since 2021, he has been working at the non-profit KIMW Forschungs-GmbH, where he contributes his expertise to innovative research projects. In addition to his main activity, he is the owner of a company that specializes in the distribution of 3D-printed components.

Non-profit KIMW Forschungs-GmbH
Lutherstrasse 7
58507 Lüdenscheid

KUNSTSTOFF INSTITUT LÜDENSCHEID
GEMEINNÜTZIGE KIMW FORSCHUNGS-GMBH

Dr. Stefan Svoboda

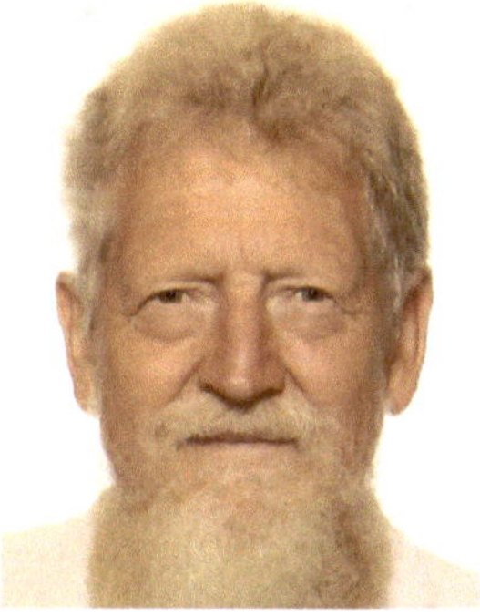

Dr. Stefan Svoboda studied materials science at the Technical University of Dresden and completed his doctorate at the Technical University of Ilmenau on the subject of "Nanostructured hard material layers on a sol-gel basis for wear protection".

He worked for several years in the tool industry in Schmalkalden and (in parallel) from 1982 at the engineering school, and from 1990 at Schmalkalden University of Applied Sciences. In addition to managing the Laboratory for Microscopy and Materials Analysis, he taught materials technology and tribology in the Faculty of Mechanical Engineering at Schmalkalden University of Applied Sciences. He also teaches materials technology, materials testing and maintenance as a private lecturer at the Eisenach Cooperative State University.

1 Introduction

Dr. Ruben Schlutter

The potential of molding markets of various countries depends on their tooling competence and their production volume, and can be separated into what can be colloquially termed “all-stars”, “established”, “rookies”, and “rising stars”.

Currently China, the USA, Japan, South Korea, and Germany belong to the all-stars. These countries are characterized by a high tooling competence and a large production volume. China is at the lower competence limit of the all-stars, but has by far the largest production volume. The USA is the second largest moldmaking market in the world and belongs to the top group for all examined tool categories in terms of production as well as export and import volumes. Germany, Japan, and South Korea, and also Switzerland, which is not an all-star, have the highest moldmaking competence in the world. A large number of organizational and technological innovations in moldmaking have their origin here. They all have a very highly developed manufacturing industry and are known for their automotive industry. The importance of these three markets will continue in the foreseeable future, not just for the respective local industries, but also for the global procurement of highly complex tools. In 2020, the mold markets of the all-stars had a volume of over €22 billion for injection molds. All all-star markets are suitable for the procurement of complex, complete tools, with the market sizes allowing tool packages to be handled quickly in large volumes [1].

Just as the quality of injection molds continues to rise, so do the demands on the plastic molded parts produced. In addition to the requirements for the design and surface of the plastic molded part, production also plays an essential role in the profile of requirements. For this purpose, the injection molds must withstand more and more physical and chemical loads, be it through the use of abrasive or corrosion-promoting plastics, fillers, and reinforcing materials or additives, or also through technological requirements, such as achieving a certain flow path or the optical lamination of a weld line. Within the injection molding process, a wide variety of defects can occur on the molded parts and the molds, which can be solved or minimized by the use of coatings in the molds.

This book summarizes the current state of science and technology in the field of coating technology. The selection of suitable coatings is always an interplay between the moldmaker, the manufacturer of the plastic molded parts, the customer of the plastic molded parts, and the coater, and we reflect this throughout the text.

1.1 Defect Patterns in Injection Molding

Gloss differences and tiger lines

The gloss of a molded part results from the fact that light shining on the molded part is reflected. The smoother and more uniform the surface of the molded part, the more uniformly the light is reflected and the greater the scattering angle of the reflected light. Structured surfaces in the mold, and also different representations of the molded part cavity by the plastic, lead to differences in the degree of gloss. In the area of cooling channels, ejectors, or wall thickness differences, gloss differences often occur because of differences in the local mold wall temperature, which lead to a different representation accuracy of the mold wall compared to the surrounding molding cavity. Coatings also influence the gloss level. Due to the different properties and modes of action of the coatings, no generally valid statement can be made here [2].

"Tiger lines" are a special case in the occurrence of gloss differences. They occur mainly when blends or multiphase systems are used. Due to the different impression of the molding cavity by the respective phase, alternating gloss impressions occur, which lead to an optical stripe formation. The main causes of tiger line formation are partial crystallization of the surface layer under high shear stresses and differences in melt elasticity. The change of the flow front velocity can also support the occurrence of tiger lines [2].

Dull spots in the gate area

In the gate area, the polymer chains of the melt are strongly stretched and oriented. Since the melt freezes immediately at the mold wall, these strains and orientations cannot be reduced by relaxation. The areas of high orientation therefore have poor mechanical properties and are very susceptible to cracking. As the melt flows under the solidified layer, the layer cracks, allowing the melt to flow into the cracks and re-solidify on the mold wall. Micro-notches are formed, which lead to a strongly scattered light reflection in the area of the gate [2].

Weld lines

When several flow fronts meet in the cavity, a weld line is formed. When they meet, the flow fronts are flattened, partially intermix, and stick together, forming a notch at the mold wall. In the case of structured surfaces, additional gloss differences can occur, resulting in optical and mechanical defects [2].

Demolding grooves

Demolding grooves occur during ejection of the molded part. Particularly with structured surfaces and moldings with large lateral surfaces, the demolding force increases sharply. Due to the structuring or the surface roughness, which lies on the side surfaces transverse to the direction of demolding, microscopic undercuts are formed, which can lead to demolding grooves [2].

Sonic plate effect

Sonic plate effects occur primarily with high-viscosity plastic melts in combination with a low injection speed. During injection, the surface layer behind the flow front solidifies. At the same time, the flow front area close to the wall also cools, making it more difficult for the melt to flow in the direction of the mold wall. The hot melt flowing towards the mold therefore cannot be conveyed as far as the flow front and contact the mold wall. Instead, it causes expansion within the flow channel. When the pressure increases, the flow front comes into contact with the mold wall again. However, since these areas of the flow front are strongly cooled, no complete contact with the mold wall can be formed [2].

Rough surface due to plaque formation

Plastics such as POM, PP, ABS, PC, PET, and PBT tend to form plaques. In addition, increased plaque formation can be observed when flame retardants, UV absorbers, and colorants or lubricants are used. When using additives, the formation of plaque is often due to a mixing incompatibility between the polymer and the additive. In some cases, the use of additives also promotes chemical reactions within the polymer or oxidative degradation of the polymer chains [2].

On the other hand, an unfavorable mold design or process control can support the formation of plaque. Especially in case of a long dwell time or high shear of the polymer melt, plaque formation in the mold cavity can occur. Poor mold venting can prevent the air and outgassing from the plastic melt from escaping from the mold cavity. The use of lubricants and release agents also leads to the formation of plaque [2].

Deformation during demolding

During demolding, forces are applied to the plastic part by the demolding system. The plastic molded part can be deformed by this demolding force, which is why the demolding force must be kept small. Shrinkage has a direct effect on the demolding force and can be favorably influenced by the process. In parallel, various plastics tend to adhere to metallic surfaces, which leads to a significant increase in the demolding force. The use of variothermal process control can also lead to an increase in the demolding force, especially with semi-crystalline plastics, since the mold cavity can be formed in greater detail by the plastic melt [2].

Ejector marks and white cracks

In addition to gloss differences in the area of ejectors, visible marks can also be caused by the ejectors in the molded part. These can have various causes, such as incorrect fitting of the length of the ejector pins or an incorrect dimensioning of the demolding system. On the process side, high demolding forces or an early demolding, or also high temperature differences within the mold or between the mold and the ejector, can lead to ejector marks [2].

White cracks occur when a maximum permissible material-dependent deformation is exceeded, and arise because they reduce the stresses that have been introduced. White cracks frequently occur during demolding under residual pressure or in the area of ejectors. In this case, the outer layers of the molded part are stretched by the inner layers [2].

Incompletely filled molded parts

In the case of an incompletely filled molded part, the cavity is not completely filled. There can be various reasons for this. In addition to insufficient dosing volume and venting difficulties, the injection pressure is often insufficient, or the flow path length is too high, so that the plastic melt freezes before it reaches the end of the flow path [2].

1.2 Preparation of a Specification Sheet and Functional Specification Sheets

In the product development environment, a three-stage development and documentation process consisting of the specification sheet, the functional specification sheet, and the requirements list has become widely accepted. Table 1.1 describes the purpose, contents, and boundaries of the documents [3–5].

The specification sheet is first prepared by the client and describes all requirements and constraints from the client's point of view. It serves as the basis for the invitation to tender and the offer.

The functional specification sheet is prepared by the supplier. It contains the specification sheet and describes the customer specifications with the corresponding requirements and how they are to be processed and solved.

The requirements list contains a systematic compilation of all data and information. It is written by the developer and is used for exact clarification of the task. After the approval by the client, the functional specification and the requirements list are binding documents [3–5].

Table 1.1 The Various Documents for Task Clarification [4]

	Specification sheet	Functional specification sheet	Requirements list
Definition	The scope of supply and services, based on the customer's requirement	A summary of all requirements	A compilation of data and information needed for product development
Author	Customer	Supplier	Designer/Developer
Task	Definition of what is to be solved, and for what purpose	Definition of how this is to be done, and the corresponding requirements	Definition of purpose and properties of requirements
Remark	The specification sheet contains all requirements and boundary conditions	The functional specification sheet contains the specification sheet with the summary of the requirements	The requirements list corresponds to an extended requirements specification

This procedure can also be transferred to the development of coatings, whereby the questions that arise from the various necessary individual aspects (including the molded part to be manufactured later, the component to be coated, and the subsequent manufacturing process) must be defined and answered. This results in the basic structure of the specifications for the selection and development of a suitable coating process:

- Requirements for the molded part
  - Part geometry
  - Part surface and relevant surfaces
  - Plastic used.

- Requirements for the mold
  - Tool material to be coated
  - Surface to be coated (roughness, texture, etc.)
  - Coating technologies to be considered
  - Principle development of the coating.
- Requirements for the functionality of the coating
  - Targeted injection molding parameters
  - Methods for mold cleaning
  - System-specific specifications and restrictions.
- Functional tests
  - Measurement and characterization methods
  - Tests to check the layer quality and layer adhesion
  - Application in the production tool
  - Efficiency tests.

From this basic structure, the methodology for developing the coatings is developed according to Figure 1.1. First, possible deposition processes are simulated in order to be able to estimate the process window and the position of the component to be coated in the reactor. In the second step, the actual coating is carried out. In addition to the component to be coated, metal coins are always positioned at various points in the reactor. The subsequent tests are carried out on these coins in order not to damage the component. Here, the coating thickness is analyzed. The structure of multilayer coatings, the presence of coating defects, and the adhesion of the coating to the substrate or the hardness of the coating can also be examined at this point. An evaluation of the coating surface is also possible. Depending on the purpose of the applied coating, further investigations are carried out to examine the properties of the coating and ensure its suitability with regard to the requirements defined in the specification sheet and the requirements list. However, the practical suitability of the coating must then always be determined in a real application. In this case, the coatings are analyzed in their original condition and then again after the coating has been applied, in order to be able to characterize any abrasion or damage to the coating.

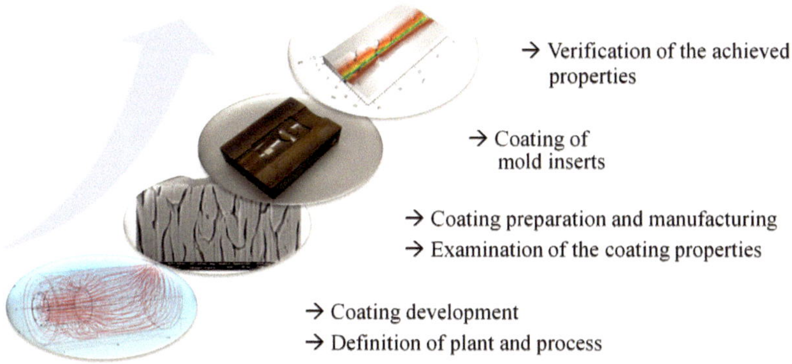

Figure 1.1 Procedure for the development of coatings (Image source: Non-profit KIMW Forschungs-GmbH)

References for Chapter 1

[1] Boos, W.; Lukas, G.; Kessler, N.; et al.: *World of Tooling 2022*, WBA Toolmaking Academy, Aachen, 2022

[2] *Guide to Defects in Moulded Thermoplastic Parts* (14th edn), KIMW NRW GmbH, 2021

[3] Conrad, K.: *Grundlagen der Konstruktionslehre* [Basics of Design Theory] (5th edn), Carl Hanser, 2010

[4] Ehrlenspiel, K.: *Integrierte Produktentwicklung* [Integrated Product Development] (3rd edn), Carl Hanser, 2007

[5] Pahl, G.; Beitz, W.; Feldhusen, J.; Grote, K.-H.: *Pahl/Beitz Konstruktionslehre* [Pahl/Beitz Design Theory] (7th edn), Springer, 2007

2 Tool Steels and Their Coatability

Markus Pothmann

2.1 Introduction

2.1.1 Definition of Tool Steels

Tool steels are a group of high-strength, highly wear-resistant steels specifically designed for use in the manufacture of molds, tools, and other components that are subject to repeated impact, abrasion, and high loads. These steels typically have high hardness, toughness, and resistance to deformation, making them ideal for use in applications that require high precision and long tool life.

The specific properties of tool steels are achieved through careful alloying and heat treatment. Alloying elements such as tungsten, molybdenum, chromium, and vanadium are added to the steel to improve its strength, wear resistance, and other important properties. The exact combination of these elements varies depending on the specific requirements of the tool to be produced.

Tool steels are typically divided into several categories based on their specific properties and intended use. For example, high-speed steels are designed for use in high-speed machining applications, while cold work steels are designed for use in applications where the tool is exposed to extremely low temperatures. Other categories of tool steels include hot work steels, steels for plastic injection molds, and impact resistant steels.

The high strength and wear resistance of tool steels make them an ideal material for use in a variety of industrial applications, including injection molding. However, selecting the appropriate tool steel for a particular application can be a complex process that requires careful consideration of factors such as the type of plastic to be molded, the desired surface finish, and the required tool life.

2.1.2 Development of Tool Steels

The development of tool steels has been driven by the need to improve the performance and durability of tools used in various industrial applications. Over time, advances in metallurgy, heat treatment, and manufacturing techniques have led to the creation of a wide range of tool steels with different properties and characteristics.

An important milestone in the development of tool steels was the invention of crucible steel in the middle of the 19th century. This new type of steel was produced by melting iron and adding various alloying elements to achieve a more homogeneous and consistent material. It was also possible to produce larger quantities of crucible steel than other types of steel, making it more accessible to manufacturers.

In the late 19th and early 20th centuries, new alloying elements were added to tool steels, including tungsten, molybdenum, chromium, and vanadium. These elements significantly improved the strength, toughness, and wear resistance of tool steels, making them more suitable for demanding industrial applications.

During the Second World War, the demand for tool steels increased dramatically, as they were used extensively in the manufacture of military equipment. This led to further advances in tool steel technology, including the development of high-speed steels that could withstand the high temperatures generated by high-speed machining.

Today, tool steels remain an important material for a variety of industrial applications, including injection molding. Technological advances continue to push the boundaries of what is possible with tool steels, and new materials and manufacturing techniques are constantly being developed to improve their performance and reliability.

2.1.3 Types of Tool Steels

Tool steels are classified into different types based on their properties and intended use. The classification is typically based on the alloying elements and heat treatment required to achieve certain properties.

Carbon tool steels

Carbon tool steels such as C75 or C100S are the oldest and simplest type of tool steel. They have a carbon content of 0.6% to 1.5%. These steels are inexpensive and easy to heat-treat, making them ideal for small tools that do not require high precision. Carbon tool steels are commonly used for chisels, knives, and hand tools.

High-speed steels

High-speed steels are designed for high-speed machining applications where the cutting speed exceeds 50 m/min. They contain tungsten, molybdenum, and vanadium as alloying elements. These steels have a high degree of hardness, wear resistance, and toughness, making them ideal for cutting and drilling tools. High-speed steels are commonly used for cutting tools in the automotive, aerospace, and medical industries.

Impact-resistant tool steels

Impact-resistant tool steels such as 1.2714 or 1.2355 are designed to withstand impact loads. They contain chromium, molybdenum, and vanadium as alloying elements. These steels are used for applications that require high toughness and resistance to cracking, such as cold chisels, hammers, and other hand-held tools.

Hot-work steels

Hot-work steels such as 1.2343 or 1.2344 are designed for use in high-temperature applications where the tool is exposed to high stress and wear. They contain tungsten, molybdenum, and chromium as alloying elements. These steels have a high degree of toughness, wear resistance, and thermal stability, making them ideal for use in forging tools, extrusion dies, and other hot work applications.

Cold-work steels

Cold-work steels such as 1.2379 or 1.2510 are designed for use in cold-work applications where the tool is exposed to high stress and wear. They contain tungsten, molybdenum, and vanadium as alloying elements. These steels have a high degree of hardness, toughness, and wear resistance, which makes them ideal for use in punching tools and other cold-work applications.

Plastic mold steels

Plastic mold steels such as 1.2311 or 1.2738 are intended for use in injection molding and other plastic molds. They contain chromium, molybdenum, and vanadium as alloying elements. These steels have a high degree of hardness, wear resistance, and thermal conductivity, which makes them ideal for use in plastic molding applications.

High-strength low-alloy (HSLA) tool steels

HSLA tool steels such as S700MC are designed for use in high-strength applications where the tool is exposed to high stress and wear. They contain molybdenum, chromium, and vanadium as alloying elements. These steels have a high degree of hardness, toughness, and wear resistance, which makes them ideal for use in highly stressed applications such as gears and shafts [1].

Powder metallurgical tool steels

Powder metallurgical tool steels are produced by mixing and pressing fine metal powders, which are then sintered at high temperatures. These steels have a high degree of density, toughness, and wear resistance, making them ideal for use in heavy-duty applications such as cutting tools, dies, and other precision components.

High-alloy tool steels

High-alloy tool steels such as 1.3247 are designed for use in applications that require high strength and wear resistance. They contain a high proportion of alloying elements such as tungsten and molybdenum.

2.2 Injection Molding Tool Steels

2.2.1 Introduction

Injection molding is a manufacturing process used to produce a wide range of molded plastic parts and products. In injection molding, plastic pellets or granules are melted and injected into a mold cavity where they cool and solidify to form the desired molded part. Injection molding is a highly efficient and cost-effective process, making it one of the most popular methods for producing molded plastic parts.

Injection molds are a critical component of the injection molding process. The tools must be designed and manufactured to withstand the stresses of the molding process. Injection molding tool steels are specifically designed for use in injection molds and must have a unique combination of properties to withstand the stresses of the molding process.

Tool steels are a group of high-strength steels used for cutting, forming, and shaping materials. They are known for their high hardness, wear resistance, and toughness. Injection molding tool steels must have similar properties to standard tool steels, but must also have specific properties to withstand the high pressures, temperatures, and abrasion associated with the injection molding process.

The properties of injection molding tool steels are influenced by their composition, microstructure, and heat treatment. The composition of the steel is crucial for determining its mechanical properties, including hardness, toughness, and wear resistance. The microstructure of the steel, which is influenced by the heat treatment process, plays a crucial role in determining the mechanical properties of the steel.

Selecting the appropriate injection molding tool steel for a particular application is critical to the success of the molding process. Factors to consider when selecting a tool steel include the type of plastic to be molded, the expected production volume, and the expected tool life.

In summary, injection molding tool steels are a critical component of the injection molding process. They must be carefully selected to ensure that they have the necessary properties to withstand the stresses of the molding process. The specific properties of injection molding tool steels and their influence on suitability for different applications are discussed in more detail in the next section.

2.2.2 Properties of Injection Molding Tool Steels

Injection molding tool steels are selected for their ability to withstand the stresses of the injection molding process. They must have certain properties to ensure that they can withstand the high pressure, high temperature, and abrasion associated with the injection molding process. This section discusses the important properties of injection molding tool steels and how these affect their suitability for use in injection molding.

Hardness

The hardness of injection molding tool steels is an important property that determines their resistance to wear and deformation. The harder the tool steel, the more resistant it is to wear and deformation under high-pressure molding conditions. However, the toughness of the tool steel decreases with increasing hardness. In order for the tool steel to withstand the stresses of the forming process, a balance must be found between hardness and toughness.

Toughness

Toughness is the ability of a material to resist cracking or breaking under conditions of high stress. During injection molding, tool steels are subjected to high stresses, and must have a high toughness to prevent cracks or fractures. Toughness is influenced by the microstructure of the tool steel as well as its alloying elements and heat treatment.

Wear resistance

Injection molding tool steels must have a high wear resistance to prevent damage of the tool surface. Wear resistance is influenced by the hardness of the tool steel as well as its microstructure and alloying elements. Tool steels with high wear resistance are typically used in applications where the tool surface is exposed to a high degree of abrasion, such as in the production of abrasive materials or parts with rough surfaces.

Corrosion resistance

Corrosion resistance is an important property for tool steels used in injection molding, as the injection molding process can be corrosive. Tool steels with high corrosion resistance are typically used in applications where the tool is exposed to corrosive materials or environments. The corrosion resistance of tool steels is influenced by their alloying elements, particularly chromium and molybdenum.

Thermal conductivity

The thermal conductivity of injection molding tool steels is important to maintain a uniform temperature throughout the mold. Tool steels with high thermal conductivity are better able to dissipate heat from the molded part surface, reducing the risk of hot spots and improving the overall quality of molded parts. The thermal conductivity of tool steels is influenced by their alloying elements, particularly copper and nickel.

Machinability

Machinability is an important property for injection molding tool steels, as they need to be machined to form the complex shapes required for injection molding. Tool steels with good machinability are easier to machine and result in less wear on the cutting tools. Machinability is influenced by the microstructure of the tool steel as well as its alloying elements and heat treatment [2].

In summary, the properties of injection molding tool steels play a critical role in determining their suitability for various applications. The properties discussed in this section, including hardness, toughness, wear resistance, corrosion resistance, thermal conductivity and machinability, must be carefully considered when selecting a tool steel for injection molding. The next section will introduce the different types of tool steels and their specific properties that are commonly used in injection molding.

2.2.3 Composition of Injection Molding Tool Steels

Injection molding tool steels such as 1.2344 or 1.2311 are high-performance materials that can withstand the extreme conditions of the injection molding process. These steels are specially formulated to provide excellent strength, toughness, wear resistance, and thermal stability, to ensure a long life and consistent performance of the injection mold. The composition of injection molding tool steels plays a crucial role in determining their properties and suitability for various applications. This chapter presents the different types of tool steels used in injection molding, and their compositions.

Low-alloy tool steels

Low-alloy tool steels are often used in the manufacture of injection molds due to their excellent combination of toughness and wear resistance. These tool steels contain a low percentage of alloying elements, typically less than 5%, including chromium, molybdenum, and vanadium. The carbon content in these tool steels is between 0.3% and 0.6%, and they are often heat-treated to achieve the desired properties.

High-alloy tool steels

High-alloy tool steels are designed to provide exceptional wear resistance, toughness, and corrosion resistance. These tool steels contain a higher proportion of alloying elements than low-alloy tool steels, often over 5%. Chromium, molybdenum, vanadium, and tungsten are commonly used alloying elements in high-alloy tool steels. The carbon content in high-alloy tool steels is between 0.7% and 1.5%, and they are typically heat-treated to achieve the desired properties.

High-speed steels

High-speed steels are used to manufacture injection molds that require high cutting speeds and temperatures, such as those used for processing thermoset plastics. These tool steels typically contain high levels of carbon, tungsten, molybdenum, and chromium. The high carbon content in high-speed steels ranges from 0.8% to 1.5%, while the alloying elements are typically present in quantities of 7% to 20%. High-speed steels are heat-treated to achieve the desired properties, including high hardness, wear resistance, and toughness.

Hot-work tool steels

Hot-work tool steels are designed to withstand the high temperatures and pressures of injection molding. These tool steels typically contain a high percentage of chromium, molybdenum, and vanadium, as well as other alloying elements such as tungsten, cobalt, and nickel. The carbon content in hot work tool steels is between 0.4% and 1.4%, and they are heat-treated to achieve the desired properties, including high hardness, toughness, and thermal stability.

Cold-work steels

Cold-work tool steels are used to manufacture injection molds that require high wear resistance, toughness, and dimensional stability. These tool steels typically contain a small percentage of alloying elements, including chromium, molybdenum, vanadium, and tungsten. The carbon content in cold work tool steels is between 0.5% and 1.5%, and they are often heat treated to achieve their desired properties.

Maraging tool steels

Maraging tool steels are high-strength, low-alloy steels that are commonly used in the manufacture of injection molds. These tool steels contain a small percentage of carbon, typically less than 0.03%, and a high percentage of nickel, cobalt, and molybdenum. Maraging tool steels are heat-treated to achieve their desired properties, which include high strength, toughness, and wear resistance.

Powder metallurgical tool steels

Powder metallurgical tool steels are produced by a process in which powdered tool steel is mixed with a binder and the mixture is then compacted into a desired shape. These tool steels offer excellent wear resistance, toughness, and dimensional stability, making them ideal for use in injection molds. Powder metallurgical tool steels typically contain high levels of alloying elements such as chromium, molybdenum, and vanadium, and their carbon content ranges from 0.4% to 2.5%. They are heat-treated to achieve their desired properties.

Stainless steels

Stainless steels are a group of corrosion-resistant steels that are commonly used in the manufacture of injection molds. These tool steels contain at least 10.5% chromium, which provides excellent corrosion resistance. They also contain varying amounts of other alloying elements such as nickel and molybdenum, which can improve their mechanical properties. Stainless steels are available in several different grades, each with their own unique combination of properties, and they are often heat-treated to improve their strength and toughness.

Specialized tool steels

In addition to the tool steels already discussed, there are also several specialized tool steels that are used in injection molding. These include high-strength tool steels, impact-resistant tool steels, and wear-resistant tool steels. These tool steels are formulated to meet specific requirements of the injection molding process, such as the ability to withstand high impact forces, or resist wear and abrasion.

In summary, the composition of injection molding tool steels plays a decisive role in determining their properties and suitability for various applications. Tool steels can vary greatly in terms of their alloying elements, carbon content, and heat treatment. Therefore, it is important to select the right tool steel for the specific requirements of the injection molding process. In the next section, the heat treatment process used to improve the properties of injection molding tool steels is presented [3].

2.2.4 Heat Treatment of Injection Molding Tool Steels

Heat treatment (see Figure 2.1) is a crucial step in the manufacturing process of injection molding tool steels. Proper heat treatment can significantly improve the mechanical properties of the tool steel, including its hardness, toughness, and wear resistance. The heat treatment process typically includes the following steps:

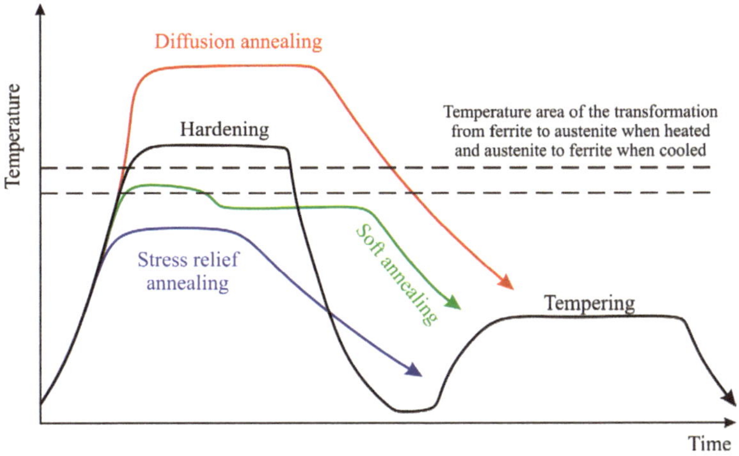

Figure 2.1 Heat treatment of steels (Image based on [4])

Annealing

During annealing, the tool steel is heated to a certain temperature and held there for a certain period of time before it is slowly cooled down. This process is used to relieve internal stresses in the tool steel and improve its machinability. Annealing also makes the tool steel softer and more ductile, which can be an advantage in certain applications.

Hardening

During hardening, the tool steel is heated to a high temperature and then quickly cooled to room temperature. This process forms a hard, wear-resistant surface layer on the tool steel while retaining a relatively soft and tough interior. Hardening can be carried out using various techniques, such as oil quenching, water quenching, or air cooling, depending on the composition and desired properties of the tool steel.

Tempering

Tempering is the reheating of the hardened tool steel to a certain temperature and subsequent cooling to room temperature. This process reduces the hardness and brittleness of the tool steel, and at the same time improves its toughness and ductility.

Tempering can also reduce residual stresses in the tool steel and improve its dimensional stability.

The specific heat treatment process used for injection molding tool steels depends on the composition of the steel and the desired properties for the application. For example, high-speed tool steels may require a different heat treatment process than high-temperature tool steels.

It is also important to note that improper heat treatment can significantly reduce the performance and lifetime of injection molding tool steels. Overheating, underheating, or rapid cooling can cause the tool steel to become too brittle or too soft, resulting in cracking, chipping, or deformation during use. Therefore, it is important to follow the correct heat treatment procedures and involve experienced professionals to achieve the best results.

In addition to heat treatment, other factors such as surface treatment, design considerations, and maintenance also play a crucial role in the performance and service life of injection molding tool steels. Proper care and maintenance, including regular cleaning, lubrication, and inspection, can extend the lifetime of the tool steel and ensure optimal performance in the injection molding process.

2.2.5 Surface Treatment of Injection Molding Tool Steels

Injection molding tool steels can be treated to improve their surface properties, such as wear resistance, corrosion resistance, and release properties. This section summarizes some common surface treatment processes for injection molding tool steels.

Nitriding

Nitriding is a surface-hardening process in which nitrogen is introduced into the surface of the tool steel to form a hard nitride layer. This process improves the wear resistance and fatigue strength of the tool steel as well as its corrosion resistance.

There are two main types of nitriding: gas nitriding and plasma nitriding. Gas nitriding is carried out at temperatures between 495–565 °C in a nitrogen-rich gas atmosphere, while plasma nitriding is carried out at temperatures between 400–580 °C in a low-pressure plasma atmosphere. Plasma nitriding provides better control over the thickness and composition of the nitride layer, resulting in better performance.

PVD coating (see also Section 3.2)

Physical vapor deposition (PVD) is a process in which a thin film of material is deposited on the surface of the tool steel. PVD coatings are typically used to improve the wear resistance, corrosion resistance, and release properties of the tool steel.

Common PVD coatings used for injection molding tool steels include titanium nitride (TiN), titanium carbonitride (TiCN) and diamond-like carbon (DLC). TiN and TiCN coatings improve wear resistance, while DLC coatings improve release properties.

DLC coating

Diamond-like carbon (DLC) is a type of PVD coating consisting of amorphous carbon with a diamond-like structure. DLC coatings are known for their excellent wear resistance, low friction, and high hardness.

DLC coatings can be applied to injection molding tool steels to improve their performance, especially in applications where high friction and wear are a problem. They can also improve the release properties of the tool steel, resulting in easier part ejection and fewer molding defects.

Galvanizing (see also Section 3.1)

Galvanizing, also known as electroplating, is a process in which a thin layer of metal is applied to the surface of the tool steel by means of an electric current. Galvanizing can be used to improve the corrosion resistance, wear resistance, and release properties of the tool steel.

Common galvanizing materials used for injection molding tool steels include chromium, nickel, and zinc. Chrome plating is particularly effective in improving corrosion resistance, while nickel plating improves wear resistance and release properties.

In summary, surface treatment processes such as nitriding, PVD coating, DLC coating, and galvanizing can be used to improve the surface properties of injection molding tool steels. These treatments can improve wear resistance, corrosion resistance, release properties, and overall performance. It is important to select the appropriate surface treatment method based on the specific application and performance requirements [5].

2.2.6 Maintenance and Care of Injection Molding Tool Steels

Proper maintenance and care of injection molding tool steels is essential to ensure their longevity and performance. Neglecting proper maintenance can lead to reduced performance, shorter lifetime, and even mold failure.

Regular cleaning and inspection

Regular cleaning and inspection are essential to detect signs of wear or damage to the tool steel. Cleaning can be done using a solvent to remove dirt or residue from the molding process. It is important to avoid the use of abrasive materials or harsh chemicals that can damage the surface of the tool steel.

After cleaning, the tool steel should be inspected for signs of wear or damage such as cracks, chips or pitting. Any problems should be addressed immediately to prevent further damage and potential tool failure.

Lubrication

Proper lubrication is essential for the longevity and performance of injection molding tool steels. Lubrication reduces friction between the tool steel and the material being molded, reducing wear and preventing the tool from seizing.

There are different types of lubricants that can be used, including oils, fats, and dry lubricants. It is important to choose a lubricant that is compatible with the specific type of tool steel being used and the material being formed.

Proper storage

Proper storage is important to prevent rust and corrosion from developing on the tool steel. Tool steels should be stored in a clean, dry environment away from dust and moisture. They should be stored on a rack or in a drawer that keeps them organized and prevents them from coming into contact with other tools.

It is also important to correctly mark and identify each tool steel to avoid confusion and ensure that the correct tool steel is used for each application.

Proper handling

Tool steels should be handled with care to avoid damaging the surface or edges. They should be lifted and moved using appropriate lifting techniques and should not be dropped or mishandled.

It is important not to use the tool steel for purposes other than injection molding to avoid unnecessary wear and damage. It should also be used within the recommended operating parameters to avoid excessive wear and premature failure.

Overhaul and repair

Over time, injection molding tool steels may need to be reconditioned or repaired to maintain their performance and longevity. Reconditioning involves restoring the surface of the tool steel to its original condition by removing damage or wear.

Repair may include welding or grinding to repair cracks or chips in the tool steel. It is important that you consult a qualified professional for any reconditioning or repair work to ensure that the tool steel retains its properties and performance.

In summary, proper maintenance and care of injection molding tool steels is essential to ensure their longevity and performance. Regular cleaning and inspection, proper lubrication, storage, handling, and overhaul or repair when needed are important steps to ensure the optimal performance and service life of injection molding steels.

2.3 Selection of Injection Molding Tool Steels

2.3.1 Factors that Influence the Choice of Material for Injection Molding Tool Steels

The selection of the right material for injection molding tool steels is influenced by several factors, including:

- Part complexity: The complexity of the part to be molded can affect the choice of tool steel. For example, molded parts with complex designs may require a tool steel with higher toughness and wear resistance to withstand the loads of the molding process.
- Part size: The size of the part to be molded can also affect the choice of tool steel. Larger molded parts may require a tool steel with higher thermal conductivity and better heat dissipation, to prevent warpage and cracking during the molding process.
- Production volume: The volume of molded parts produced can also influence the choice of material. Mass production may require a tool steel that can withstand the high heat and stress of continuous use, while production in smaller quantities may allow the use of less expensive tool steels.
- Mold material: The mold material can also influence the choice of tool steel. Different materials have different melt flow characteristics and thermal properties, which can affect the wear resistance, thermal conductivity, and overall performance of the tool steel.

2.3.2 Challenges in the Selection of Injection Molding Tool Steels

Selecting the right tool steel for injection molding can be challenging due to the numerous factors that can affect material selection. Some of the challenges include:

- Balancing performance and cost: High-performance tool steels often have a higher price, which makes it difficult to reconcile the desired performance with the material costs.
- Compatibility with the injection molding process: The tool steel must be compatible with the specific injection molding process being used. This includes the ability of the tool steel to withstand the temperature and pressure changes during the molding process.
- Customized tool steels: In some cases, custom tool steels may be required to meet the specific requirements of the application. This can increase the complexity and cost of the tool steel selection process.

2.3.3 Selecting Materials for Injection Molding Tool Steels

The selection of injection molding tool steels is a critical factor in ensuring the performance, quality, and efficiency of the injection molding process. There are several factors to consider when selecting tool steels for injection molding applications, including:

- Material properties: The material properties of the tool steel must be carefully considered to ensure that it meets the requirements of the application. Factors such as hardness, toughness, wear resistance, corrosion resistance, and machinability are important considerations.
- Molded part material and geometry: The material and geometry of the molded part to be produced affect the choice of tool steel. For example, if the part is made of abrasive or corrosive material, a tool steel with high wear or corrosion resistance may be required. If the molded part has a complex geometry or requires high precision, a tool steel with good machinability may be required.
- Production volume: The expected production volume of the injection molding process will also affect the choice of tool steel. For low-volume production, a tool steel with lower wear resistance may be acceptable, while for high-volume production, a tool steel with high wear resistance is required to ensure that the molds will last long enough.
- Cost: Cost is always a factor when selecting tool steels. While high-performance tool steels can offer significant advantages in terms of performance and tool life, they can also be considerably more expensive than inferior tool steels. The cost of tool steel must be weighed against the performance and efficiency requirements of the application.
- Availability: The availability of tool steel must also be considered. Some specialized tool steels may be difficult to source, which can affect production times and costs.

The German Institute for Standardization (DIN) and the International Organization for Standardization (ISO) have established a classification system for tool steels based on their properties and applications (DIN EN ISO 4957). ISO tool steel grades are often used in injection molding applications due to their high quality and durability.

Some examples of injection-molded tool steels according to ISO standards are presented below:

- 1.2311 (P20): This is a low-alloy carbon tool steel that is widely used in injection molding applications due to its good machinability, toughness, and wear resistance. It is suitable for the production of large and complex parts with high surface quality requirements.

- 1.2343 (H11): This is a hot-work tool steel with high heat resistance and excellent toughness. It is often used in injection molding applications to produce parts with high dimensional accuracy and surface finish requirements.
- 1.2738 (P20+Ni): This is a low-alloy carbon tool steel that contains nickel for improved toughness and machinability. It is suitable for the production of large and complex parts with high surface quality requirements, as well as for mold structures and core and cavity inserts.
- 1.2083 (420): This is a stainless tool steel with good corrosion resistance and polishability. It is often used in injection molding applications to produce parts with high aesthetic requirements such as consumer electronics and medical devices.
- 1.2312 (P20+S): This is a low-alloy carbon tool steel with good machinability and toughness. It is suitable for the production of large and complex parts with high surface quality requirements, as well as for mold structures and core and cavity inserts.
- 1.2379 (D2): This is a tool steel with a high carbon and chromium content that has good wear resistance and toughness. It is often used in injection molding applications to produce parts with high precision and dimensional stability, as well as for ejector pins and inserts.
- 1.2767: This is a high-strength, low-alloy tool steel with excellent toughness and wear resistance. It is suitable for the production of parts subject to high stress and pressure and for molds that require frequent maintenance and repair.
- 1.2085 (420F): This is a stainless steel with a high carbon content that has good corrosion resistance and high polishability. It is often used in injection molding applications to produce parts with high aesthetic requirements, such as products in the medical and food industries.
- 1.2764 (X19NiCrMo4): This is a tool steel with a high nickel and chromium content that has good wear resistance and toughness. It is suitable for the manufacture of parts subject to high stress and pressure and for molds that require frequent maintenance and repair.
- 1.2316 (X38CrMo16): This is a low-carbon, chromium-rich stainless steel with good corrosion resistance and high polishability. It is often used in injection molding applications to produce parts with high aesthetic requirements such as automotive and appliance components.
- 1.2436 (D6): This is a tool steel with a high carbon and chromium content that has good wear resistance and toughness. It is often used in injection molding applications to produce parts with high precision and dimensional stability, as well as for ejector pins and inserts.

- 1.2842 (O2): This is a low-alloy carbon tool steel that has good machinability and toughness. It is suitable for the production of large and complex parts with high surface quality requirements, as well as for mold structures and core and cavity inserts.
- 1.2378 (X210CrW12): This is a tool steel with a high carbon and chromium content that has good wear resistance and toughness. It is often used in injection molding applications to produce parts with high precision and dimensional stability, as well as for ejector pins and inserts.
- 1.2310 (P20+S): This is a low-alloy carbon tool steel with good machinability and toughness. It is suitable for the production of large and complex parts with high surface quality requirements, as well as for mold structures and core and cavity inserts.

These ISO tool steel grades offer a range of properties and characteristics that highlight their suitability for various injection molding applications. The selection of the appropriate steel grade depends on factors such as the molding conditions, the material to be molded, and the required tool life. It is important to work with a tool steel supplier or material engineer to select the best tool steel for a particular injection molding application [6].

2.3.4 Future Development of Injection Molding Tool Steels

As the manufacturing industry continues to evolve and demands for higher quality, efficiency, and sustainability increase, the need for advanced materials and technologies in injection molding becomes more and more important. The development of new and improved injection molding tool steels is essential to meet the demands of the industry.

In recent years, significant progress has been made in the development of tool steels for injection molding, with a focus on improving properties such as wear resistance, toughness, and corrosion resistance. However, there is still room for improvement, and the future of injection molding tool steels is exciting.

One of the key areas of development for injection molding tool steels is the incorporation of advanced alloying elements. For example, boron is known to improve the toughness and wear resistance of tool steels, while nitrogen can improve corrosion resistance and fatigue strength. Other elements such as copper and nickel can also be incorporated to improve specific properties.

In addition to new alloying elements, advances in microstructure technology are also being researched. The use of advanced heat treatment techniques such as quenching and tempering can produce a more uniform microstructure and improve the mechanical properties of the tool steel. The incorporation of nanoscale particles such as carbides can also increase the wear resistance of tool steel.

Another area of development is the use of surface coatings or treatments to improve the performance of injection molds. For example, coatings with diamond-like carbon (DLC) can increase the wear resistance of the tool steel, while surface treatments such as nitriding can improve surface hardness and corrosion resistance.

The increasing demand for sustainability in manufacturing is also driving the development of more environmentally friendly injection molding tool steels. The use of recycled materials such as steel scrap and other metals is being investigated to reduce waste and minimize the environmental impact of the manufacturing process.

Overall, the future of injection molding tool steels is focused on improving the properties and performance of these materials to meet the evolving needs of the manufacturing industry. The integration of advanced alloying elements, microstructure technology, surface coatings, and sustainability initiatives will all play a crucial role in the development of the next generation of injection molding steels.

References for Chapter 2

[1] Tirler, W.: *Europäische Stahlsorten* [European Steel Grades], Beuth, 2017

[2] Roos, E.; Maile, K.: *Werkstoffkunde für Ingenieure: Grundlagen, Anwendung, Prüfung* [Materials Science for Engineers: Basics, Applications, Testing], Springer, 2005

[3] Schatt, W. E.: *Konstruktionswerkstoffe des Maschinen- und Anlagenbaues: 131 Tabellen* [Construction Materials in Mechanical and Machine Engineering: 131 Tables], Wiley-VCH, 1998

[4] Schlegel, J.: Wärmebehandlung von Stählen [Heat treatment of steels], in: *Die Welt des Stahls* [The World of Steels]. Springer, 2021, *https://doi.org/10.1007/978-3-658-33916-6_6*

[5] Liedtke, D.: *Wärmebehandlung von Eisenwerkstoffen II* [Heat Treatment of Ferrous Materials II], Expert Verlag, 2018

[6] Bleck, W.; Moeller, E.: *Handbuch Stahl: Auswahl, Verarbeitung, Anwendung* [The Steel Handbook: Selection, Processing, Applications], Hanser, 2017

3 Fundamentals of Coating Technologies

3.1 Electrolytically Deposited Metallic Layers and Hybrid Systems

Dr. Orlaw Massler

3.1.1 Background and Challenges

Tools for injection molding and extrusion are becoming increasingly complex in terms of their functionality, tolerances, and load requirements. A wide variety of additives are used in modern plastics, including glass fibers, flame retardants, colorants, antioxidants, antistatic additives, and heat stabilizers. Some of these are highly abrasive, which places considerable demands on the surfaces of the tools used. This can lead to the failure of critical components, downtime, and production losses. Widely used glass fiber reinforcements, for example, wear out tools and other critical components early on in operation. Corrosive and chemically aggressive additives also exacerbate the problem, as they can release vapors that can lead to corrosion and wear on the tool, which also has a negative impact on surface quality and performance. This also increases the risk of molded parts or raw material sticking on the tool surface, which in turn can lead to costly interruptions in production. This is a serious problem in view of the price competitivity compared to low-cost countries, where simple tools and spare parts are produced at a fraction of the cost.

In order to remain competitive, technology-oriented plastics processors must therefore set themselves apart from low-cost producers by making their systems significantly more productive, more resistant to wear and corrosion, and with lower operating costs. Surface technology plays a key role here, not only in terms of sustainability and resource conservation, but also in giving manufacturers a competitive advantage. The technological approach of a functionalized surface involves the application of a surface coating that counteracts the specific destruction mechanism without causing a significant modification to the base material. This means that the manufacturing process chain can be largely retained and the performance can still be significantly increased by the applied surfaces.

The functionalization of surfaces is usually achieved using technologies such as bath-supported processes, for example:

- Electroplating (also known as galvanization)
- Electroless (chemical) coating
- Physical vapor deposition (PVD) (see Section 3.2)
- Plasma-assisted chemical vapor deposition (PACVD) processes (see Section 3.3.3)
- Thermo-chemical processes
- A combination of these (hybrid processes) [1, 2].

3.1.2 Galvanic Layers

Bath-supported coating processes have clear advantages over many other processes. The coating process generally does not interfere with the mechanical properties of common base materials because the coating temperatures tend to be low. Thus, for example, thermal destabilization of the structure of steels due to loss of hardness or grain coarsening can be excluded. This means that even more sensitive steels can be coated easily and reliably. The nature of the bath processes enables the deposition of thicker layers, which primarily enables good coverage and corrosion resistance [3].

The basic principle of electroplating processes dates back to the 19th century. The coating process consists of a sequence of various pre-treatment and rinsing steps (see Table 3.1) before the plating bath (Figure 3.1) is approached. Each active process medium must be removed cleanly before the next step in order to guarantee the coating quality and prevent carryover of the process media.

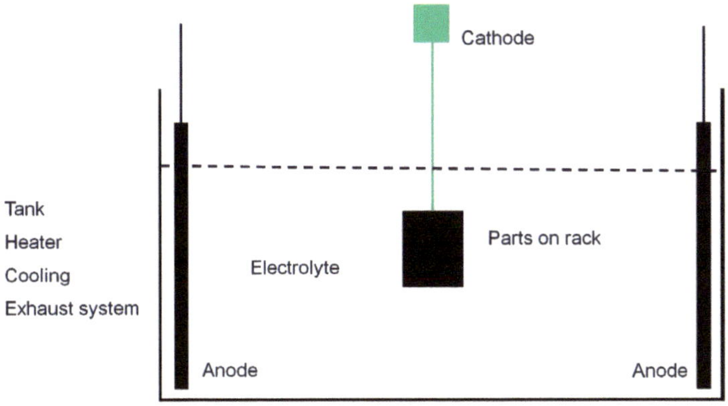

Figure 3.1 Structure of a coating tank for electroplating deposition

Table 3.1 Process Steps and Their Function in Electroplating

Process step	Goal	Remark
Batching of rack or bulk material	Make secure contact with the bracket Establish electrical contact	Contact point must be present; drum goods are possible for many processes
Cleaning, degreasing	Remove corrosion protection residues Produce a clean surface	
Pickling	Remove oxide layers	Produce metallic surface for good adhesion/bonding between coating and coating material
Coating	Apply coating and supply power	Depending on the process
Rinsing	Produce a clean surface after each process step, to avoid carryover.	
Drying	Produce a stain-free surface	

The electroplating process itself, also known as electrolytic deposition or electroplating, is a process of coating a metal object with a thin layer of another metal. This process is based on the use of electrolysisto transfer metal ions from a solution to the surface of the metal object.

The parts are immersed as a cathode in an electrolytic solution containing, among other things, the metal to be used for the coating. An anode material made of the same metal as the solution is also placed in the bath as an electrode. When current is passed through the solution, the metal ions are dissolved at the anode and attracted to the cathode. During this process, the metal ions are deposited on the surface of the cathode and form a layer that is dense and uniform. The thickness of the coating depends on the duration of the process and the current intensity. By controlling the current intensity and time, the thickness and quality of the coating can be precisely controlled. The current density distribution over the geometry of the coating material is often controlled by the appropriate design of an auxiliary electrode system on the component, to improve (or even enable) development of a well-distributed coating of the desired thickness.

Common coatings for tools are limited to hard or corrosion-resistant coatings (Table 3.2). Many coatings are not suitable due to their limited hardness.

Table 3.2 Common Electroplated Coatings for Use in Tools

Coating system	Field of application
Electroplated nickel, matte nickel, sulfamate nickel	Corrosion protection, often as a base or intermediate layer
Hard chrome	Wear protection, avoidance of plastic adhesion
Tin	Corrosion protection, conductivity
Electroplated nickel, matte nickel, sulfamate nickel	Corrosion protection, often as a base or intermediate layer

3.1.2.1 Challenges and Actions

Hydrogen embrittlement

Hydrogen is produced in the galvanic and also in the electroless plating process, which can lead to the phenomenon of hydrogen embrittlement in the microstructure of some high-strength steels. The hydrogen causes cracks at the grain boundaries of the microstructure, which can spread further under tensile stress and lead to the failure of the corresponding components.

This effect is particularly well-known (and undesirable) in high-strength steels. In addition to the presence of hydrogen, a material susceptible to cracking and a state of tensile stress are required for hydrogen embrittlement to occur. If all these criteria are met, damage can occur.

To mitigate this effect, one measure that can be carried out after the coating process is hydrogen de-embrittlement. This involves carrying out a heat treatment immediately after the coating process, whereby the hydrogen is expelled before the damage can occur at the grain boundaries.

Pre-treatment and adhesion issues

Electroplating and chemical coating processes rely on the electrical conductivity of metallic surfaces. If this is compromised by surface deposits or other such interference, the deposition process cannot proceed as desired. Therefore, in order to produce the components, any potentially disruptive surface layer must be avoided or removed before the actual coating process. This includes oxides and scale layers, as well as surface coatings caused by erosion and thermo-chemical processes.

These interfering layers are either avoided using suitable processes or removed prior to the coating process. A range of processes are available, from conventional abrasive blasting treatment and chemical processes to modern technologies, such as ablative laser cleaning or blasting processes with CO_2 ice pellets or "snow".

Edge structure of galvanic processes

It is known that in the electroplating process, an edge build-up of the coating occurs due to an inhomogeneous current density on the geometric features of many components and tools. This must be taken into account if tight tolerances are to be maintained. In special cases, a significant improvement is achieved by suitable veneering in the coating rack of these parts. However, this is rather complex and can possibly be solved more efficiently by selecting suitable tolerances or the coating system.

Chemical or thin-film processes, which do not exhibit this edge build-up effect, have a significant advantage here.

Burns on the tips of the components

Particular attention must be paid to the positioning of tools with geometries with peaks. If these are aligned directly with the anodes in the bath, very high local current densities can lead to the formation of buds, burns, or thermal effects due to the direct current flow, which can result in the parts becoming unusable.

3.1.3 Electroless Separation

3.1.3.1 Electroless Nickel Plating

Electroless nickel plating is a process for coating a metal object with a thin alloy layer of nickel-phosphorus. In contrast to the electroplating process, electroless nickel plating does not require an electrical power source but is based on chemical reactions [4–6].

3.1.3.2 Principle of Electroless Nickel Plating

The pre-treated metal object to be plated is placed in an electrolyte to cause nickel deposition. The electrolyte usually contains complexing or chelating agents to stabilize the nickel ions, a reducing agent to reduce the nickel ions to produce the nickel, and a pH regulator to control the acidity of the solution (Figure 3.2).

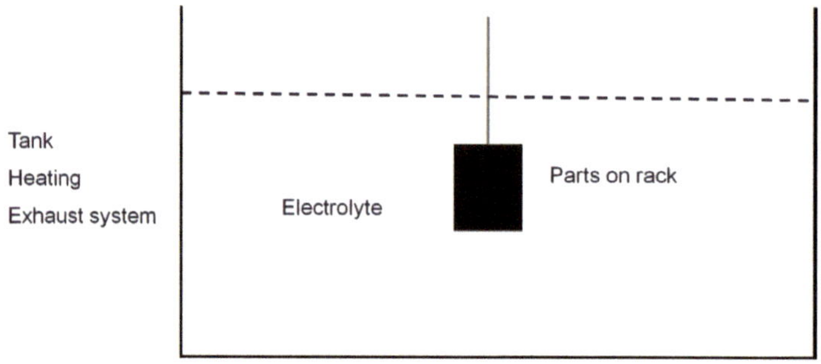

Figure 3.2 Principle of a coating basin for electroless deposition

When the metal object is immersed in the bath, the reducing agent diffuses onto the surface of the object, where it reacts with the nickel ions in the solution. The nickel ions are reduced and deposited on the surface of the metal object, forming a uniform layer of nickel. Over time, this layer thickens and a hard, dense coating is formed.

There are different types of electroless nickel coatings that are used depending on specific requirements and areas of application. In general, electroless nickel-phosphorus coatings are characterized according to their phosphorus content. These are differentiated into:

- "Lowphos" layers with 1–5% phosphorus content
- "Midphos" layers with 5–9% phosphorus content
- "Highphos" layers up to 12% phosphorus content.

3.1.3.3 Electroless Nickel Coatings with Solid Additives (Dispersion)

Electroless nickel coatings can also be produced with various additives to provide specific properties such as increased hardness, corrosion resistance, low friction, or electrical conductivity. Examples of such additives are particles of diamond, PTFE, or silicon carbide. These are discussed in more detail in Section 3.1.3.4.

The choice of electroless nickel coating type depends on the specific application in order to meet the requirements for hardness, abrasion resistance, corrosion resistance, and other properties [4].

3.1.3.4 Dispersion Layers as a Special Case

A dispersion layer consists of a metallic matrix (binding phase) and uniformly distributed solid particles (not soluble in the electrolyte) bound in it (Figure 3.3). These dispersoids can be solids as well as encapsulated solids or liquids, which thus contribute their properties to the resulting coating [5, 6]. This opens up the possibility of synergistically combining the properties of the matrix material and embedded particles. In particular, wear properties and friction behavior can be significantly changed as a result.

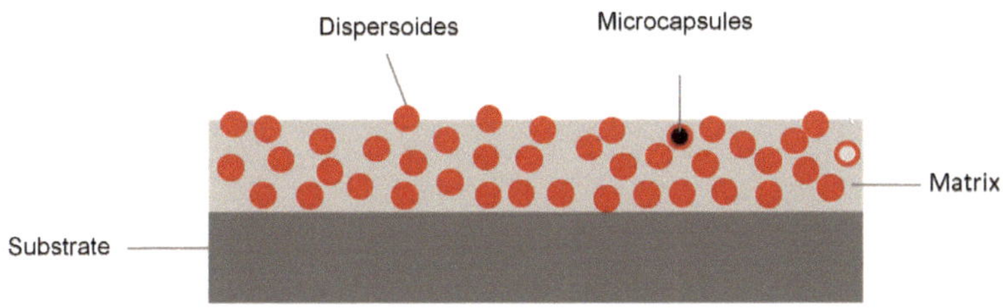

Figure 3.3 Basic structure of dispersion layers

The resulting coefficient of friction of the tribological system depends strongly on the corresponding sliding partners and intermediate materials [7, 8].

3.1.3.5 Friction-Increasing Layers

In many applications, a secure connection between two contact surfaces of detachable connections or a defined high coefficient of friction is important. The use of coatings with large particles, which ensure an additional form fit in addition to the frictional connection, leads to micro-interlocking (flat interlocking) and significantly better protection against slipping or a reduction in the required tension of a detachable connection. In particular, large inorganic particles such as diamond in a nickel matrix are used to achieve this goal. These coating systems can either be applied to connecting elements such as friction disks or directly to one of the components.

3.1.3.6 Sensor- and Indicator Layers

Indicator layers are a special type of dispersion layer. Here, special inorganic particles are incorporated into a chemical nickel matrix, which can then be identified using a spectroscopic method. The great advantage here is the ability to clearly determine the origin of the coating material. Especially for safety-relevant components that cannot otherwise be protected against product piracy, this method is a way of ensuring reliable product identification. Another application is to incorporate such indicator elements into a coating, which can then be identified by sensors in the system, for example to determine the wear condition of a system in situ. Table 3.3 provides an overview of common dispersion coatings.

Table 3.3 Overview of Common Dispersion Layers

Field of application	Coating system	Function
Friction reduction	Electroless nickel plus solid particles (PTFE, hBN)	Incorporation of solids with friction-reducing behavior Reduction of system friction coefficient
Anti-adhesive surfaces	PTFE	Non-stick properties
Improved wear resistance	Electroless nickel plus solid particles (SiC, BC, cBN, diamond)	Dispersion hardening through hard material incorporation
Increased friction	Electroless nickel plus solid particles (diamond, large grains)	Tight fit due to large particles in the contact area
Indicator layers	Electroless nickel with built-in inorganic oxide particles	Spectroscopic detection of the additives to determine the origin of the coating or identify the state of wear

3.1.4 Loading Types

For all bath processes, the method of introduction into the bath must be considered. Depending on the process, rack fixing or bulk coating are possible. Partial coating and partial covering of the parts to be coated are also possible.

For rack coating, the components to be coated are mounted with defined electrical contacting on special coating racks that serve as mechanical supports. This requires that the components have a designated position for attachment to the rack. No coating is applied at this position. This mounting point must be provided for in the design phase. A lack of this mounting point often leads to significant quality or functional losses and additional costs, which can often be easily avoided.

Some processes and components allow coating in bulk. The surface quality that can be achieved here is usually limited, but the costs are very advantageous due to simplified handling.

Partial coating, in which only part of the tool is coated, is particularly important in many cases. This can be achieved by covering with paint, covering material, sealing sleeves, or masking tape, or by partial immersion in the bath. Partial coverage can also be achieved by so-called reactor coating, where the component is partially sealed by a special construction and the bath medium is pumped through this vessel. This is used in particular for very large quantities of identical parts, because this process can be automated very well and therefore leads to a significant reduction in costs for the corresponding quantities.

3.1.5 Coating-Compatible Design

Coating plays an important role in the design of products and components in order to achieve the desired performance. However, the importance of careful planning and coating-compatible component design is often underestimated. Coating-compatible design is an important step in achieving optimum results.

The suitability of the coating for the component to be coated is particularly important in terms of cost-effectiveness and suitability for use. In addition to the design with regard to tolerances and coatings, aspects such as material selection, delivery conditions, manufacturing processes, and surface condition are also critical here. Coating-compatible design means that the coating requirements are taken into account from the outset when planning products and components. An important factor here is the selection of the material from which the component is manufactured. The material must be suitable for the desired coating and ensure sufficient adhesion. Table 3.4 shows an overview of common coatable materials for the various process types. A list of suitable surface conditions is shown in Table 3.5.

Table 3.4 Coatable Materials

Material	Electroplating	Currentless	PVD/PACVD
Steel, hard metals	Surfaces should be shiny (individual exceptions), note hydrogen embrittlement		
Iron/steel casting	Surfaces should be shiny (individual exceptions) graphite phase must be taken into account, outgassing is an undesired effect and can occur		
Non-ferrous metals, Al	Surfaces should be shiny (individual exceptions)		
Sintering, 3D printing	Closed porosity required, no foreign residues		
Brass	Yes	Yes	No
Sn, Zn, Pb, Cd phases	—	—	No
Ceramics	With activation		Individual applications
Titanium			Yes
Plastics			Decoration

Table 3.5 Material Condition for Coating

Favorable conditions	Unfavorable conditions
▪ Shiny metals, rust-free, paint-free, oxide-free ▪ Grinding, polishing, fine blasting, lapping, finish-eroded ▪ Burr-free ▪ Dense, no pores ▪ Sufficient load-bearing strength due to base material ▪ Low surface roughness – run-in effect ▪ Hard metals, sintered parts, MIM ▪ No sintered skins, leaching, cracks, fingerprints	▪ Abrasive skins, abrasive martensite ▪ Non-metallic layers from nitriding, boriding, chromalizing, burnishing, steam tempering, phosphating, eroded surfaces, compound layers ▪ Drilling residues, salt residues, grinding residues, polishing agents, blasting agents, cooling lubricant residues ▪ Color markings in the coating area ▪ Eggshell effect

If some basic aspects of the component to be coated are considered, many problems can be avoided in advance (Table 3.6).

Table 3.6 Practical Tips and Coating Guidelines

Material	▪ Steels with a high carbon content (and in particular cast materials) can impair the adhesion of the coating ▪ There is a risk of embrittlement with high-strength steel ▪ Welded components are critical (oxidation, blowholes, inclusions, outgassing, cleaning) ▪ A continuous V-joint is more favorable than an overlap joint or a spot-welded joint ▪ Combinations of different materials on one workpiece can lead to problems, e. g. during pre-treatment.
Processing status	▪ Remove machining residues ▪ Surface must be shiny-metallic.
Construction, design	▪ Functional areas are to be defined ▪ Functional surfaces to be coated as simply as possible ▪ Provide option for easy pick-up for coating ▪ External surfaces are generally easier to coat ▪ Through holes are cheaper than blind holes ▪ Rounded contours are easier to coat ▪ Small components can be coated more cost-effectively than large ones ▪ Differential design is to be aimed for ▪ Ensure good cleanability of the component (narrow gaps, holes, cooling channels, scooping geometries).

Coating-appropriate design requires close cooperation between the designer and the coating expert. The coating expert can make recommendations on the selection of the appropriate coating and the pre-treatment of the surface, while the designer ensures that the component meets the necessary requirements. It is also important that the design is adapted to the subsequent processing and assembly in order to prevent damage to the coating.

References for Section 3.1

[1] Moller, P.; Nielsen L.: *Advanced Surface Technology* (1st edn), M & M, 2013

[2] Hofmann, H.; Spindler, J.: *Verfahren in der Beschichtungs- und Oberflächentechnik* (3rd edn), Hanser, 2015

[3] Kanani, N.: *Galvanotechnik*, Hanser, 2009

[4] DIN EN ISO 4527: Autocatalytic (electroless) nickel-phosphorus alloy coatings – Specifications and test methods

[5] Sörgel, T.; Meyer, J.: Chemische und elektrochemische Dispersionsbeschichtung, *WOMag*, 2013, issue 09

[6] Meyer, J.: Eigenschaften und Anwendungen von Chemisch Nickel Dispersionsschichten, *Materialwissenschaft und Werkstofftechnik*, 2008, 39: 958–962

[7] Massler, O.; Meyer, J.; Melidis, S.: Tribologisches Verhalten von autokatalytisch abgeschiedenen Nickelschichten gegen ausgewählte funktionale Oberfläche, *WOMag*, 2019, issue 07–08

[8] Massler, O.: Reibverhalten von Dichtelastomeren im Kontakt mit funktional beschichteten Oberflächen, *WOMag*, 2019, issue 09: 44

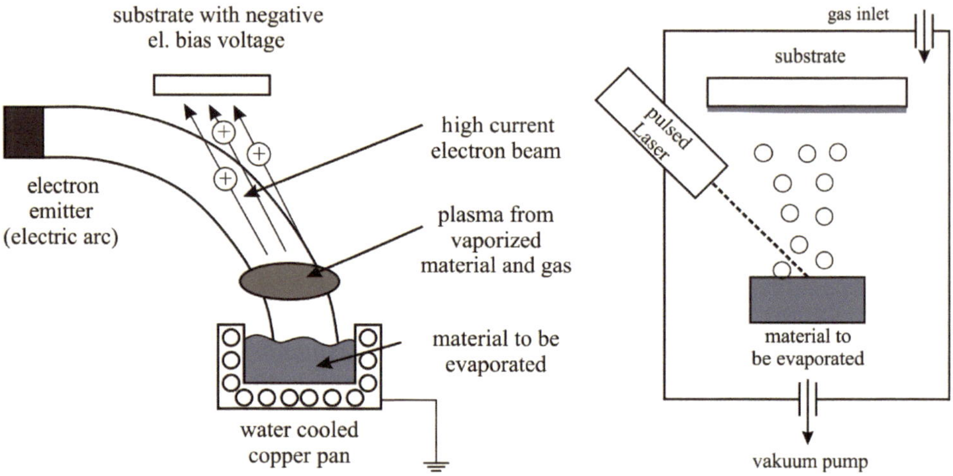

Figure 3.5 Evaporation by means of electron beam and pulsed laser (Image based on [2] and [6])

As the mean free path lengthof the atoms increases in a vacuum, the evaporated atoms do not interact with the residual gas or with each other, but hit the substrate in a straight line. There they condense as adatoms and form the coating.

High coating rates and an even distribution of the coating thickness can be achieved by evaporation deposition, which provides sufficient adhesion. However, because the atoms hit the substrate in a straight line, complex-shaped components can only be coated poorly. Temperature control must be precise, as even small temperature fluctuations lead to significant differences in the evaporation rate and therefore also in the deposition rate. The energy supply to the evaporator must be constantly adjusted, as the heat absorption of the coating material depends, among other things, on the filling level of the coating tank. The necessary parameters must be determined in preliminary tests [1, 4].

3.2.2.2 Sputtering

While the coating material is thermally evaporated during evaporation deposition, evaporation during sputtering takes place through the momentum transfer of atomic particles. For this reason, sputtering requires an inert process gas, usually argon, and a high-voltage source. The sputtering process does not require a high vacuum. The mean free path length of the argon atoms is a few millimeters at the process pressure of 0.1–1 Pa. The coating material is subject to constant bombardment during the coating process, which is why it is called the target. Collisions of the argon atoms on the surface of the coating material cause further collisions of the particles within the target, whereby particles are released from the target. These are then accelerated towards the negatively charged substrate [1, 7].

Various methods are used to carry out sputtering processes. In direct current sputtering (DC sputtering), a DC voltage is applied between the target and the substrate so that the target is negatively charged and the substrate is positively charged (Figure 3.6). Voltages between 1 kV and 5 kV are often used. A glow discharge plasma is ignited between the target and the substrate, which ionizes the process gas (argon). The argon ions now trigger particle collisions in the target. The dissolved coating particles are deposited on the substrate and form the coating [1, 7].

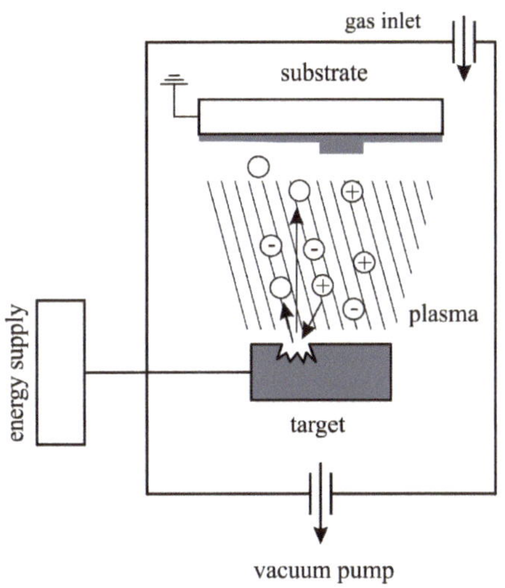

Figure 3.6
Direct current sputtering (Image based on [5])

High-frequency sputtering (HF sputtering) is characterized by the application of a high-frequency alternating current instead of a DC voltage. The ionized particles are accelerated alternately in both directions. Above a frequency of 50 kHz, the inertia of the ions is so great that they no longer follow the alternating electric field. A negative offset voltage is superimposed on the alternating current so that the ions are attracted to the target. Due to the electrons remaining in the alternating field, the degree of ionization increases further [2, 7].

With magnetron sputtering a permanent magnet is placed behind the target. The magnetic field acts within the process gas and causes the electrons to rotate on a spiral path, which increases the dwell time of the electrons in the process gas. The longer dwell time of the electrons increases the number of ions, the necessary process pressure decreases, and the deposition rate increases. As the magnetic field is static, the target is not ablated homogeneously, but, influenced by the magnetic field, “ablation trenches” form in the target [8].

If the inert process gas is exchanged for a reactive process gas, such as oxygen, nitrogen, or acetylene, chemical reactions with the target material can be achieved, enabling the deposition of oxide, nitride, or carbide coatings [7].

Only low deposition rates for the coating can be achieved by sputtering, and the coating thickness is comparatively irregular. However, the coating adhesion is good, and the interaction of the detached particles with the process gas improves the scattering power [1].

3.2.2.3 Ion Plating

With the help of ion plating (Figure 3.7), the properties of thin films can be specifically modified by the introduction of ions. For this purpose, additional anodes are placed in the gas discharge chamber, and electric fields are generated in front of the substrate and behind the evaporator. The electric fields also expose the substrate to high-energy particles, which leads to an improvement in the adhesive strength and structure of the coating, and an increase in the growth rate of the coating. The temperature of the substrate can be lowered, and the use of reactive gases for further functionalization of the coating is also possible [1].

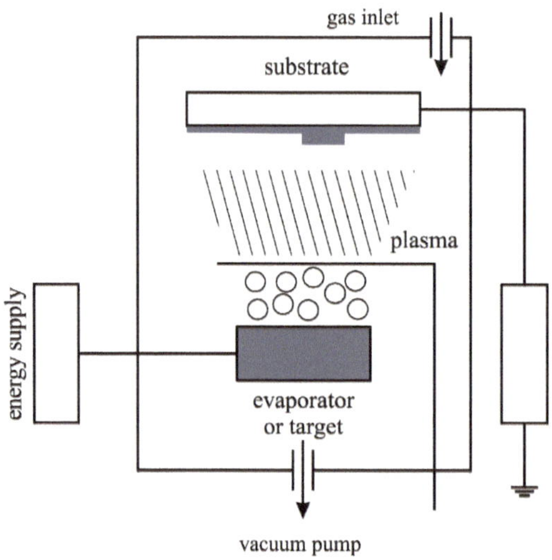

Figure 3.7
Ion plating (Image based on [5])

3.2.3 Layer Growth and Adhesion Mechanisms for PVD Coatings

The processes described below are equally valid for the deposition of PVD coatings and chemical vapor deposition (CVD) coatings. Therefore, in the subsequent sections (Sections 3.3.1–3.3.3) the deposition processes will not be further discussed [1, 9].

The growth of the coating can be divided into four phases. When atoms hit a solid surface, they are either reflected or they release energy to the surface so that they become loosely bound as adatoms. The adatoms then diffuse across the surface and can leave it again by desorption, or they condense with other atoms to form a stable nucleus on the surface. The movement of the adatoms is controlled by the temperature of the substrate, the kinetic energy of the adatoms on impact, and the strength of the interactions between the adatoms and the atoms of the substrate surface [1]. What occurs depends on the kinetic energy of the atom. If the kinetic energy is too high, the atom is reflected and cannot attach. If the temperature of the substrate is too high, the atom evaporates, so that no attachment takes place either. In addition, the bond between the attached adatom and the substrate must be so strong that the impact of further atoms does not overload the bond. A cluster is formed by the addition of further adatoms [2, 7].

In the second step, nuclei form on the condensed atoms. To this end, the adatoms migrate on the surface and form the first crystalline structures. The layer formation processes – desorption, condensation, and surface diffusion – are most important here. The nature of the coating and its surface structure can be significantly influenced by the deposition temperature. The rate at which the crystals grow depends on the temperature of the substrate and the crystal level of the nucleus. The nucleus can grow at different rates in different directions. The adatoms migrate across the substrate surface until they have found an energetically favorable position, such as surface defects or changes in the chemical composition of the substrate, and attach themselves [2, 7, 10].

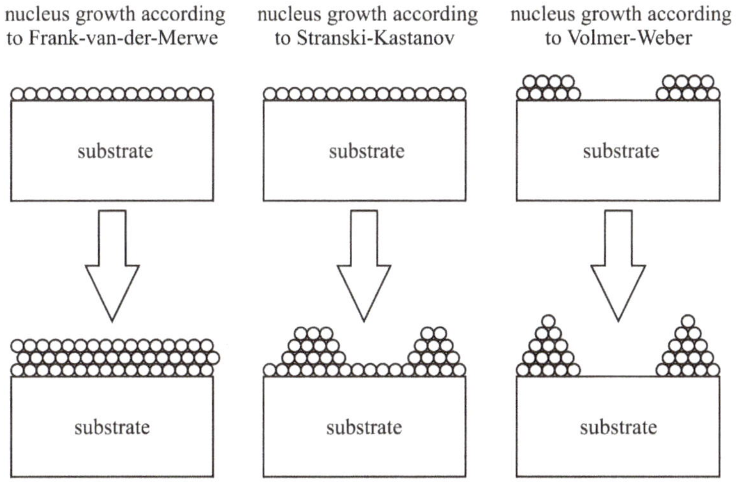

Figure 3.8 The types of nucleus growth (Image based on [7, 11])

In the third step, the nuclei grow and form a closed layer on the substrate surface. A distinction can be made between three different types of nucleus growth, which influence the growth of the layer (Figure 3.8). In the Frank–van der Merwe model the layer grows homogeneously [12]; this type of layer growth can be observed at lower deposition temperatures [11]. Taking into account that the individual nuclei have a different sublimation pressure or solubility, homogeneous growth no longer takes place. After the formation of a thin top layer, the layer grows heterogeneously according to the

Stranski–Krastanov model [13]. If no closed top layer is formed at the start of the coating process, the Volmer–Weber model can be used to describe the growth process. In this case, small coating islands are formed [14]. At higher deposition temperatures and with more lattice defects during deposition, deposition increasingly follows the Stranski–Krastanov model and the Volmer–Weber model [11].

The type of nucleation growth depends, among other things, on the roughness and the chemical composition of the substrate, the surface temperature of the substrate, the mobility of the atoms, any reactive gas used, and the complexity of the geometry of the substrate. As a rule, columnar structures are formed, regardless of whether the layer is crystalline or amorphous [2, 7].

In the final step, the layer grows over the entire surface. The growth depends strongly on the temperature and pressure (Figure 3.9, left). At low temperatures, a closed coating forms from the columnar structures. This is a few nanometers thick and forms a rough surface. Atoms cannot be placed between the columns, so that geometric imperfections occur, and the layer grows in a columnar and inhomogeneous manner (zone 1). This layer growth results in high residual stresses and low strength. As the substrate temperature increases, the adatoms can migrate further along the surface, resulting in a fine columnar structure with few defects (zone 2). The density increases and the surface roughness decreases. With a further increase in temperature, the layer can recrystallize so that stresses are relieved (zone 3). The surface roughness of this coating is very low, and the mechanical properties of this coating are the best [2, 7].

Taking into account the gas pressure during sputtering, a transition zone is formed (zone "T" in Figure 3.9, right) is formed. Due to the gas pressure, the adatoms have greater mobility and can compensate for the shadowing, so that a closed layer of thick fibrous grains is formed, which has a very smooth surface [10].

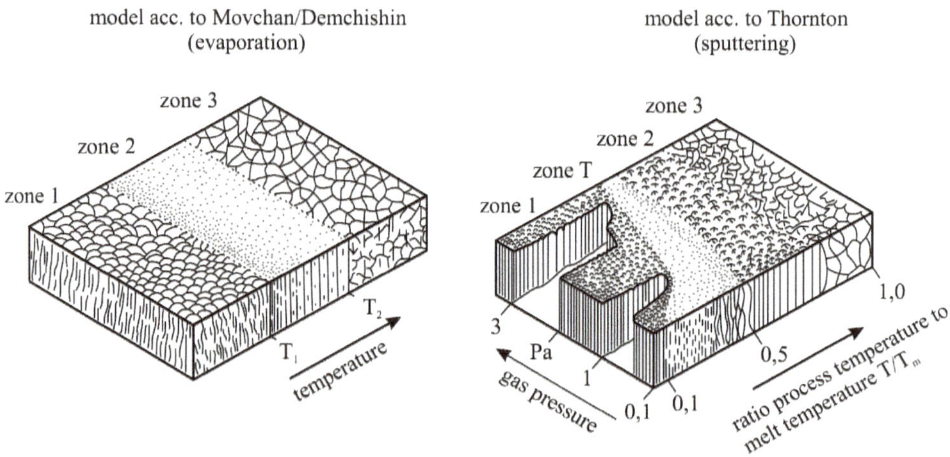

Figure 3.9 The layer zone models for deposition via evaporation and sputtering (Image based on [1, 7, 10])

With regard to the adhesion of the coating, a fundamental distinction is made between whether the coating is bonded to the substrate through mechanical, chemical, or physical interactions. Mixed forms are possible. In the case of mechanical interactions, the roughness of the substrate surface is used. As tiny undercuts form in the roughness, these can be used to anchor and hook the coating. With chemical bonding, the coating is firmly bonded to the substrate by covalent bonds, metallic bonds, or ionic bonds. The type of bond used for the adhesion of a coating depends on the substrate and the coating. If both partners are metallic, metallic bonds are formed. If the coating is doped with non-metallic elements, such as oxygen, carbon, or nitrogen, covalent bonds are increasingly formed. If very large differences in electronegativity occur, ionic bonds are formed between the substrate and the coating. Physical bonds involve dipole–dipole bonds, hydrogen bonds, and van der Waals forces, which are weaker than chemical bonds and mostly occur in plastic substrates [7, 15, 16].

Pre-treatment of the substrate is an additional option for improving the adhesion of the coating. The coating and the substrate must be considered as a composite in which both partners must be optimized in terms of coating adhesion. During pre-treatment, the surface, mechanical properties, and chemical properties of the substrate can be modified. The surfaces are often hardened by nitriding or boriding in order to change the mechanical and chemical properties of the surface. Nitriding or boriding causes the elements to diffuse into the surface. The restructuring of the metal lattice also changes the surface roughness. In addition, a brittle ceramic surface layer forms, which must be removed before coating using PVD. Other pre-treatment methods include shot peening or roller burnishing, in which the surface is plastically deformed. The aim of pre-treating the substrate is to equalize the hardness of the substrate surface and the coating in order to prevent the coating from peeling off later, due to delamination or the "eggshell effect" [17, 18].

In addition, the surface of the substrate can be structured to improve the topography of the surface. By modifying the topography, growth defects can be minimized during the coating process. The load of the coating is also distributed more evenly on the substrate surface [19].

3.2.4 Multilayer Coating Systems

The deposition of coatings using PVD is not limited to a single layer. By depositing different coatings, the properties of the coating can be better adapted to the requirements of the coating. Figure 3.10 shows a selection of coating systems that can be produced.

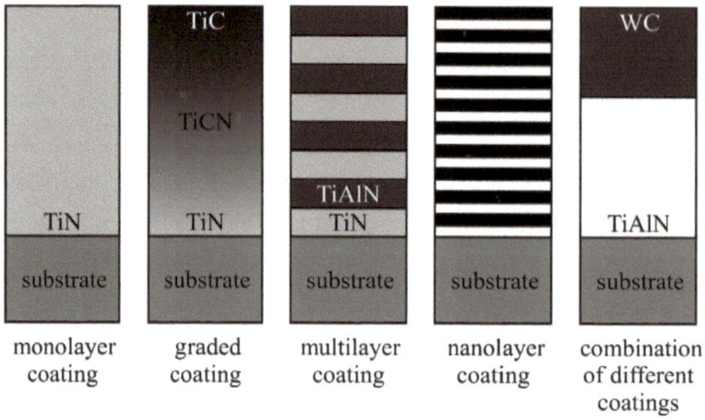

Figure 3.10 Schematic representation of selected PVD coating systems (Image based on [20])

In the simplest case, a single-layer coating is deposited. Various intermediate layers are often added to increase the functionality of the coating. A material boundary can be formed between the individual layers to create a multi-layer coating. However, the transition can also be seamless. In this case, a graded coating is created. If the thickness of the individual layers is in the nanometer range, this is referred to as a nanolayer coating. Several hundred individual layers can be deposited to form the coating system. Combinations of different layer systems are also possible [20, 21].

The combination of individual layers into multi-layer systems leads to a number of advantages. For example, the residual stresses and susceptibility to cracking in the coating are reduced. As growth defects do not spread across several layer boundaries, they also decrease with multilayer coatings. The thickness of individual layers is limited by the residual stresses in the respective layer. By using multi-layer coating systems, the thickness of the individual layers can be reduced, which also reduces the residual stresses in the layer. The layer thickness can be further increased by building up multi-layer coating systems. In addition, the functional properties of the individual layers can be combined by using and combining different layer systems. In this way, diffusion barriers and wear protection layers can be integrated in a multilayer coating [22, 23, 24].

References for Section 3.2

[1] Bach, F.-W.; Möhwald, K.; Laarmann, A.; Wenz, T.: *Moderne Beschichtungsverfahren* (2nd edn), Wiley-VCH, 2005

[2] Mattox, D. M.: *Handbook of Physical Vapor Deposition (PVD) Processing* (2nd edn), Elsevier, 2010

[3]Zimmer, O.: Harte Schichten > 20 µm – Neue Möglichkeiten für die Dünnschichttechnik [Fraunhofer IWS Annual Report], 2008, p. 74

[4] Sree Harsha, K. S.: *Principles of Physical Vapor Deposition of Thin Films*, Elsevier, 2006

[5] Pantke, K.: Entwicklung und Einsatz eines temperatursensorischen Beschichtungssystems für Zerspanwerkzeuge [Dissertation], Technical University of Dortmund, 2012

[6] Schematic representation of a PVD evaporation process, *https://de.wikipedia.org/wiki/Physikalische_Gasphasenabscheidung#/media/Datei:PVD-CVD.jpg*, created on 04 August 2010, accessed on 10 March 2023

[7] Buttler, T. A.: Verschleißuntesuchungen zu Stegoberkanten dünnfilmbeschichteter Einschnecken-extruder bei der Verarbeitung ungefüllter Polymere [Ph. D. thesis], Technical University Clausthal, 2021

[8] Kelly, P. J.; Arnell, R. D.: Magnetron sputtering: A review of recent developments and applications, *Vacuum*, 2000, 56: 159–172

[9] Unrecht, B.: Chemische Gasphasenabscheidung: Ein Verfahren zur Erzeugung heterogenkatalytisch aktiver Oberflächen [Ph. D. thesis], University of Hamburg, 2001

[10] Thornton, J. A.: High rate thick film growth, *Annual Review of Materials Science*, 1977, 7: 239–260

[11] Khazaka, R.: From atomic level investigations to membrane architecture: An in-depth study of the innovative 3C-SiC/Si/3C-SiC/Si heterostructure [Ph. D. thesis], Université François-Rabelais, Tours, 2016

[12] Frank, F. C.; van der Merwe, J. H.: One-dimensional dislocations. I. Static theory, *Proceedings of the Royal Society A*, 1949, 198: 205–216

[13] Stranski, I. N.; Krastanow, L.: Zur Thoerie der orientierten Ausscheidung von Ionenkristallen aufeinander, *Monatshefte für Chemie*, 1937, 71: 351–364

[14] Volmer, M.; Weber, A.: Keimbildung in übersättigten Gebilden, *Zeitschrift für Physikalische Chemie*, 1926, 119U: 277–301

[15] Riedel, E.; Meyer, H.-J.: *Allgemeine und Anorganische Chemie* (12th edn), De Gruyter, 2019

[16] Holleck, H.: Basic principles of specific applications of ceramic materials as protective layers, Surface and Coating Technology, 1990, 43/44: 245–258

[17] Tönshoff, H. K.; Blawit, C.; Rie, K. T.; Gebauer, A.: Effects of surface properties on coating adhesion and wear behavior of PACVD-coated cermets in interrupted cutting, *Surface and Coatings Technology*, 1997, 97: 224–231

[18] Tönshoff, H. K.; Mohlfeld, A.: Surface treatment of cutting tool substrates, *International Journal of Machine Tools and Manufacture*, 1998, 38: 469–476

[19] Neves, D.; Diniz, A. E.; Fernandes Lima, M. S.: Microstructural analyses and wear behavior of the cemented carbide tools after laser surface treatment and PVD coating, *Applied Surface Science*, 2013, 282: 680–688

[20] Wijngaard, J. H.; Schütze, A.: Entwicklungen und Trends in der Beschichtungsindustrie, *Vakuum in Forschung und Praxis*, 2003, 15: 194–197

[21] Zha, X.; Jiang, F.; Xu, X.: Investigating the high frequency fatigue failure mechanisms of mono and multilayer PVD coatings by the cyclic impact tests, *Surface and Coatings Technology*, 2018, 344: 689–701

[22] Sander, T.: Ein Beitrag zur Charakterisierung und Auslegung des Verbundes von Kunststoff-substraten mit harten Dünnschichten [Ph. D. thesis], University of Erlangen-Nuremberg, 2020

[23] Koch, R.: Stress in evaporated and sputtered thin films – A comparison, *Surface & Coatings Technology*, 2010, 204: 1973–1982

[24] Panjan, P.; Drnovšek, A.; Kovač, J.; et al.: Oxidation resistance of CrN/(Cr,V)N hard coatings deposited by DC magnetron sputtering, *Thin Solid Films*, 2015, 591B: 323–329

3.3 Chemical Vapor Deposition

3.3.1 Metalorganic Chemical Vapor Deposition

Vanessa Frettlöh

3.3.1.1 Classification of the Technology

There are various types of chemical vapor deposition (CVD) technology, which are categorized on the basis of the process conditions used. The coating material is produced by a chemical reaction from the precursor or precursor compound on the surface of the substrate during the coating process. For this purpose, the precursor is transferred into the gas phase and brought into the coating area, the so-called reactor, with a carrier gas flow and, if necessary, with the additional introduction of a reaction gas. In cold-wall reactors only the heated substrate is coated, whereas in hot-wall reactors the coating is also deposited on the inner walls of the reactor and all surfaces that come into contact with the gas.

The classic thermal CVD coating requires process temperatures between 720 °C and 1100 °C [1]. Metal salts (e.g. chlorides such as $TiCl_4$), metal hydrides (e.g. SiH_4, GeH_4) [2], silanes and acetylene etc. are usually used as precursors. This process is used to apply classic hard material coatings such as titanium nitride (TiN) or titanium carbide (TiC) to tool surfaces. The CVD process technology offers the advantages of uniform layer deposition, even in bores and holes, and very good adhesion of the coating to the substrate can be achieved [2]. The technology can also be used to fill defects, and can produce coatings with excellent wear protection and corrosion protection [3].

However, classic CVD technology at high (900–1050 °C) and medium (720–900 °C) temperatures [1] also has some disadvantages, primarily due to the increased process temperature. For example, the coating of tools requires subsequent vacuum hardening and post-treatment of the surfaces. Furthermore, there is a risk of cracking and distortion of the tools, and dimensional changes can occur [3]. Although the tools can be subsequently restored to their original dimensions, the hardness and dimensions of the tool cannot both be restored to their original state. Post-hardening changes the tool dimensions again.

By using a plasma, the coating temperature can be significantly reduced. The energy for the chemical reaction to form the coating material is supplied to the molecules by a plasma rather than thermally. PACVD (plasma-activated CVD) enables coating deposition at temperatures between room temperature and a few hundred degrees Celsius. However, the plasma ensures a directional process, meaning that the 3D capability of the coating and the coating of holes and gaps are only possible to a limited extent.

In order to utilize the advantages of thermal CVD and at the same time be able to coat at lower process temperatures, chemical compounds must be used that can be converted into the gas phase at significantly lower temperatures and activated with less energy in order to undergo the chemical reactions required to form the desired coating material. Organometallic compounds have proven to be ideal precursor substances. In contrast to classic CVD, in which the metal halides or oxides used form multi-component precursors with the reaction gases used, in metalorganic CVD (MOCVD),[1] single-component organometallic precursors are often used [5]. A large variety of chemical structures and material types are available [4], which can be adapted and tailored to the required reaction conditions by chemical synthesis [6].

By using these precursor compounds, it is possible to achieve layer deposition in the CVD process at temperatures below 500 °C, and thus also to coat injection molding tools to achieve the necessary size and hardness without changing the mechanical and geometric properties due to the temperatures in the process [7].

3.3.1.2 Processes during MOCVD-coating

During the MOCVD process, the substrate is exposed to one or more gaseous substances in a heated reactor. These accumulate on the surface, undergo chemical reactions, and finally form the coating. The processes involved can be divided into the following steps [8, 9] (Figure 3.11):

1. In the evaporator the precursor is transferred from the solid or liquid state to the gas phase.
2. With the help of a carrier gas stream the gaseous precursor is transported into the reactor.
3. Formation of gaseous intermediate products that:
   a) at temperatures above the decomposition temperatures of the precursors, react in the gas phase to form particles (gas phase nucleation/particle formation/generation of nanoparticles) (Figure 3.11, upper path)
   b) at temperatures below the decomposition temperature of the precursors, diffuse through the boundary layer (shown transparent white in Figure 3.11) above the substrate surface and pass through the subsequent process steps.
4. The gaseous intermediates adsorb on the substrate surface (blue) and react to form layered material (gray) and by-products (white).
5. The deposited atoms diffuse along the substrate surface and coating growth takes place.
6. The resulting by-products (organic residues) desorb from the substrate surface and are removed by diffusion and convection from the boundary layer.

[1] The terms "metalorganic chemical vapor deposition" (MOCVD) and "organometallic chemical vapor deposition" (OMCVD) can both be found in the literature. Within this book, MOCVD will be used for consistency [4].

By-products, particles, and unreacted precursor are removed from the reactor in the transport gas and by the negative pressure applied in the direction of the vacuum pump from the reactor.

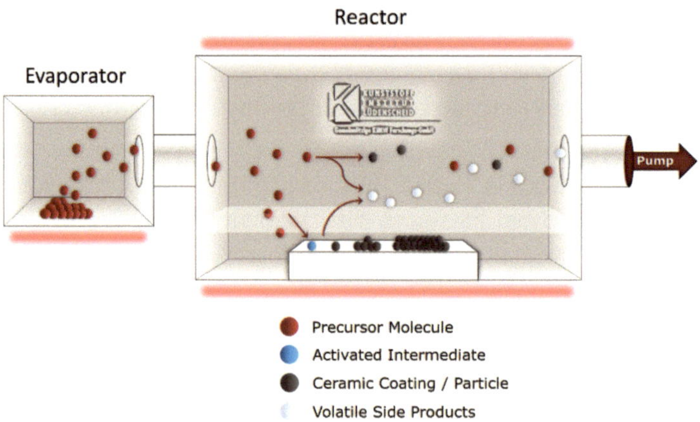

Figure 3.11 Schematic sequence of an MOCVD coating process (Image source: Non-profit KIMW Forschungs-GmbH)

Chemical reactions that take place in the gas phase can lead to increased layer purity and homogeneity if the reaction already forms an activated precursor compound. However, as shown above, gas-phase reactions can also lead to impurities if the particles formed are incorporated into the coating (see Figure 3.12 for example). The probability of such undesirable side-reactions increases with increasing precursor partial pressure and increasing process temperature due to the increased collision frequency of the molecules. By running the process at lower pressures, the mean free path length of the molecules can be increased and the probability of undesired gas-phase reactions reduced. However, this also leads to a reduction in the deposition rate [4].

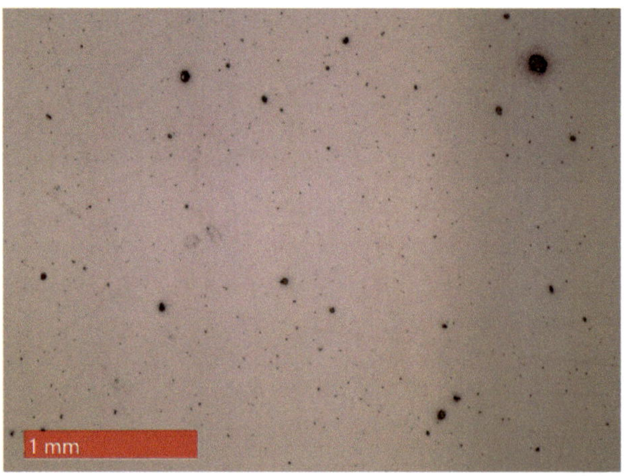

Figure 3.12 Light microscopy image of a tool surface coated with zirconium oxide; the dark gray particles have been incorporated into the coating and cause an unclean tool surface (Image source: Non-profit KIMW Forschungs-GmbH)

3.3.1.3 Requirements for Organometallic Precursors

Precursors used in industrial applications have to meet a wide range of requirements. In order to achieve stable, pure functional layers using chemical compounds, the precursors must be liquid at room temperature or easily soluble in an organic solvent. As they have to be converted into the gas phase, the vapor pressure should be sufficiently high, and evaporation should be possible without decomposition. Decomposition in the reactor should be possible at low temperatures so that the coating can also be applied to temperature-sensitive substrates. Long-term stability during storage and low sensitivity to oxygen and moisture are also important for handling in an industrial environment. In order to simplify the process sequences and occupational health & safety requirements for coating, the precursor compounds should be non-toxic, non-pyrophoric, not spontaneously flammable, and should not release any hazardous by-products during the coating process. Economic aspects also play an important role. The precursor compounds should be inexpensive and commercially available. A complex synthesis of the precursor upstream of the process is not relevant for industrial applications.

Acetylacetonates, which are available for many metal elements, fulfill most of the points mentioned above. They are available as powdery solids and can be easily dissolved in organic solvents, or conveyed directly as solids and vaporized.

Due to the metal–oxygen compound in acetylacetonates, metal oxide layers such as aluminium oxide, zirconium oxide, copper oxide, or chromium oxide can be deposited very well with this substance class (Figure 3.13). These processes are very robust, and compressed air or oxygen can be used as a carrier gas. In addition, they can also be used to deposit pure metals, e.g. in the case of nickel and copper [10, 11]. The coating process takes place in the absence of oxygen and with the addition of ethanol.

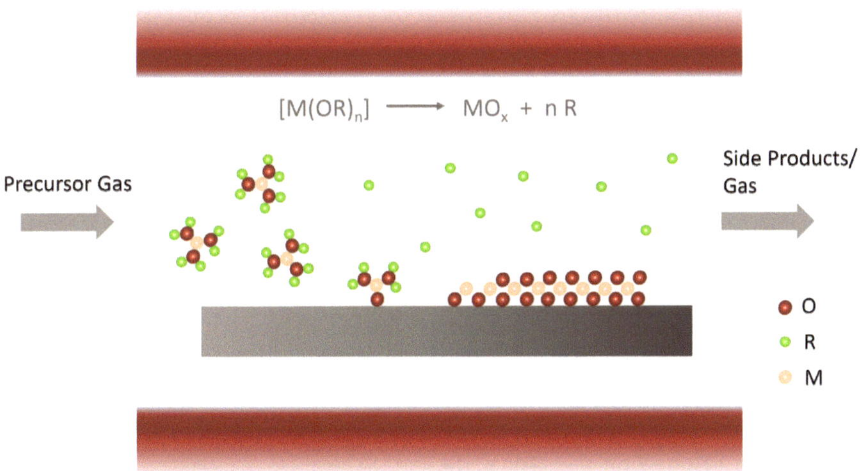

Figure 3.13 The coating deposition of metal oxides from metal acetylacetonates using MOCVD

With the liquid precursor tetraethyl orthosilicate (TEOS), SiO_2 coatings can be deposited at very low process temperatures (200–250 °C) if oxygen enriched with ozone is used as the reaction gas [12, 13]. These coatings can be used as surface protection layers for oxidation-sensitive tool steels if they are to be equipped with a functional layer that requires higher process temperatures. Elemental nickel coatings can also fulfill this purpose, as they can also be deposited at lower temperatures and in the absence of oxygen in the MOCVD process.

An overview of possible precursors used at the Non-profit KIMW Forschungs-GmbH for the deposition of various functional layers is shown in Table 3.9 (Section 3.3.2.2). Table 3.8 provides an overview of the liquid precursors used at the Non-profit KIMW Forschungs-GmbH.

Table 3.8 Overview of Liquid Precursors used at the Non-profit KIMW Forschungs-GmbH

Precursor	Isolated layer
Tetraethyl orthosilicate (TEOS)	Silicon oxide (SiO_2)
Zirconium acetylacetonate ($Zr(acac)_4$) and yttrium acetylacetonate ($Y(acac)_3$) are dissolved in a mixture of acetylacetone and benzyl alcohol (50:50 wt%)	Yttrium-doped zirconium oxide ($Y:ZrO_2$)
Zirconium acetylacetonate ($Zr(acac)_4$) and triphenyl phosphate ($(C_6H_5)_3PO_4$) are dissolved in a mixture of acetylacetone and benzyl alcohol (50:50 wt%)	Phosphorus-doped zirconium oxide ($P:ZrO_2$)

3.3.1.4 Structure of an MOCVD System

The reactors used for CVD coating deposition are as varied as the process itself. Horizontal process control is just as possible as vertical. Gas flow and sample positioning differ depending on the reactor type and coating and substrate material [14]. In order to achieve optimum coating deposition, the design of the CVD reactor is adapted to the substrates to be coated. In Figure 3.14 and Figure 3.15, a CVD hot-wall reactor for the deposition of coatings from organometallic precursors is shown schematically. In this case, a liquid precursor (or a mixture of solid precursors dissolved in solvent) is used (see Table 3.8). The precursor is fed into the reactor using liquid flow controllers (LFC), where it is converted into the gas phase and fed into the reactor with the carrier gas flow. With the help of mass flow controllers (MFCs) the amount of gas conveyed for the reaction gas and the carrier gas can be controlled. By alternately controlling two LFCs coupled with two different precursor solutions, alternating layers can be deposited and thus a multilayer coating structure can be achieved. The substrates to be coated are positioned on a metal plate in this reactor. To protect the vacuum pump, a condensation trap is installed on the system to trap the reaction products and solvents.

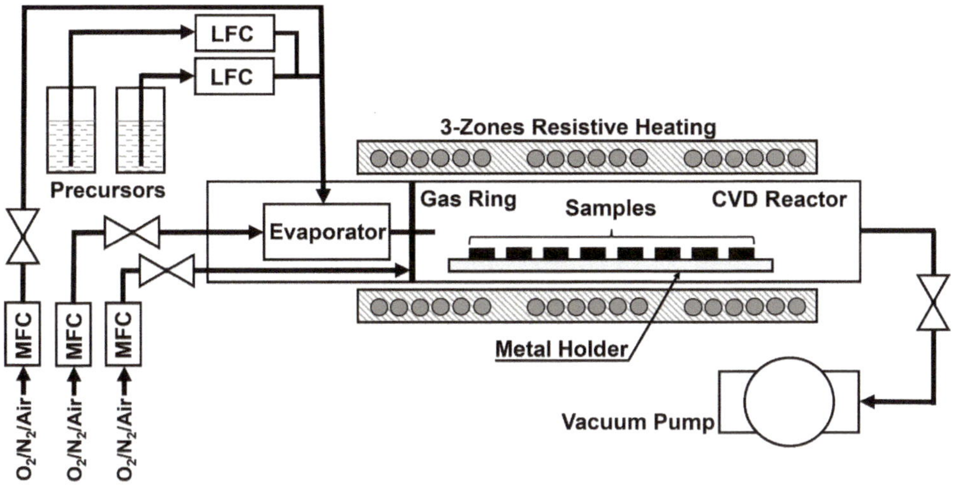

Figure 3.14 Schematic structure of a CVD hot-wall reactor used for MOCVD coating at the Non-profit KIMW Forschungs-GmbH

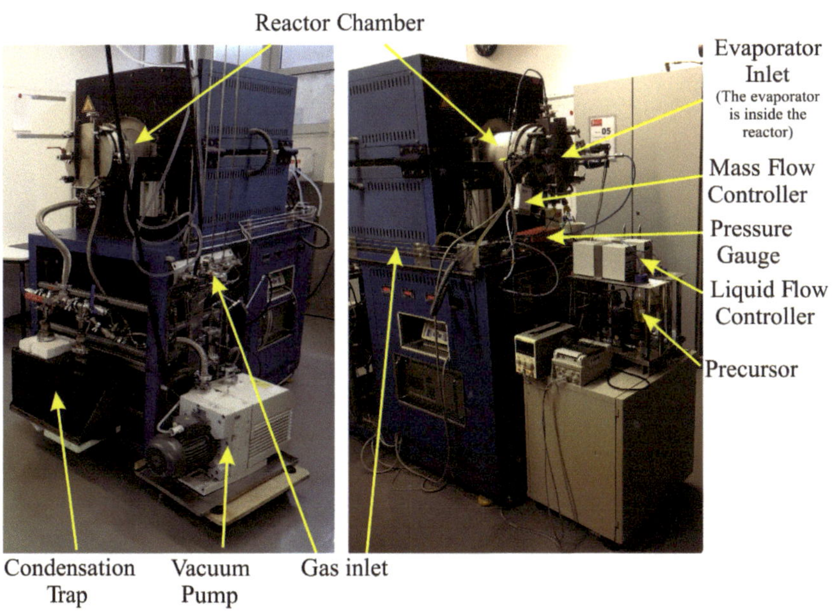

Figure 3.15 Hot-wall MOCVD system used at the Non-profit KIMW Forschungs-GmbH: Exhaust gas side (left) and precursor reservoir and reactor inlet (right)

3.3.1.5 Gap mobility and 3D capability of the coatings

The advantage of all thermal CVD processes is the very uniform layer deposition. This means that the technology is also suitable for coating tools and components with complex 3D surfaces or undercuts, or for the internal coating of hollow bodies and tubes.

Intensive and detailed investigations at the Non-profit KIMW Forschungs-GmbH have shown that a remarkable 3D capability of the coating can be achieved through a targeted choice of CVD process parameters. A homogeneous coating of non-flat tool surfaces as well as recesses and gaps with aspect ratios (opening of the gap/depth of the gap) of up to 1:60 and beyond can be achieved with this technology [15, 16]. There is a clear dependence of the 3D capability on the selected process pressure and the deposition temperature. As part of investigations with zirconium oxide-based coating systems, the 3D capability was demonstrated using different demonstrator components with various gap geometries (Figure 3.16 and Figure 3.17). The tests were carried out at temperatures between 445 °C and 540 °C and process pressures of 1–5 mbar. The gap mobility is characterized using the achievable aspect ratio. Especially in application areas with complex surfaces, gaps, and holes, such as in plastic injection molding, the gap mobility and the 3D conformal coating of the surface play a decisive role.

The demonstrator in Figure 3.16 has a thread which can be made accessible by inserting a slot and into which a metal specimen can be inserted. This then forms a part of the internal thread and thus enables measurement of the coating thickness inside the thread. With this demonstrator, aspect ratios of 1:2 can be mapped, which can be achieved by coating under the conditions mentioned. However, at a lower temperature and lower pressure, a more homogeneous layer deposition occurs on the component in the gap.

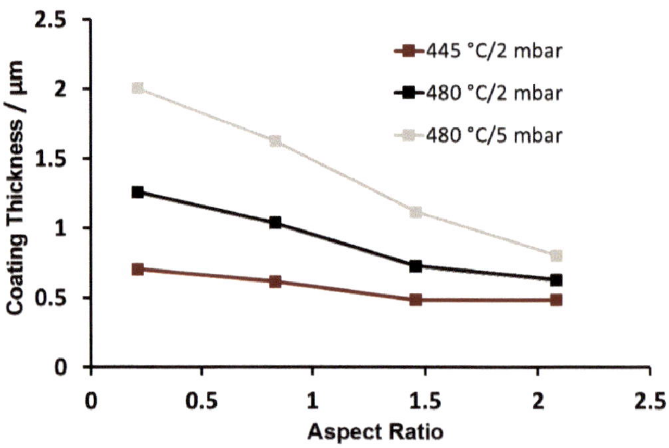

Figure 3.16 Left: Demonstrator with internal thread into which a flat test specimen can be inserted for coating thickness measurement. Right: Coating thickness achieved in the same time with different process parameters as a function of the aspect ratio in the component (Image source: Non-profit KIMW Forschungs-GmbH)

On a cube-shaped device (see Figure 3.17), test specimens can be clamped in front of indentations with defined gap widths (100, 150, 200, or 300 µm). Due to the slit depth of 10 mm, aspect ratios of up to 1:100 can be achieved with this demonstrator.

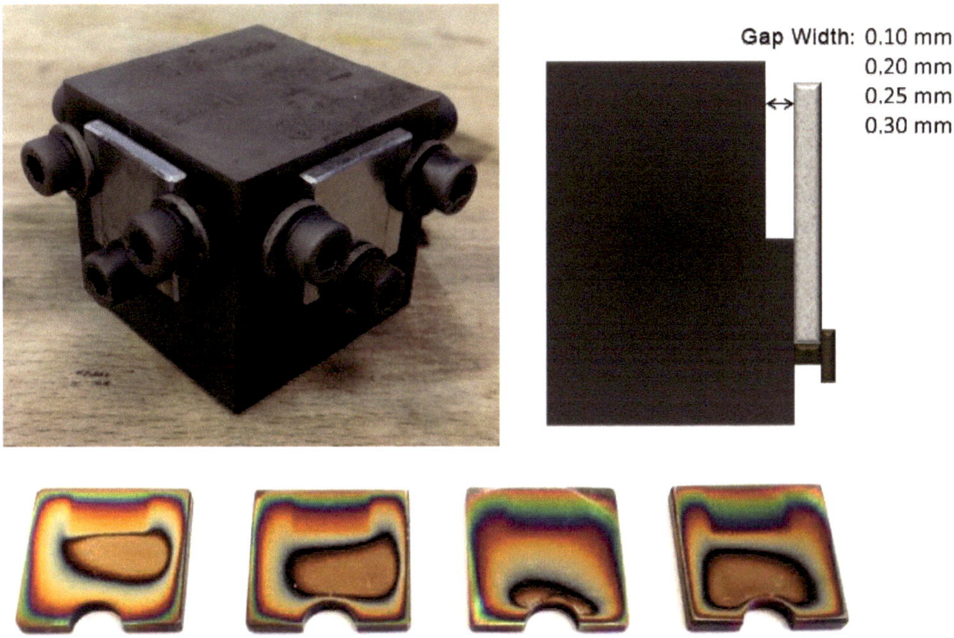

Figure 3.17 Left: Cube-shaped demonstrator component with which the gap mobility of the coating can be verified as a function of the gap width. Right: The coating thickness is determined after coating on the mounted metal test specimens as a function of the distance from the surface (Image source: Non-profit KIMW Forschungs-GmbH)

By measuring the coating thickness on the coated metal specimens as a function of the distance from the upper edge, the achievable aspect ratio and the speed of growth at the various points can be determined. The results are shown in Figure 3.18 graphically.

The comparison of the coating tests shows that a complete coating of the inner wall of the gap can be achieved up to an aspect ratio of 1:60. At higher temperatures in the range of lower aspect ratios, stronger coating growth can be seen, but the coating growth rate, i.e. the coating thickness, drops sharply with increasing gap depth. The gaps can therefore no longer be completely coated.

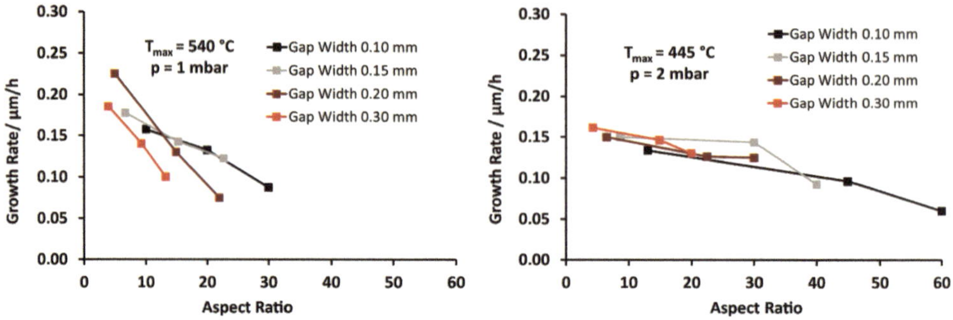

Figure 3.18 Growth rate as a function of gap width, coating temperature, and pressure, determined on the demonstrator from Figure 3.17. The achievable aspect ratio can be seen in the diagrams

The investigations show that at lower coating temperatures, which also result in a lower layer growth rate, a more uniform layer deposition can be achieved within narrow cavities. This is due to the fact that a lower temperature in the reactor and on the substrate prevents excessive reaction of the precursor molecules at the edge of the cavity. The precursor also reaches deeper areas in the gas phase and only reacts there. Lower process pressures also favor higher aspect ratios and more uniform layer deposition. Due to the longer free path of the molecules at lower pressure, they can make more collisions with the wall of the cavity before they react, which makes it possible to coat deeper gaps. Even complex components and tools can be coated to conform to contours using MOCVD technology (see Figure 3.19).

Figure 3.19 Gear tool inserts coated using the MOCVD process (Image source: Non-profit KIMW Forschungs-GmbH)

References for Section 3.3.1

[1] Erkens, G.; Vetter, J.; Müller, J.; auf dem Brinke, T.; Fromme, M.; Mohnfeld, A.: Plasmagestützte Oberflächenbeschichtung, Verfahren, Anlagen, Prozesse, Anwendungen, Süddeutscher Verlag onpact GmbH, 2010

[2] Bach, F.-W.; Möhwald, K.; Laarmann, A.; Wenz, T.: *Moderne Beschichtungsverfahren* (2nd edn), Wiley-VCH, 2005

[3] Schubert, A.: CVD: Vor- und Nachteile anhand von Anwendungsbeispielen, *Symposium on Tool Surfaces*, Kunststoff-Institut Lüdenscheid, 2019

[4] Unrecht, B.: Chemische Gasphasenabscheidung: Ein Verfahren zur Erzeugung heterokatalytisch aktiver Oberflächen [Dissertation], Hamburg, 2001

[5] Schneider, A. W.: Metallorganische Chemische Gasphasenabscheidung (MOCVD) von Übergangsmetallen am Beispiel von Eisen, Ruthenium und Wolfram [Dissertation], Erlangen, 2006.

[6] Fornalczyk, G.: Alternative Vorstufenbibliotheken der Metalle Eisen, Mangan, Cobalt und Nickel sowie deren Einsatz in der Materialsynthese [Dissertation], Cologne, 2014

[7] Fornalczyk, G.; Sommer, M.; Mumme, F.: Yttria-stabilized zirconia thin films via MOCVD for thermal barrier and protective applications in injection molding, *Key Engineering Materials*, 2017, 742: 427–433

[8] Choy, K.: Chemical vapor deposition of coatings, *Progress in Material Science*, 2003, 48: 57–170

[9] Hamden-Smith, M. J.; Kodas, T. T.: Chemical vapor deposition of metals: Part 1. An overview of CVD processes, *Chemical Vapor Deposition*, 1995, 1: 8–23

[10] Bahlawane, N.; Premkumar, P. A.; Reilmann, F.; Kohse-Höinghaus, K.; Wang, J., Qi, F.; Gehl, B.; Bäumer, M.: CVD of conducting ultrathin copper films, *Journal of The Electrochemical Society*, 2009, 156: D452–D455

[11] Bahlawane, N.; Premkumar, P. A.; Tian, Z.; Hong, X.; QI, F.; Kohse-Höinghaus, K.: Nickel and nickel-based nanoalloy thin films from alcohol-assisted chemical vapor deposition, *Chemistry of Materials*, 2010, 22: 92–100

[12] Lubnin, A.; Kudriavtsev, V.: SiO2 TEOS-ozone CVD: Mechanism, kinetic scheme and process implications, *Dumic Symposium* (Session 2F), 1998

[13] Juárez, H.; Pacio, M.; Días, T.; Rosendo, E.; Garcia, G.; García, A.; Mora, F.; Escalante, G.: Low temperature deposition: Properties of SiO2 films from TEOS and ozone by APCVD system, *Journal of Physics: Conference Series*, 2009, 167: 012020

[14] Hitchman, M. L.; Jensen, K. F. (ed.): *Chemical Vapor Deposition: Principles and Applications*, Academic Press, 1993

[15] Fornalczyk, G.: 3D-Konforme CVD-Beschichtungen von Werkzeugformeinsätzen, Kunststoffverarbeitung Deutschland, 2016, 170–171

[16] Fornalczyk, G.: Anwendungsbereiche deutlich erweitert – CVD-Beschichtung dreidimensionaler Konturen schützt Formeinsätze, *VDWF im Dialog*, 2017, issue 2: 38–39

3.3.2 Solid-Based Chemical Vapor Deposition

Dr. Ruben Schlutter

3.3.2.1 Basics of CVD with Solid Precursors

In chemical vapor deposition (CVD), the coating material is only produced during the process. This takes place through a chemical reaction. The precursor, which contains the molecules required for coating deposition in liquid or solid form, is converted into a gaseous state by an evaporator. This reactive gas is fed into the reactor together with a carrier gas, where it reacts with the substrate surface. This reaction takes place at elevated temperatures (up to 2000 °C). Depending on the process parameters, the precursor can also react before deposition, producing particles instead of a coating. In a well-adjusted process, the substrate surface acts as a catalyst so that the chemical reaction takes place close to the substrate. The reaction products are then deposited in atomic layers on the surface. The remaining reaction products are removed from the reactor by the carrier gas. Temperatures of over 700 °C are required to thermally excite and maintain the reaction. In the CVD process, several processes take place, consisting of diffusion and convection as well as chemical reactions. The most common reactions are chemosynthesis, pyrolysis, and disproportionation [1].

The chemical reactions are subject to the reactor temperature (Figure 3.20). If this is low, the chemical reaction for layer deposition takes place exclusively on the surface of the substrate (region I). The temperature dependence of the deposition rate in this range can be described by the Arrhenius approach. With increasing temperatures, the reaction rate increases until the gas flow is the limiting factor, as this is responsible for transporting the precursor. The temperature-independent plateau occurs in region II, since in this region only the gas flow influences the reaction rate and thus the deposition rate. If the reactor temperature continues to rise, the reaction is increasingly moved to the gas phase. The deposition rate decreases, as powder particles are increasingly formed that are not deposited on the substrate [2].

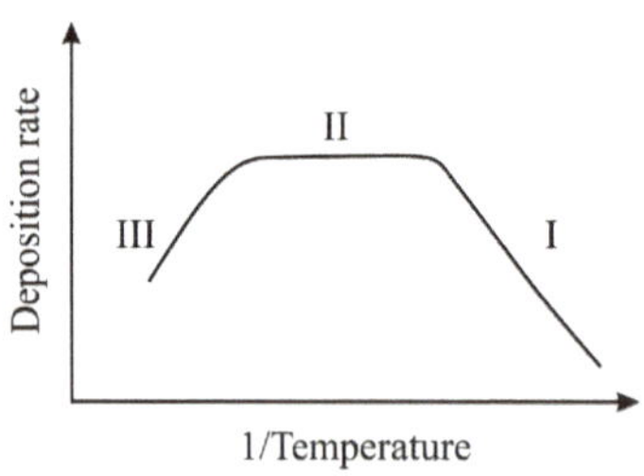

Figure 3.20
Temperature-dependent CVD characteristics (Image based on [2])

The choice of operating conditions for the chemical reaction depends primarily on the part geometry and the coating structure. If internal areas are to be coated, the kinetically controlled region I is preferred, as this ensures that the precursor also reaches the inner areas and does not react completely beforehand. If flat areas are to be coated, a high deposition rate is usually preferred, so that region II is selected as operating point. In this range, the coating is dependent on the flow of the process gas, so that no homogeneous layer is deposited. The components to be coated must therefore be rather small [2].

In almost all CVD processes, the precursor is either gaseous or liquid at room temperature. Efforts to improve and expand the use of such processes through the development of new precursors share a common problem, namely that the aggregate state of the molecular precursor changes to solid at standard temperature and pressure as the size and complexity of the precursor increases. Solid precursors are known to deliver low and temporally fluctuating concentrations in the gas phase, which causes problems in CVD processing. In addition, lab-scale solid precursor delivery systems do not achieve complete saturation of the carrier gas flowing over or through the powder's thermoregulated reservoir [3]. This situation is particularly disadvantageous for solid starting materials with relatively low vapor pressure. Numerous inorganic compounds, for example, result in mass flows through classic conveying systems that are even below the thermodynamically predicted values. For this reason, among others, they are used as starting compounds for the synthesis of molecular precursors, which are, as expected, more volatile but also more expensive [4].

A well-known solution to this problem is the injection of droplets or vapors of a liquid solution containing the solid precursor into the deposition zone [5]. However, these technologies have disadvantages such as a lower solubility of the precursor, contamination of the deposits, a limited application range (mainly related to oxide layers), and the need to handle the solvents and the effluents of the process [4].

As the material transport within the reactor is is non-directional, there is no preferred direction for coating. This means that even more complex shaped parts can be coated more or less evenly. The coating thickness is therefore largely constant. The CVD coatings also have good adhesive strength, and are generally dense.

The microscopic form of the coating can be influenced by varying the process parameters. The process temperature and the deposition time as well as the gas pressure of the process gas and the power of the evaporator have the greatest influence on the growth rate of the coating and therefore the formation and size of the individual zones. In order to avoid shading effects during the coating process, the surface roughness of the substrate can be reduced by grinding or polishing [1].

In contrast to MOCVD, the metal component to be deposited does not have to be converted into a complex compound and the precursor dissolved in order to ensure transport into the CVD reactor. As a result, fewer impurities resulting from the decomposition of the complex compound and the degradation of the solvent can be incorporated into the coating and contaminate them. In dissolved precursors, the proportion of solvent is up to 90%, which makes a condensation trap at the outlet of the CVD reactor necessary in order to collect the solvents. The decomposition of the precursor and the solvent can also lead to oiling inside the CVD reactor [6].

Despite these general advantages, the tuning of CVD processes must overcome numerous difficulties related to either the complex chemistry or technical aspects of the process. The latter include the precise control of the gaseous reactive feedstock introduced into the process chamber to produce the desired coating [7]. The use of complex metallic alloys in CVD processes intensifies these difficulties for the reliable and robust deposition of such coatings. Since these coatings consist of several elements, the starting materials must be compatible in terms of their physical, chemical, and physicochemical properties, which requires a special reactor design and an optimized technology for feeding the starting materials. If the starting materials are liquids or solids with low evaporation pressure (which is often the case with complex metallic alloys), one of the difficulties is generating the vapors from them at high velocity and in a reproducible and stable manner throughout the process [8].

Various liquid precursors are successfully used in CVD processes by feeding the precursor vapor to a carrier gas. A common method is to bubble a carrier gas through the liquid at a controlled temperature to saturate the gas phase with the evaporated precursor. Such a bubbler requires that the compound in question is sufficiently volatile at the bubbling temperature to allow a compatible and high flow rate of precursor vapors into the process chamber for the process to be viable. The flow rate of the precursor is proportional to the flow rate of the carrier gas through the precursor and the saturation vapor pressure of the precursor at the bubbling temperature, and it is inversely proportional to the difference between the total pressure in the bubbler and the saturation vapor pressure of the precursor at the bubbling temperature [9]. However, it is difficult to precisely control this mass transfer rate [10]. Therefore, when evaporating different liquid feedstocks, uncontrolled fluctuations in mass transfer in the process streams can lead to fluctuations in product stoichiometry.

Analogous attempts have been made to feed vapors of solid powder reactants into a CVD reaction chamber using the sublimator/bubbler method. This method is often problematic because it is not possible to feed a reproducible and stable stream of evaporated solid feedstock at a controlled rate [11]. Several factors, including inefficient heat and mass transfer and the changing specific surface area of the powdered feedstock, reduce the efficiency of the sublimation process.

A common problem with CVD coating is a low deposition rate, which is either due to a low vapor pressure of the liquid or solid feedstock or to transport problems associated with such low vapor pressures [12]. This relationship is straightforward if the reactor operates in the diffusion-limited region, where the growth rate is proportional to the availability of reactants in the gas phase. However, it can also affect operation in the kinetically limited regime when the availability of reactants in the gas phase is lower than their consumption at the surface.

The relationship between the saturation vapor pressure and the temperature for pure components is described by the Antoine equations. A low saturation vapor pressure of the precursor can be a thermodynamic consequence of the equilibrium between solid or liquid and gas. It can also be a consequence of early, partial, or complete decomposition of the precursor in the evaporation system during heating, which limits the evaporation temperature.

In the development of new evaporation technologies based on direct liquid injection, attempts have therefore been made to overcome the problems of uniform precursor feed by introducing a solution of the solid precursor into the reactor. Ideally, the solvent evaporates rapidly as it enters the evaporation chamber, allowing the gas phase of the precursor to deposit on the substrate. An advantage of this approach is that the precursors are kept at a lower temperature, at which they are stable, until they reach the evaporation chamber.

The equilibrium vapor pressure limits the process, but it is now higher due to the higher temperature of the evaporation chamber. Problems can arise if supersaturation leads to the formation of precursor particles. In addition, the introduction of a large amount of solvent into the reactor can lead to contamination of the layer and environmental problems. In addition, several inexpensive and potentially interesting solid CVD precursors cannot be vaporized with direct liquid injection, because it is not easy (and sometimes even impossible) to dissolve them in an organic solvent [8].

3.3.2.2 Transport of the Solid Precursor

Various approaches to conveying and dosing are described in detail in [13]. The dosing device shown in [6] is described below.

Various hurdles have to be overcome when conveying the solid precursors. As a rule, the coating time for for CVD processes ranges from several hours to several days. During this time, coatings with a thickness of a few micrometers to approx. 30 µm are built up. This results in the need for a process-reliable, constant supply of solid precursors into the CVD reactor. Figure 3.21 shows a possible dosing device for feeding the solid precursors in the direction of the carrier gas for the CVD system. The solid precursor is stored as a powder in a vertically aligned storage container. A mixing tool ensures constant conveying in the direction of the conveyor screw, and also prevents bridging over the feed area into the conveyor screw by breaking up any clumps that may form.

Conveying the small quantities required poses a further challenge. The evaporation of the solid particles should take place spontaneously in the evaporator, which is located on the carrier gas line below the nozzle. This results in a particle size of the powder between 30 µm and 300 µm. During storage in the storage container, the powder must not clump together so as not to negatively influence further conveying and thus the coating process. Compression must also not take place during conveying in the screw, as this leads to caking of the powder within the conveyor screw. This prevents further conveying. In the worst-case scenario, the caking leads to the rotary movement of the conveyor screw being blocked. Due to the compression of the powder, continuous conveying cannot be guaranteed, which means that the proportion of precursor in the carrier gas can fluctuate greatly, meaning that the layer build-up is not controlled [6].

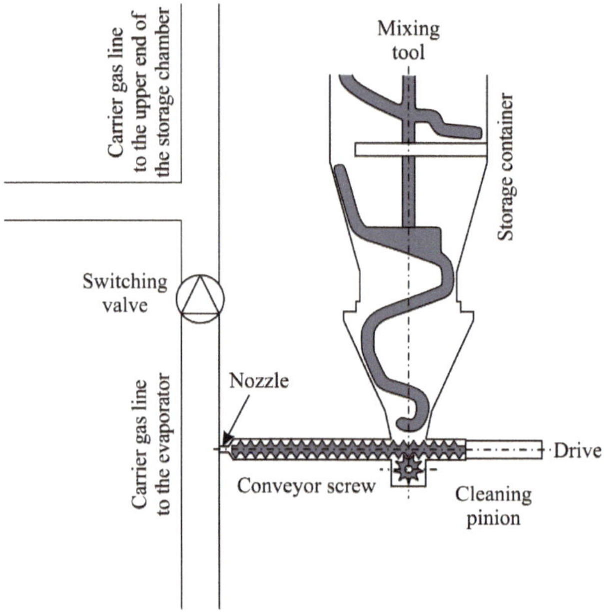

Figure 3.21 The system for conveying solids (Image based on [6])

To overcome these problems, a ventilation line is installed at the top of the storage tank. The inside of the storage tank is therefore also pressurized with carrier gas. By switching the switching valve, the gas flow can be directed through the storage container in such a way that it carries the powdery precursor in the nozzle as suspended matter. To loosen the powder in the storage container, the switching valve is closed briefly so that the gas pressure in the storage container increases. Opening the switching valve causes the powder to swirl in the storage container and in the conveyor screw, which leads to loosening. Closing the switching valve briefly interrupts the coating process. However, as the switching time is comparatively short, this interruption has no negative effect on the deposition of the coating. In addition, mixing elements are introduced into the storage container and the area of the screw feed that support the loosening of the powdery precursor. The mixing elements must extend to the lower end of the feed hopper to prevent clumping directly above the feed zone. The shape of the mixing elements, their rotary movement, and the switching intervals of the switching valve can be adapted to the requirements of the powder [6].

To prevent clumping within the conveyor screw, a cleaning pinion is installed below the feed zone. This is rotatably mounted and combs the helix of the conveyor screw to keep it clear. Ideally, the diameter of the feed zone corresponds to the width of the cleaning pinion, so that it can clean the entire width of the feed zone. The use of the cleaning pinion also ensures that the subsequent screw flights are only partially filled.

The measures described can be used to achieve the necessary low conveying rates of powdery solids, which are in the range 0.1–2 g/h. The deposition of boride, carbide, and oxide layers is therefore possible. As the dosing device requires very little installation space, several devices can also be arranged next to each other and connected in parallel in order to deposit several successive, different layers. To extend the coating process, two dosing devices can also be equipped with the same precursor so that the switch-over to the second dosing device can take place when the precursor supply of the first dosing device is exhausted. It is then possible to refill the first dosing device so that the coating process does not have to be interrupted or stopped when the storage container is completely empty [6].

The dosing device can be used to convey different solid precursors so that different coatings can also be built up. Table 3.9 summarizes various solid precursors and the coatings that can be deposited with them.

Table 3.9 Possible solid precursors (Non-profit KIMW Forschungs-GmbH)

Precursor	Deposited coating	Remark
Aluminium acetylacetonate ($Al(acac)_3$)	Aluminium oxide (Al_2O_3)	
Copper acetylacetonate ($Cu(acac)_2$)	Copper oxide (CuO)	
	Elemental copper (Cu)	
Zirconium acetylacetonate ($Zr(acac)_4$)	Zirconium oxide (ZrO_2)	
Yttrium acetylacetonate ($Y(acac)_3$)	Yttrium oxide (Y_2O_3)	Only for doping ZrO_2
Triphenyl phosphate ($(C_6H_5)_3PO_4$)	Phosphorus oxide (P_4O_6)	Only for doping ZrO_2
Lanthanum phenanthroline acetylacetonate ($La(acac)_3(phen)$)	Lanthanum oxide (La_2O_3)	
Chromium acetylacetonate ($Cr(acac)_3$)	Chromium oxide (Cr_2O_3)	

Table 3.9 Possible solid precursors (Non-profit KIMW Forschungs-GmbH) *(continued)*

Precursor	Deposited coating	Remark
Nickel acetylacetonate ($Ni(acac)_2$)	Elemental nickel (Ni)	
	Nickel oxides	With the addition of an oxygen source
Tungsten hexacarbonyl ($W(CO)_6$)	Tungsten carbide (WC, W_2C)	
	Tungsten sulfide (WS_2)	With the addition of a sulfur source
	Tungsten nitride (W_2N, WN, WN_2)	
Chromium hexacarbonyl ($Cr(CO)_6$)	Chromium carbide (Cr_3C_2)	Other chromium carbides possible
	Chromium nitride (CrN)	

References for Section 3.3.2

[1] Bach, F.-W.; Möhwald, K.; Laarmann, A.; Wenz, T.: *Moderne Beschichtungsverfahren* (2nd edn), Wiley-VCH, 2005

[2] Kienel, G.; Röll, K.: *Vakuumbeschichtung 2 – Verfahren und Anlagen*, Springer, 1994

[3] O'Brien, P.; Pickett, N. L.; Otway D. J.: Developments in CVD delivery systems: A chemist's perspective on the chemical and physical interactions between precursors, *Chemical Vapor Deposition*, 2002, 8: 237–249

[4] Vahlas, C.; Caussat, B.; Senocq, F.; Gladfelter, W. L.; et al.: Fluidized bed as a solid precursor delivery system in a chemical vapor deposition reactor, *15th European Conference on Chemical Vapor Deposition (EUROCVD-15)*, Bochum, 2005

[5] Kreisel, J.; Lucazeau, G.; Dubourdieu, C.; Rosina, M.; Weiss, F.: Raman scattering study of $La_{0.7}Sr_{0.3}MnO_3/SrTiO_3$ multilayers, *Journal of Physics: Condensed Matter*, 2002, 14: 5201–5210

[6] Non-profit KIMW Forschungs-GmbH, Dosing device for dosing a powdery substance, Patent specification EP000003699320A1, 2019

[7] Ganguli, S.; Ku, V. W.; Chung, H.; Chen, L.: Method and apparatus for monitoring solid precursor delivery, Patent specification US6772072, 2004

[8] Vahlas, C.; Guillon, H.; Senocq, F.; et al.: Solvent-free method for intense vaporization of solid molecular and inorganic compounds, *Gases & Instrumentation*, 2009, 3: 8–11

[9] Hersee, S. D.; Ballingall, J. M.: The operation of metalorganic bubblers at reduced pressure, *Journal of Vacuum Science & Technology A: Vacuum, Surfaces, and Films*, 1990, 8: 800–804

[10] Paz, D. A. C. A.; McMillan, L. D.; Solayappan, N.; Bacon, J. W.: Method and apparatus for fabrication of thin films by chemical vapor deposition, Patent US6511718, 2003

[11] Wang, L.; Baum, T. H.; Xu, C.: Delivery systems for efficient vaporization of precursor source material, Patent US6909839, 2005

[12] Vahlas, C.; Caussat, B.; Senocq, F.; Gladfelter, W. L.; et al.: An original device for the delivery of vapors to a deposition apparatus based on the sublimation in a fluidized bed, *Chemical Vapor Deposition*, 2007, 13: 123–131

[13] Vahlas, C.; Caussat, B.; Gladfelter, W. L.; et al.: Liquid and solid precursor delivery systems in gas phase processes, *Recent Patents on Materials Science*, 2015, 8: 91–108

3.3.3 Plasma-Based Chemical Vapor Deposition

Patrick Engemann

3.3.3.1 Plasmas

If energy is added to a gas, the gas atoms separate into ions and electrons. This state is often referred to as the fourth state of matter. The splitting of the atoms into ions and electrons results in the gas having different properties. As the separation can also occur sporadically due to external influences, it is only referred to as a plasma when the ionized components of the gas dominate the properties. The proportion of the ion density (n_i) to the neutral gas density (n_n) in a gas can be expressed by the degree of ionization. This can be calculated using the Saha equation (Equation 3.1). The temperature (T), the ionization energy (W_{ion}) in eV, and the density (n) in m^{-3} must be used for the calculation [1].

$$\frac{n_\mathrm{i}}{n_\mathrm{n}} \approx 3 \cdot 10^{27} \frac{T^{3/2}}{n_\mathrm{i}} \exp\left(-\frac{W_\mathrm{ion}}{k_\mathrm{B} \cdot T}\right) \tag{3.1}$$

where

n_i = ion density ($1/m^3$)

n_n = neutral gas density ($1/m^3$)

W_{ion} = ionization energy (eV)

T = temperature (K)

k_B = boltzmann constant (J/K)

In order to maintain the plasma, new ions and electrons are constantly being generated, as ions and electrons are continuously recombining to form atoms in the plasma. The energy released in this process is the cause of the characteristic glow. The energy required to maintain the plasma can be generated in different ways.

Among other things, plasmas are characterized by their ability to conduct electrical energy. Another property of plasmas is that they emit light, which is why plasmas are also used for lighting purposes, for example in the form of neon tubes. Under certain circumstances, they can also have high thermal conductivity. Furthermore, plasmas can be confined and manipulated by magnetic fields [1]. This property is used, for example, in the construction of new fusion power plants to prevent the plasma from coming into contact with the reactor wall [2].

One way to cause ionization is to apply a high voltage. This type of plasma is also known as a glow discharge. The difference in charge causes free electrons to be accelerated towards the anode. If there is sufficient free path length, the electron can absorb enough kinetic energy to release another electron from the atom when it collides with it, ionizing the atom. In order to achieve the necessary free path length, the process takes place under low pressure. The electrons that are released are accelerated towards the anode again, allowing them to ionize new atoms. The ions, on the other hand, move towards the cathode due to their charge. Due to their low mass, electrons can have a much higher temperature than ions and therefore contribute more to impact ionization than positively charged ions. Due to their special properties, plasmas are used in both CVD coating technology and PVD technology.

3.3.3.2 Plasma-Activated Chemical Vapor Deposition

PACVD (plasma-activated chemical vapor deposition, or plasma-assisted chemical vapor deposition) is a special form of CVD technology. By using a plasma, the required process temperature can be lowered to such an extent that plastics can be coated with ceramic coatings. PACVD technology is therefore used in the packaging industry on PET bottles, for example, to reduce the permeability of the plastic [3].

The lowering of the process temperature is made possible by the high temperature of the electrons in the plasma. The temperature of the electrons in the plasma can rise to several 10,000 K. However, as they have a low mass in relation to the ion, the temperature of the process gas is 50 °C. The degree of ionization for coating PET bottles with silicon oxide is less than 1%. A pulsating microwave is used to generate the plasma. is used to generate the plasma. Coating thicknesses of between 5×10^{-8} and 5×10^{-7} m are sufficient to achieve the required effect [4].

Coating systems for tool technology use pulsed and non-pulsed direct currents as well as alternating current to generate a glow discharge. In order to accelerate the charge carriers in the gas, a vortex field can also be generated by means of induction. The structure of the systems is shown in Figure 3.22. The PACVD reactor has similarities to plasma nitriding systems. The reactor wall serves as an anode for the coating process, whereas the workpieces are isolated from the wall and used as a cathode [5].

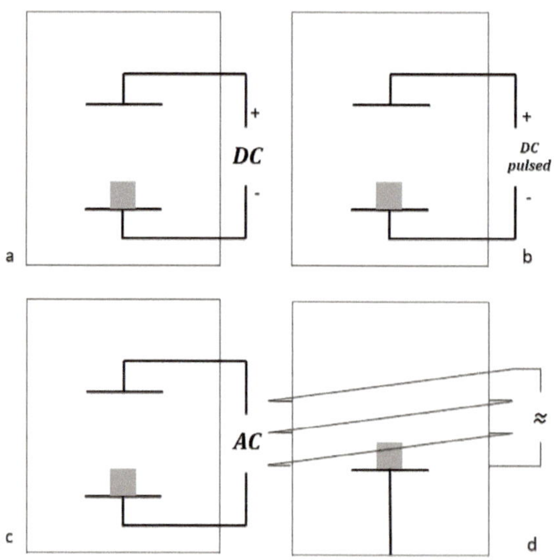

Figure 3.22
Schematic representation of PACVD: (a) DC voltage; (b) pulsed DC voltage; (c) AC current; (d) induction (Image based on [5])

With remote plasma, the gas is ignited outside the actual coating chamber. Among other things, ionization using microwaves is also used as a RPS (remote plasma source). The corresponding plasma sources have become established for cleaning and etching surfaces [6].

In the industrial environment, hard coatings such as titanium nitride (TiN), titanium carbonitride (TiCN) and aluminium oxide (Al_2O_3) are used, which can be applied at temperatures of 400–600 °C using PACVD. The necessary process energy is applied both via the temperature in the reactor and via the plasma. These deposition processes should therefore be regarded as a combination of a thermal CVD process and a plasma-activated CVD process. In addition, the application of a glow discharge enables the deposition of a carbon coating at 200 °C [7]. The amorphous carbon coatings are usually also referred to as DLC (diamond-like carbon) coatings.

The composition of the process components for the application of a TiN coating are identical to those of a classic CVD process. These are nitrogen (N_2), hydrogen (H_2), and titanium tetrachloride ($TiCl_4$) (Equation 3.2). However, argon is added to the process in order to achieve a stable plasma [5].

$$2TiCl_4 + N_2 + 4H_2 \longrightarrow 2TiN + 8HCl \tag{3.2}$$

The stabilizing effect of argon in the coating process was proposed by Archer as early as 1981 [8]. Aluminium chloride ($AlCl_3$) or trimethylaluminium $Al(CH_3)_3$ can be used for the deposition of Al_2O_3. In combination with molecular oxygen or carbon, these suppliers of aluminium form an Al_2O_3 coating on the substrate surface in the process [9].

References for Section 3.3.3

[1] Stroth, U.: *Plasmaphysik*, Vieweg+Teubner Verlag, 2011

[2] Wagner, F.: Auf den Wegen zum Fusionskraftwerk, *Physik Journal*, 2009, 8: 35–41

[3] Behle, S.; Neuhäuser, M.; Neuhäuser, M.: Damit die Qualität stimmt, *Plastverarbeiter*, 2002, 54: 50–51

[4] Kalinowski, R.: Ein Urknall in jeder Flasche, *Getränkeindustrie*, 2007, 4: 21–24

[5] Morlok, O.: *Die Kombination von Plasmanitrierung und plasmagestützter Schichtabscheidung der Gasphase (PACVD) in einem Verfahrensablauf*, Springer-Verlag Berlin Heidelberg GmbH, 1998

[6] Hellriegel, R.: Ätzen von Titannitrid mit Halogenverbindungen – Kammerreinigung mit externer Plasmaquelle [Doctoral thesis], Dresden (Books on Demand GmbH), 2009

[7] Plasmagestütze Oberflächenbeschichtung, Sulzer Metaplas GmbH, Süddeutscher Verlag onpact GmbH, 2010

[8] Archer, N. J.: The plasma-assisted chemical vapor deposition of TiC, TiN and TiCxN1–x, *Thin Solid Films*, 1981, 80: 221–225

[9] Kyrylov, O.: Abscheidung und Charakterisierung von PECVD-Aluminiumoxidschichten [Doctoral thesis], Aachen, 2003

3.3.4 Precursors – Molecular Sources for Functional Materials

Prof. Dr. Sanjay Mathur, Dr. Veronika Brune, and Dr. Thomas Fischer

Molecular precursors suitable for coating technologies are solid, liquid, or gaseous compounds from which solids can be obtained through targeted chemical transformations. These molecular species, also known as precursors, represent a variety of chemical categories which, subject to their different chemical composition, physico-chemical properties, and reactivity, can be transformed into solids with defined functional properties. Tailored organometallic molecules are therefore essential starting materials in chemical material synthesis for the controlled synthesis of functional materials (Figure 3.23) [1–5]. In general, the term "precursor" is used for reactants in chemical synthesis or biochemical processes that serve as starting materials or compounds for further subsequent syntheses or metabolic steps [6–8].

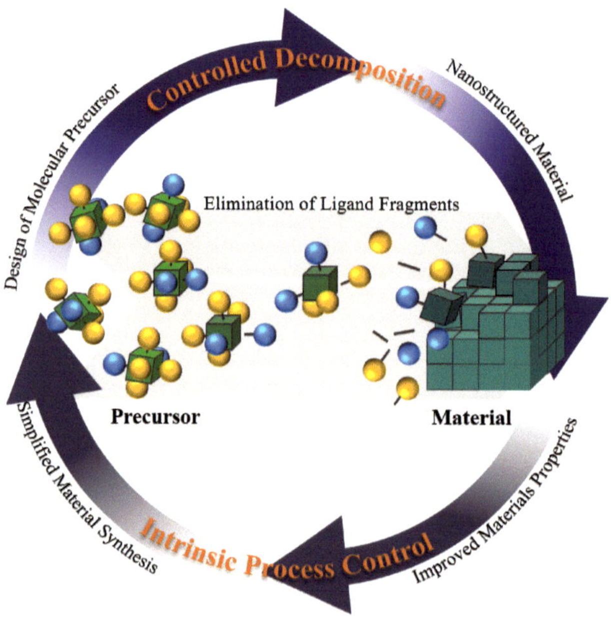

Figure 3.23 Schematic illustration of defined material preparation from single molecular precursor using gas-phase methods

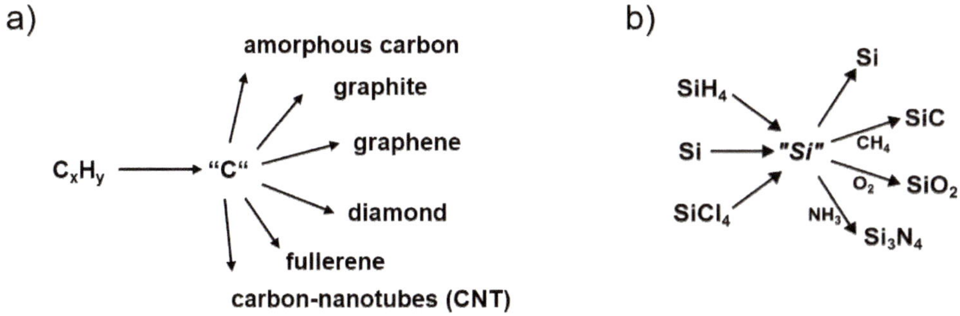

Figure 3.24 Generalized scheme of molecules as precursors for materials using the example of (a) a carbon precursor for amorphous carbon, graphite, graphene, diamond, fullerenes, or carbon nanotubes (CNTs) and (b) a silicon precursor for silicon, silicon carbide, silicon dioxide, or silicon nitride

For example, simple molecular precursors such as hydrocarbons (C_xH_y) can react under different process conditions via hydrogen elimination to form various carbon allotropes such as amorphous carbon, graphite, graphene, diamond, fullerenes, or carbon nanotubes (CNTs) (Figure 3.24a). Silicon (Si), silicon carbide (SiC), silicon dioxide (SiO_2), or silicon nitride (Si_3N_4) can be obtained from the reaction of silane (SiH_4) or silicon tetrachloride ($SiCl_4$) with hydrogen (H_2), methane (CH_4), oxygen (O_2), or ammonia (NH_3) (Figure 3.24b). However, not every chemical compound can serve as a precursor, and a number of essential criteria need to be met in identifying a suitable molecular starting material. The salient requirements and qualifies for molecular precursors are:

1. **Controlled conversion into the desired target material**

   Under given reaction conditions, only the desired material with a defined composition, structure, and morphology should be accessible in a reproducible manner without formation of undefined product mixtures that are difficult to purify. This can be facilitated by pre-formed structural units in the precursor (M–C for metal carbides, M–O for metal alkoxides, M–N for metal nitrides, etc.), that remain stable during the process conditions (Figure 3.25).

   The type of functional groups in molecular precursors can determine the process parameters. For example, the oligomeric or polymeric structures of the precursor can change the physicochemical properties such as vapor pressure (volatility) or solubility.

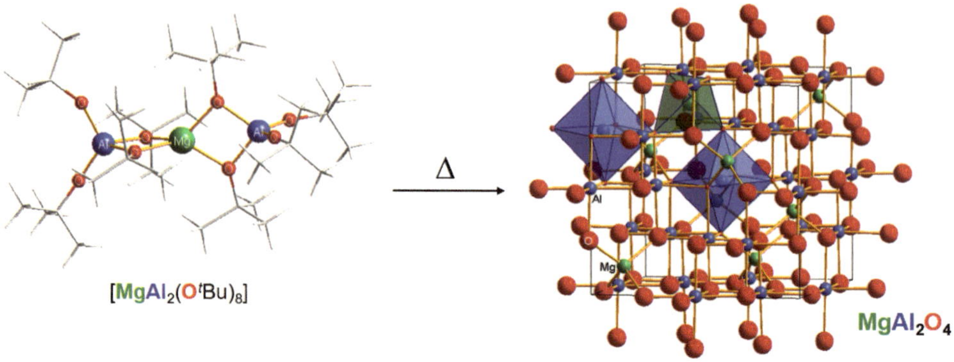

Figure 3.25 [$MgAl_2(O^tBu)_8$] as precursor for the synthesis of the spinel $MgAl_2O_4$ [9]. The preformed structural units ((R)O–Al–O(R)–Mg–O(R)–Al–O(R)) are already present in the precursor, and are retained during the decomposition process due to the stable metal–oxygen bonds

2. **Specific molecular properties and reactivity for a chosen material synthesis method**

   Depending on the requirements of the processing technique, the precursors are required to possess good solubility in the solvent used for liquid-phase processing (e.g., sol–gel process, solvothermal synthesis, etc.) or exhibit sufficiently high vapor pressure for gas-phase processes (e.g. CVD, ALD, etc.). In addition, the reactivity should be matched to the process conditions in order to avoid premature decomposition or undesirable side reactions during the transport and activation steps.

   In this context, the electronic or steric stability of a suitable precursor must be considered. The electronic stability describes a stable redox behavior of the precursor, whereas the steric shielding of the metal center in combination with repulsive intermolecular interactions among precursor molecules determines whether they have an appropriate volatility. The organometallic compounds in which the metal centers are shielded or stabilized with sterically demanding ligand[2] scaffold usually lead to increased fragility, which results in premature decomposition of the precursor molecule.

3. **Formation of stable and easy-to-separate by-products**

   The by-products formed during molecular precursors must either have a high vapor pressure or good solubility in the solvent used, to ensure easy separation from the targeted solid material. In addition, the by-products formed during the

2 A ligand is an atomic or molecular fragment that is bound to a central metal atom via a covalent or coordinative bond, or (in the case of bridging or chelating ligands) that can coordinate to more than one metal center.

decomposition process should have a low reactivity towards the target material, in order to prevent side-reactions or contamination. For example, intramolecular C–C or C–H bond activation in the organic ligand framework of suitable precursors favors the formation of by-products that are easy to remove from the reaction zone (e. g. β-hydride eliminations during the thermal decomposition of metal alkoxides, Figure 3.26) [10]. Additionally, the formation of stable, non-toxic, and non-corrosive by-products also simplifies process control and post-treatment.

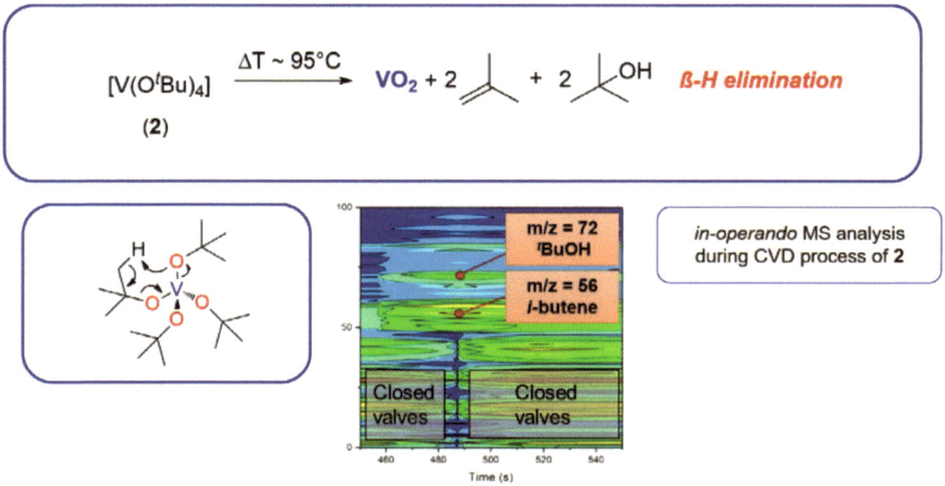

Figure 3.26 Thermal decomposition of a vanadium alkoxide ([V(O^{t}Bu)$_4$]) to VO_2 with mass spectrometric characterization of the generated by-products (isobutene and butanol) (Image source: [10], with permission)

4. **Minimum necessary complexity in structure, and synthetic access**

   Precursors should be synthetically accessible by simple chemical reactions from readily available starting materials with high atom economy and high yields. A mononuclear molecular structure allows a better understanding of the conversion reactions to the target material and reduces unwanted side-reactions. In addition, straightforward chemical pathways also reduce the effort and costs for the production of precursors.

Considering the above-mentioned general characteristics of an “ideal” precursor, real molecular precursors always need to be evaluated in relation to their clean and direct transformation to the desired target material, as well as boundary conditions for material processing. In a typical inorganic “retrosynthesis”, suitable precursors can be identified for the target material and specific process conditions, as shown in Figure 3.27.

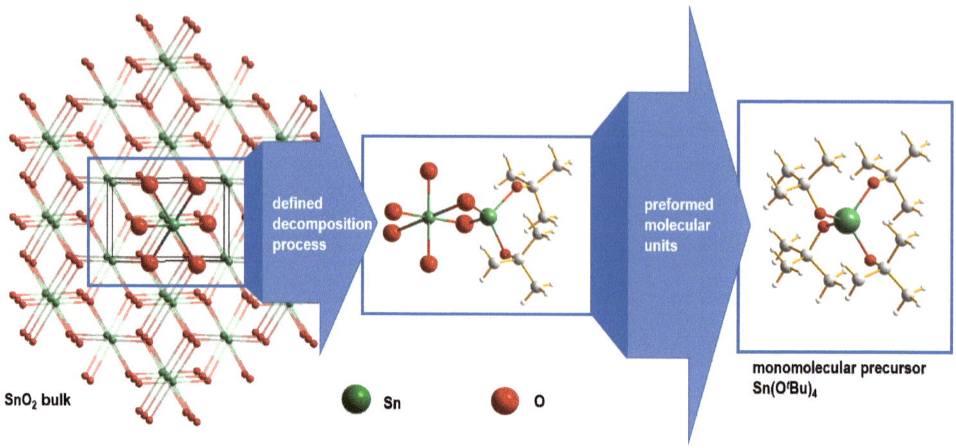

Figure 3.27 Inorganic retrosynthesis for the preparation of SnO_2, starting from the monomolecular precursor tin *tert*-butoxide ($[Sn(O^tBu)_4]$)

In describing the precursor concept, a general distinction is made between single and multi-source precursors. In *multi-source precursors* (MSPs), several chemical compounds react to form the target material, whereas in *single-source precursors* (SSPs), all necessary material constituting components are pre-organized in a single molecular unit.

3.3.4.1 Chemical Strategies in Material Synthesis

The reaction parameters are specified by the choice of precursor and its properties. If the selected precursor has good solubility in common organic or inorganic solvents, the desired target material is accessible by a liquid-phase synthesis (sol–gel, solvothermal, etc.), and the formed by-products can be easily removed by washing with the corresponding solvents. If the selected precursor possesses suitable volatility at appropriate temperatures and pressures, good stability in the gas phase, and a clean ligand elimination mechanism, it is suitable for material processing from the gas phase.

By using a tailored chemical precursor, the properties of the target material can already be influenced at the molecular level.

MSPs (also known as multi-component systems) (Figure 3.28) consist of sources of at least two elements present in the target material (e.g. $SnCl_4$ as the tin source and H_2O as the oxygen source in the synthesis of SnO_2). A multi-component synthesis is typically diffusion-driven, and so usually requires higher temperatures and longer reaction times to form the thermodynamically favored solid-state structure. The bonds and building units required for the final material have to be formed in the thermodynamic equilibrium state by diffusion-controlled processes (in the liquid or gas phase).

In traditional synthesis routes using MSPs, the different properties of the starting materials (solubility, volatility, and stability) can lead to the formation of inhomogeneous products, after which subsequent processing steps are necessary, such as fine grinding, calcination, or further decomposition steps. In addition, the entropy-favored segregation of the starting materials counteracts the formation of thermodynamically metastable phases in particular.

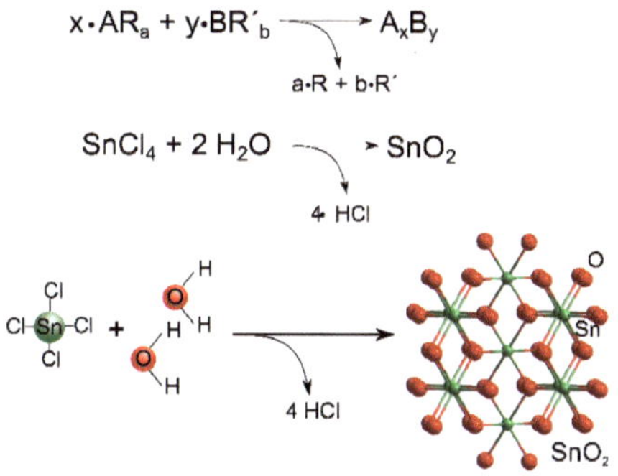

Figure 3.28 MSP synthesis to the target material SnO_2 using $SnCl_4$ and H_2O as starting components

In contrast, organometallic compounds that already contain all the components of the target material at the molecular level belong to the class of single-source precursors (e.g. $Sn(OR)_4$ as a precursor for SnO_2, or $Ti(OR)_4$ as a precursor for TiO_2, where R = alkyl). These molecules are attracting increasing attention in the synthesis of specific materials. These SSPs consist of a metal core surrounded by organic ligands (e.g. $[Sn(O^tBu)_4]$, Figure 3.29).

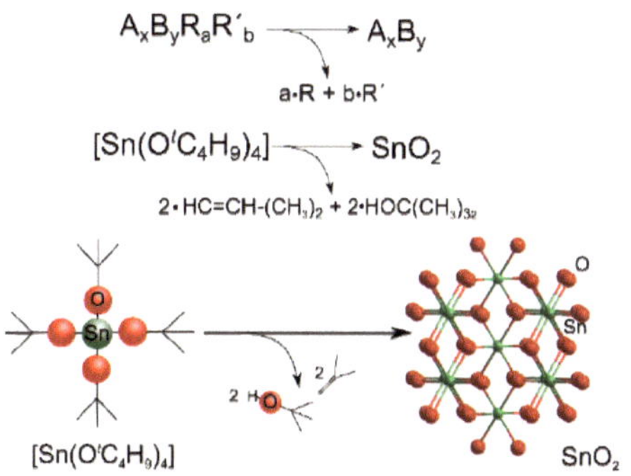

Figure 3.29 Single-source precursor (SSP) synthesis to the target material using the example of SnO_2 starting from $[Sn(O^tBu)_4]$

This predefined metal–ligand interaction enables the formation of the solid material phase under mild conditions, since the decomposition of the precursor requires a low activation energy due to specifically pre-constructed molecular building units that contain the material-specific interatomic bonds. The transformation of a molecular precursor into a defined solid state involves the structural rearrangement and elimination of the organic ligands. The metal–ligand framework must already be constructed in a balanced way at the atomic level so that the corresponding ligand fragments can be completely eliminated after the desired reactions, and can be easily removed from the reaction process, all while maintaining the pre-constructed components. This is easier to achieve if the ligand fragment to be eliminated itself forms molecular by-products (e.g. CO, alkanes, alcohols). Specially designed organometallic precursors allow a high degree of control over the composition, structure, and functional properties of the target material. In particular, the molecular design of precursors enables kinetically controlled material synthesis in order to be able to produce metastable phases in a targeted manner [9–19].

The production of materials and substances using molecules as a preliminary stage is based on the **transformation of molecules into materials**, which means that chemical or physical methods can be used to exert a targeted influence on the decisive building units on a molecular scale.

A comparison of the properties of commercially available precursors for the production of SnO_2 materials reveals significant differences (Table 3.10). The oxygen-to-tin ratio, the carbon content, and the physical properties (solubility, volatility, and aggregate state) vary in the individual precursors.

Table 3.10 Overview of Some Commercially Available Precursors for the Preparation of SnO_2

	$Sn^{IV}(CH_3)_4$	$Sn^{IV}Cl_4$	$Sn^{II}(acac)_2$	$Sn^{IV}(OAc)_4$	$Sn^{IV}(O^tBu)_4$
Category	Metal alkyl	Metal halide	Metal acetyl-acetonate	Metal carboxylate	Metal alkoxide
Structure	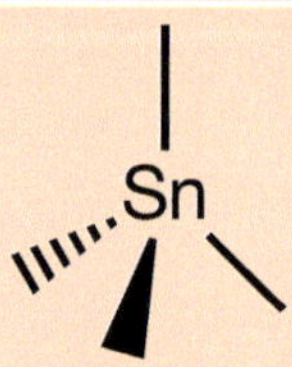	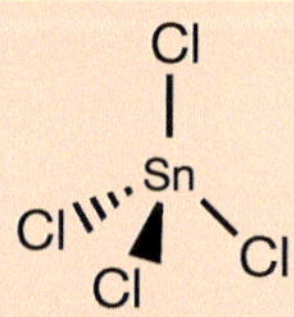	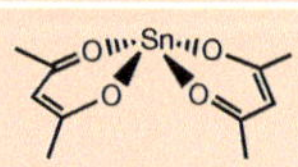	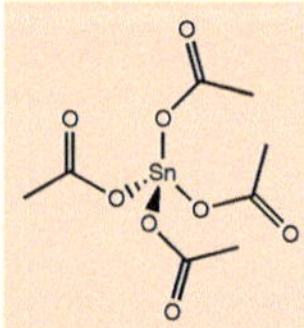	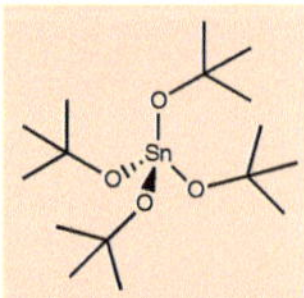
Mass (g/mol)	180.00	259.78	317.99	355.96	412.16
Sn content (%)	66.37	45.57	37.46	33.45	28.87
Sn:O ratio	1:0	1:0	1:4	1:8	1:4
Use as a precursor	Multi-source	Multi-source	Single-source	Single-source	Single-source
Binding type	Covalent	Ionic	Covalent/ coordinative	Covalent/ coordinative	Covalent

Ionic compounds, for example, have a high solubility in polar solvents, which favors their use in solution-based manufacturing processes, but their use in gas-phase methods is severely limited due to their low vapor pressure. In contrast, covalent molecular compounds usually have less solubility in organic solvents, which limits their use in solution processing, but favors their application in gas-phase synthesis, due to their increased volatility.

For an appropriate volatility of the molecular precursor, the degree of molecular aggregation should be minimized. This is achieved in particular by using (poly)dentate ligands (Figure 3.30a) or by introducing sterically demanding groups[3] (Figure 3.30b) in a favorable position, which prevent an attack on the metal center or oligomerization of the precursor.

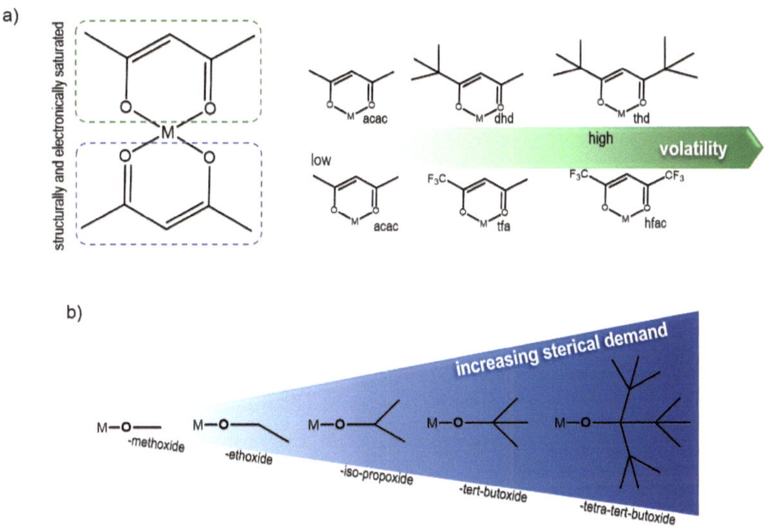

Figure 3.30 Influence on the volatility of precursors by (a) donor ligands with different functional groups (-CH_3, -^tBu, -CF_3; acac = acetylacetonate; thd = 2,2,6,6-tetramethylheptane-3,5-dionate, dhd = 2,2-dimethylhexane-3-5-dionate, tfa = trifluoroacetylacetonate, hfac = hexafluoroacetylacetonate) and (b) by increasing the steric demand of the alcoholate ligands

Although huge research efforts in the field of organometallic chemistry using β-diketonates as donor ligands for novel precursors have been undertaken, many CVD processes based on this precursor class still result in unfavorable ligand fragmentation during decomposition, leading to impurities in the resulting materials [20, 21].

3 Steric demand means that the substituent or molecule takes up a certain amount of space. For example, a methyl group (-CH3) is sterically less demanding (i.e. takes up less space) than a tert-butyl group (-C(CH3)3). The higher the steric requirement, the more likely it is that the metal center, for example, will be shielded from "attacks", as the functional group's size prevents it from getting close enough to the metal center.

The substitution of -H, $-CH_3$ or $-CR_3$ groups by fluorine-containing groups generates repulsive inter- and intramolecular interactions of the precursor molecules due to the increased electron density at the perfluorinated groups. The influence of perfluorinated groups on the volatility of precursor molecules has been extensively studied in the chemistry of β-diketonates [22], and could be transferred to the heteroaryl-alkenolate ligands that were introduced by Mathur et al. and that have been continuously expanded in recent years (Figure 3.31) [10, 15, 17, 23–29]. Sterically demanding ligands, or steric influence and electronic effects (through the introduction of electron-withdrawing fluorine atoms), support the shielding of the positively charged metal ion, thus allowing intermolecular interactions to be reduced. In addition, the polarizing forces of the metal ion can be minimized by charge transfer using donor ligands, thus further lowering the intermolecular interactions and increasing the volatility. If the electronic and steric shielding is combined in one ligand, the volatility and thermal stability of the molecular precursors can be optimized [23, 24, 30].

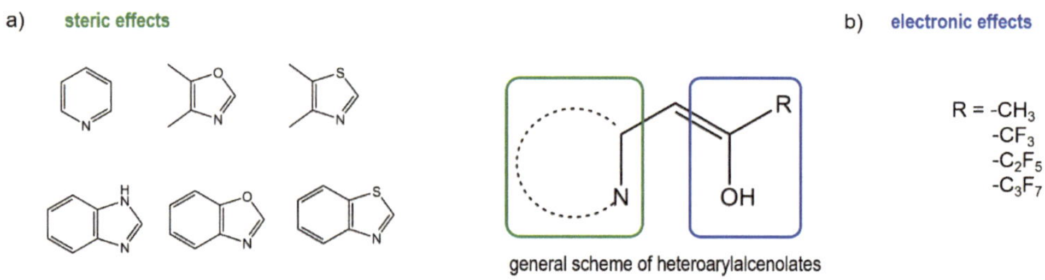

Figure 3.31 Heteroarylalkenolates and the variable influences of (a) the steric effect and (b) the electronic effect

Synthetically accessible molecular compounds allow the modification of physical and chemical properties by optimizing the ligand framework (e.g. β-diketonates, alkoxides, and heteroarylalkenolates). For example, the complexes shown in Figure 3.32 show increasing volatility through the introduction of perfluorinated units to the donor ligands, despite some high molecular masses. The boiling points of the alcohols used for the synthesis of these various donor ligands already exhibit significantly different physical properties. $HOC(CF_3)_3$ (45 °C/760 Torr, M = 236 g/mol) has a lower boiling point than $HOC(CMe_3)_3$ (85 °C/1.5 Torr, M = 200 g/mol), which implies a better volatility of the fluorine-modified ligands.

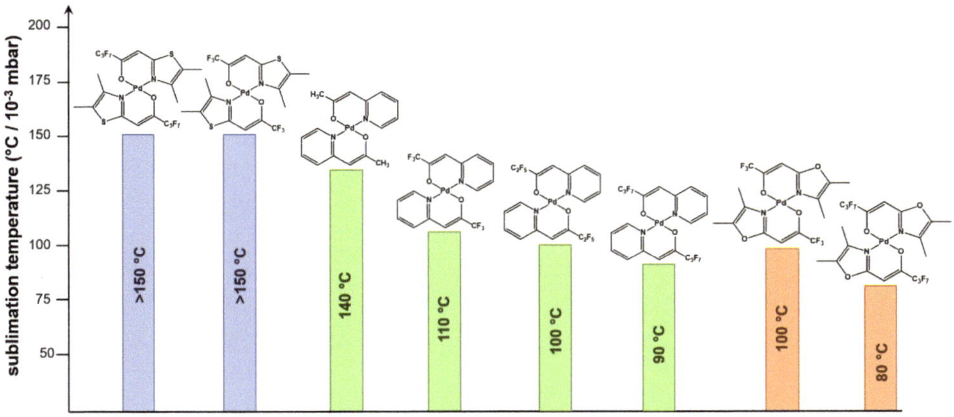

Figure 3.32 Influence of the sublimation temperatures of modified precursor molecules with perfluorinated functional groups, exemplified by palladium complexes [31]

The nuclearity (degree of molecular association) of the precursor molecules also influences their volatility. High nuclearity is associated with a high molecular mass, which in turn results in an increased sublimation temperature and thus a negative effect in gas-phase processing [32]. The nuclearity of the precursor compound can be influenced either by introducing sterically demanding ligands coordinated to the metal center, or alternatively by the introduction of ligands with additional (neutral) donor sites. If, for example, the steric requirement of the alkoxo ligands is reduced by decreasing the size of alkyl chains, the molecular mass of the monomeric unit of the precursor molecule decreases, but at the same time, the reduced shielding of the metal center leads to an increase in the nuclearity of the molecular compound. This results in the formation of dimeric, trimeric, or polymeric molecular units, thus enhancing the molecular weight of the precursor unit (Figure 3.33a). To counteract this undesirable effect, specific donor ligands can be introduced into the molecule, which help to increase the steric shielding and repulsive effects and thus enable the formation of a stable monomolecular compound (Figure 3.33b) [33]. These donor-stabilized monomolecular compounds usually exhibit good volatility and high stability in the gas phase.

By influencing the molecules through special design, it is possible to optimize the process parameters and material formation at the molecular level. The targeted pre-structuring of bonds and the controlled embedding of functional groups is an area of research that contributes to the continuous development of (functional) materials as well as the optimization of process conditions, handling, and efficiency. The enormous interest in the development and improvement of molecular complexes as precursors for functional materials is demonstrated by the large number of publications in this field [11, 12, 16, 19, 25, 35–41].

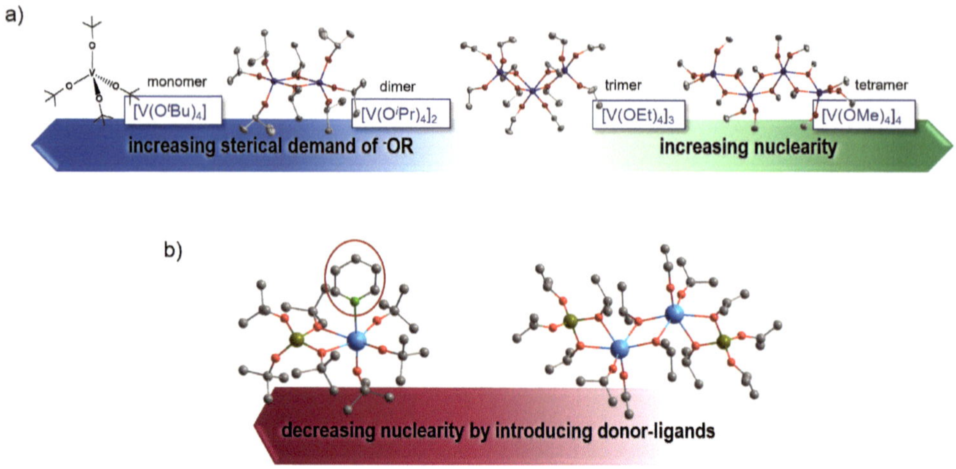

Figure 3.33 Influence of precursor nuclearity through (a) increased steric requirement of ligand scaffolds and (b) introduction of donor ligands (here pyridine, marked with a red circle) [33, 34]

However, which processing route is the most efficient for which material depends on a variety of parameters, such as the availability of resources, available methodologies, previous synthetic knowledge and infrastructural requirements, and the desired quantity and morphology of the target material.

Using the example of tin(IV) oxide (SnO_2), the approach of analyzing commercially available molecules (Table 3.10) as a precursor for material synthesis will be explained in more detail below.

SnO_2 crystallizes in the rutile structure, where the tin atom (with the formal oxidation state +4) is 6-fold coordinated by oxygen atoms and each oxygen atom is 3-fold coordinated by tin atoms; in a first approximation, the first coordination sphere of tin consists of oxygen atoms. An ideal precursor must therefore provide tin atoms already coordinated with oxygen for phase formation (Figure 3.27–Figure 3.29).

Different routes can be selected for the production of SnO_2, depending on the requirements of the precursor used:

- For the production of SnO_2 materials using CVD methods, volatility (atmospheric-pressure CVD (APCVD), low-pressure CVD (LPCVD) and atomic layer deposition (ALD)), as well as the decomposition profile of the precursors, has to be considered. Therefore, decision of suitable precursor must be based on the MSP or SSP approach, respectively [8].
- Most examples for the deposition of tin oxide materials from the gas phase are reported using dual-source routes using O_2 or H_2O as oxygen sources in combination with inorganic tin halides (e.g. $SnCl_4$), organometallic compounds of tin (involving a direct Sn–C bond, e.g. tin tetra-alkyls (SnR_4, where R = alkyl)) and

organometallic compounds (compounds in which no direct Sn–C bond exists, but organic ligands are bound to the tin center, e.g. tin acetylacetonate ($Sn(acac)_2$)) as a tin source).

- Tin tetrachloride ($SnCl_4$) is a liquid and has a sufficiently high volatility to produce SnO_2 from the gas phase. The targeted deposition of SnO_2 thin films by APCVD has been performed using $SnCl_4$ and atmospheric oxygen at 350–520 °C [42]. SnO_2 layers were obtained by ALD methods at 180–300 °C using the precursors $SnCl_4$ and H_2O [43]. Thin films of SnO_2 were obtained in CVD experiments using $SnCl_2$ and O_2 at 900–1400 °C [44, 45].
- Organotin compounds exhibit a high volatility, due to their weak intermolecular interactions, which underlines their preferred application in dual-source gas-phase processes [46–49]. Using tetramethyl tin ($Sn(CH_3)_4$, TMT), nanostructured SnO_2 thin films were produced at 550 °C in an oxygen-enriched nitrogen gas stream following an aerosol-assisted CVD process [49]. In ALD processes, SnO_2 thin films have been produced at 250–400 °C using the precursors $Sn(C_2H_5)_4$ (tetraethyl tin, TET), oxygen plasma, and H_2O_2, respectively [50, 51].
- Tin alkoxides, tin β-diketonates, and tin carboxylates belong to the class of organometallic precursors in SnO_2 material synthesis.
- $Sn(O^tBu)_4$ acts primarily as an SSP in gas phase methods. VLS (liquid–vapor–solid) growth resulted in one-dimensional SnO_2 structures using liquid gold droplets as templates (Figure 3.34) at 650–800 °C [52–54].

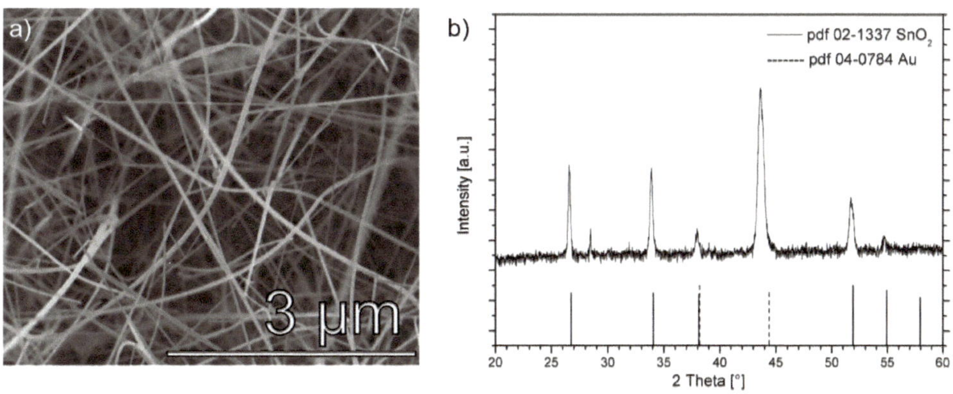

Figure 3.34 Left: SEM (scanning electron microscope) images of SnO_2 wires produced from SSP [$Sn(O^tBu)_4$] using VLS growth method. Right: The corresponding X-ray diffraction characterization method (Image source: [52], with permission)

- Furthermore, SnO_2 structures have been obtained on non-catalytic substrates using tin *tert*-butoxide as a single-source precursor (Figure 3.35) [55–57].

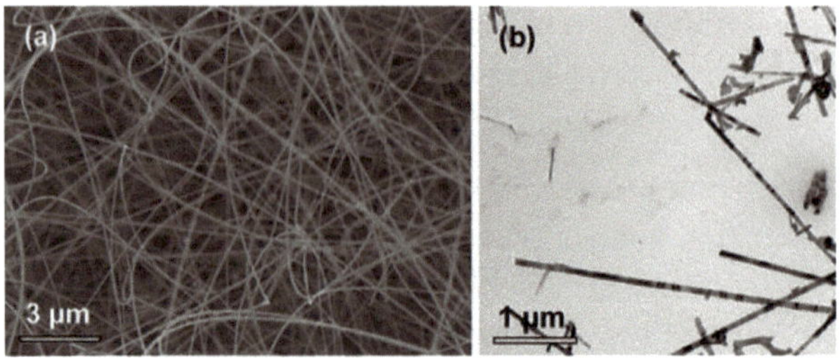

Figure 3.35 (a) SEM and (b) TEM (transmission electron microscope) images of SnO_2 wires obtained from CVD process (Image source: [55], with permission)

- Tin acetylacetonate ($Sn(acac)_2$) was decomposed to SnO_2 layers at 100–600 °C in the presence of oxygen [58].
- Tin acetate has been used as tin source in UV-catalyzed gas phase synthesis using atmospheric oxygen to form SnO_2 materials [59].
- The hydrolysis of tin alkoxides (e.g. $Sn(O^tBu)_4$) to SnO_2 material is a well-investigated procedure. The included hydrolysis with subsequent condensation reaction is, e.g. in connection with the sol–gel method, predominantly known in bulk (macroscopic) material syntheses.
- The hydrolysis of $Sn(O^tBu)_4$ leads to various particulate SnO_2 structures (Figure 3.36) [60, 61].

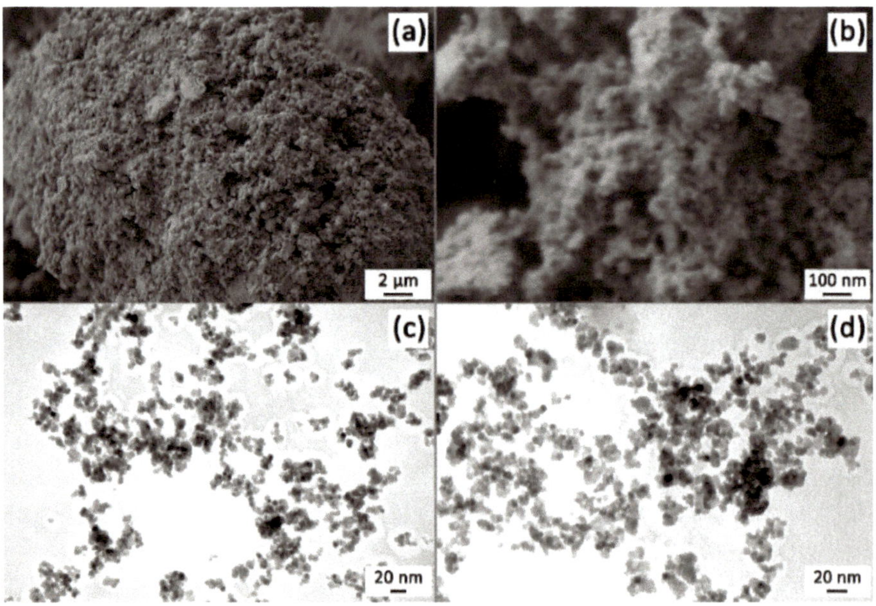

Figure 3.36 SEM (a,b) and TEM (c,d) images of the SnO_2 particles using tin acetylacetonate as precursor (Image source: [62], with permission)

- Solvothermal reactions of tin acetylacetonate in a solvent mixture of water and alcohol lead to the formation of nanoscaled SnO_2 material (Figure 3.36) [64].

A suitable molecular precursor must always be selected carefully, considering the processing method and the requirements of available conditions (liquid phase, gas phase, solubility, stability, etc.). Each method imposes different demands on the respective precursor. Conversely, not every molecule fulfills the above-mentioned requirements of a precursor in material synthesis.

The complexity of available molecular starting compounds and their challenges in material synthesis will be explained in detail using the example of titanium nitride (TiN).

Titanium nitride (TiN) is used for hard coating in various fields of application (including tool coating). In addition to physical vapor deposition (PVD, e.g. sputtering) using TiN as a material source, various chemical transformation methods are listed below, which demonstrate the advantages and disadvantages of chemical precursors in material synthesis. In the following, only gas-phase coatings will be evaluated (material preparation using nitration of molten salts is discussed elsewhere [65]). The reaction equations shown below are simplified, and cannot describe the complex reaction networks that take place in the gas phase and on the substrate surface; instead, they focus on the final product formation.

- The chemical production of TiN directly from the elements (Equation 3.3) requires high activation energies in order to activate the N–N triple bond of nitrogen [66], demanding high temperatures (exceeding 1000 °C) under inert conditions to avoid oxidation of the titanium by oxygen and/or water [67].

$$2\,\mathrm{Ti} + \mathrm{N_2} \longrightarrow 2\,\mathrm{TiN}(T > 1000^\circ\mathrm{C}) \tag{3.3}$$

- Using more reactive and volatile metal halides ($TiCl_4$) in combination with dinitrogen (N_2) enables the formation of TiN (Equation 3.4), but reactive and corrosive by-products are formed (in this case, HCl from the reaction of Cl· + H· ↔ HCl (radicals in the plasma)), which must be handled according specific safety procedures (separation and disposal) [68].

$$2\,\mathrm{TiCl_4} + \mathrm{N_2} + 4\mathrm{H_2} \longrightarrow 2\,\mathrm{TiN} + 8\,\mathrm{HCl}(\mathrm{RF\ Plasma}) \tag{3.4}$$

- By using ammonia (NH_3) instead of N_2 and H_2 in a Lewis acid ($TiCl_4$)–base (NH_3) reaction, product formation can be improved by lowering the activation energy (Equation 3.5) [69, 70]. However, the formation of ammonium chloride from the reaction of NH_3 + HCl ↔ NH_4Cl represents a significant side-reaction [71]. This either results in an increased consumption of NH_3, or demands increased process temperatures (approx. >500 °C) to avoid side-product formation (Equation 3.6) [72].

$$4\,TiCl_4 + 22\,NH_3 \rightarrow 4\,TiN + 16\,NH_4Cl + N_2 + H_2 \quad (3.5)$$

$$4\,TiCl_4 + 6\,NH_3 \rightarrow 4\,TiN + 16\,HCl + N_2 + H_2 (> 500^\circ C) \quad (3.6)$$

- Due to the challenging handling of corrosive by-products – especially at high temperatures – chloride-free molecular precursors were introduced in material synthesis [73]. Tetrakis(dimethylamino)titanium (TDMAT; $Ti(N(CH_3)_2)_4$) already contains direct Ti–N bonds and decomposes to TiN at temperatures of around 350 °C in thermal CVD processes [72–74], or at around 200 °C in ALD processes [75, 76]. Next to the formation of TiN, the only by-products are volatile trimethylamine ($N(CH_3)_3$) and dinitrogen (N_2) (Equation 3.7) [70, 77, 78].

$$6\,Ti(N(CH_3)_2)_4 \rightarrow 6\,TiN + 16\,N\,(CH)_3)_3 + N_2 (350^\circ C) \quad (3.7)$$

While the complexity of described precursors increases continuously, the process temperatures and formation of corrosive educts (NH_3) or by-products (HCl) have simultaneously decreased. More complex organometallic molecular precursors ($Ti(N(CH_3)_2)_4$) are often more expensive (both financially and in the life cycle assessment) compared to corresponding metal chlorides ($TiCl_4$), but the complete process chain must always be considered. Even "simple" anhydrous reactants (including NH_3) are cost-intensive and furthermore, the complete coating process must always be taken into account, due to reduced machine lifetime through corrosive by-products or having to deal with unwanted by-products. A life cycle analysis can provide more precise insights into the respective benefits of molecular precursors in a complete process analysis.

3.3.4.2 Outlook

Through targeted chemical modification of the molecular structure, the physicochemical properties of molecular precursors (e.g. vapor pressure, decomposition temperature) can be tuned to suit the coating process in question. The key criteria employed for the development of new precursors involves chemical design concepts enabling high thermal stability and resistance towards adventitious hydrolysis.

More recent work based on the use of a tailored ligand periphery [11, 13, 15–17, 28, 79] allows optimal saturation of the coordination sphere using neutral Lewis base functions, which is important for high volatility and solubility – even in non-polar solvents.

In the development of precursors for gas-phase deposition processes, the major challenge is to assemble phase-forming elements completely or partially into a single molecular source, which simplifies the delivery of essential materials constituents. By adapting the structure of the compounds and selecting the ligands and co-ligands, the conversion into the end product can be optimally adapted for material design at a molecular level.

The choice of ligands influences the stability of the compounds, and when there is an optimal balance between steric and electronic factors, efficient conversion of the molecule into the target material can be achieved (Figure 3.37). The steric requirement can be varied by specifically selecting the alkoxo-ligands used (-O^iPr vs. -O^tBu) (Figure 3.37a). The sterically less demanding isopropoxide (-O^iPr) units do not shield the metal centers completely, so that isopropyl alcohol molecules additionally coordinate to the compound for the purpose of coordinative saturation. This has the disadvantage that these alcohol molecules, which are weakly coordinated compared to the covalently bound alkoxo-units, are eliminated prior to the actual decomposition process; this leads to premature fragmentation and reorganization of the molecular fragments to create species that do not possess the initial Mg:Al stoichiometry.

The use of sterically more demanding *tert*-butoxide (-O^tBu) units leads to a complete saturation of the coordination sphere of the complex by covalently bound ligands, resulting in a stable compound. However, the increased carbon content in [$MgAl_2(O^tBu)_8$], due to the introduction of the more sterically demanding alcoholate ligands, can lead to increased carbon impurity in the target phase. By substituting the terminal -O^tBu units with hydrides (Figure 3.37 b), the shielding of the metal centers is maintained while the carbon content in the precursor compound is reduced, which effectively reduces the molar mass of the precursor compound, resulting in a positive effect on the volatility. In addition, the hydrides initiate an intramolecular ligand stripping mechanism, leading to stable by-products *tert*-butanol (HO^tBu) and hydrogen (H_2) that are easy to remove from the reaction, thus favoring the formation of a phase-pure spinel deposit [9]. This example of spinel synthesis from three different precursors demonstrates that crucial role precursors can play in governing the compositional purity and phase of the target material. Further development of new chemical concepts and better understanding of mechanistic steps involved in the transformation of the molecular precursors into functional materials is a promising avenue for advancing materials technologies.

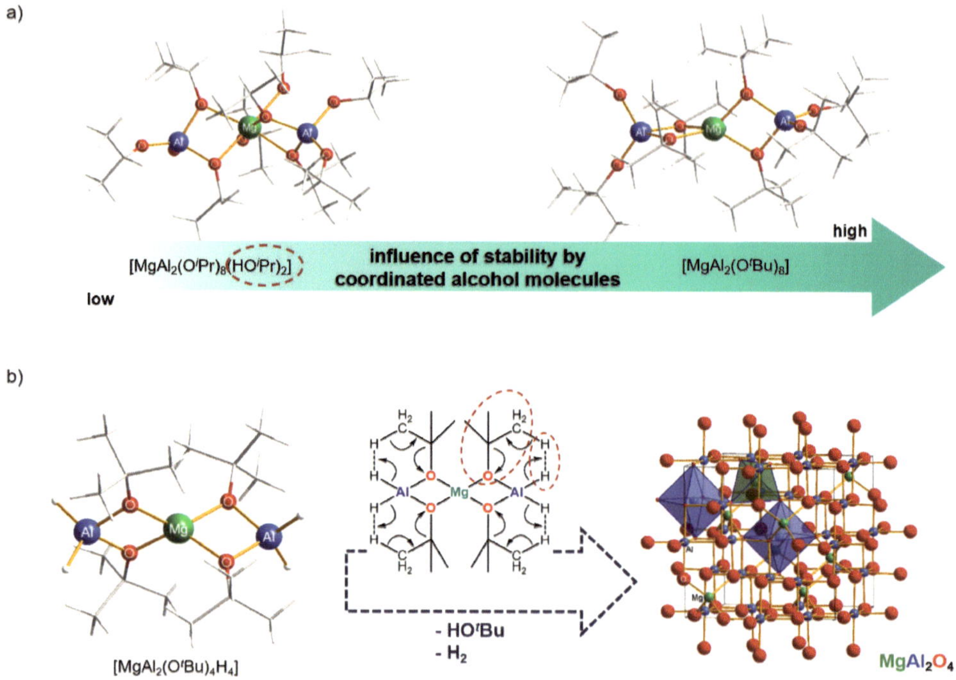

Figure 3.37 Influence of the ligands used in the precursor synthesis on (a) the stability of the compounds and (b) the clean decomposition to the target material $MgAl_2O_4$ [9]

Acknowledgments

The authors would like to thank the University of Cologne (UzK) for providing the infrastructure and the German Research Foundation (DFG) for financial support within the framework of the priority program “Manipulation of matter controlled by electric and magnetic field: Towards novel synthesis and processing routes of inorganic materials” (SPP 1959).

References for Section 3.3.4

[1] Mendes, R. F.; Almeida Paz, F. A.: Transforming metal-organic frameworks into functional materials, *Inorganic Chemistry Frontiers*, 2015, 2: 495–509

[2] Möller, F.; Piontek, S.; Miller, R. G.; Apfel, U.-P.: From enzymes to functional materials – Towards activation of small molecules, *Chemistry – A European Journal*, 2018, 24: 1471–1493

[3] Beaujuge, P. M.; Fréchet, J. M. J.: Molecular design and ordering effects in π-functional materials for transistor and solar cell applications, *Journal of the American Chemical Society*, 2011, 133: 20009–20029

[4] MacInnes, A. N.; Power, M. B.; Barron, A. R.: Chemical vapor deposition of gallium sulfide: Phase control by molecular design, *Chemistry of Materials*, 1993, 5: 1344–1351

[5] Mathur, S.; Shen, H.; Donia, N.: Chemical nanotechnology: From molecules to applications, *ECS Transactions*, 2006, 3: 3–13

[6] Nalivaeva, N. N.; Turner, A. J.: The amyloid precursor protein: A biochemical enigma in brain development, function and disease, *FEBS Letters*, 2013, 587: 2046–2054

[7] Veith, M.: Molecular precursors for (nano) materials - A one step strategy, *Journal of the Chemical Society, Dalton Transactions*, 2002: 2405–2412

[8] Molloy, K. C.; Precursors for the formation of tin(IV) oxide and related materials, *Journal of Chemical Research*, 2008: 549–554

[9] Mathur, S.; Veith, M.; Ruegamer, T.; Hemmer, E.; Shen, H.: Chemical vapor deposition of $MgAl_2O_4$ thin films using different Mg–Al alkoxides: Role of precursor chemistry, *Chemistry of Materials*, 2004, 16: 1304–1312

[10] Graf, D.; Frank, M.; Ojelere, O.; Gessner, I.; Juergensen, L.; Grosch, M.; Mathur, S.: Structure-decomposition relationships of homoleptic and heteroleptic vanadium(IV) alkoxides for defined gas phase depositions of VO_2 and V_2O_3 thin films, *Materials Today Proceedings*, 2019: 2445–2450

[11] Brune, V.; Hegemann, C.; Wilhelm, M.; Ates, N.; Mathur, S.: Molecular precursors to group IV dichalcogenides MS_2 (M=Ti, Zr, Hf), *Zeitschrift für Anorganische und Allgemeine Chemie*, 2022, 648: e202200049

[12] Fiz, R.; Hernandez-Ramirez, F.; Fischer, T.; Lopez-Conesa, L.; Estrade, S.; Peiro, F.; Mathur, S.: Synthesis, characterization, and humidity detection properties of Nb_2O_5 nanorods and SnO_2/Nb_2O_5 heterostructures, *Journal of Physical Chemistry C*, 2013, 117: 10086–10094

[13] Atamtürk, U.; Brune, V.; Mishra, S.; Mathur, S.: Vapor phase synthesis of SnS facilitated by ligand-driven "launch vehicle" effect in tin precursors, *Molecules*, 2021, 26: 5367

[14] Sekhar, P. K.; Graf, D.; Ojelere, O.; Saha, T. K.; Riheen, M. A.; Mathur, S.: Electrochemical gas sensor integrated with vanadium monoxide nanowires for monitoring low concentrations of ammonia emission, *Journal of the Electrochemical Society*, 2020, 167, 027548

[15] Büyükyazi, M.; Hegemann, C.; Lehnen, T.; Tyrta, W.; Mathur, S.: Molecular Co(II) and Co(III) heteroarylalkenolates as efficient precursors for chemical vapor deposition of Co_3O_4 nanowires, *Inorganic Chemistry*, 2014, 53: 10928–10936

[16] Brune, V.; et al.: Single source precursor route to nanometric tin chalcogenides, *Dalton Transactions*, 2021, 50: 17346–17360

[17] Jürgensen, L.; Frank, M.; Pyeon, M.; Czympiel, L.; Mathur, S.: Subvalent iridium precursors for atom-efficient chemical vapor deposition of Ir and IrO_2 thin films, *Organometallics*, 2017, 36: 2331–2337

[18] Leduc, J.; Pacold, J. I.; Shuh, D. K.; Dong, C. L.; Mathur, S.: Uranium oxide nanocrystals by microwave-assisted thermal decomposition: Electronic and structural properties, *Zeitschrift für Anorganische und Allgemeine Chemie*, 2018, 644: 12–18

[19] Brune, V.; Hegemann, C.; Mathur, S.: Molecular routes to two-dimensional metal dichalcogenides MX2 (M = Mo, W; X = S, Se), *Inorganic Chemistry*, 2019, 58: 9922–9934

[20] Pousaneh, E.; et al.: Iron(III) β-diketonates: CVD precursors for iron oxide film formation, *Inorganica Chimica Acta*, 2019, 487: 1–8

[21] Pollard, K. D.; Puddephatt, R. J.: Chemical vapor deposition of tantalum oxide from tetraethoxo (β-diketonato) tantalum(V) complexes, *Chemistry of Materials*, 1999, 11: 1069–1074

[22] Gardiner, R.; Brown, D. W.; Kirlin, P. S.; Rheingold, A. L.: Volatile barium β-diketonate polyether adducts. Synthesis, characterization, and metalloorganic chemical vapor deposition, *Chemistry of Materials*, 1991, 3: 1053–1059

[23] Giebelhaus, I.; Müller, R.; Tyrra, W.; Pantenburg, I.; Fischer, T.; Mathur, S.: First air stable tin(II) β-heteroarylalkenolate: Synthesis, characterization and application in chemical vapor deposition, *Inorganica Chimica Acta*, 2011, 372: 340–346

[24] Appel, L.; Leduc, J.; Webster, C. L.; Ziller, J. W.; Evans, W. J.; Mathur, S.: Synthesis of air-stable, volatile uranium(IV) and (VI) compounds and their gas-phase conversion to uranium oxide films, *Angewandte Chemie International Edition*, 2015, 54: 2209–2213

[25] Brückmann, L.; Tyrra, W.; Stucky, S.; Mathur, S.: Novel air-stable and volatile bis (pyridylalkenolato) palladium(II) and-platinum(II) derivatives, *Inorganic Chemistry*, 2012, 51: 536–542

[26] Heidemann, T.; Mathur, S.: Air-stable and volatile bis (pyridylalkenolato) germanium(II), -tin(II), and -lead(II) complexes, *European Journal of Inorganic Chemistry*, 2014: 506–510

[27] Büyükyazi, M.; Hegemann, C.; Lehnen, T.; Tyrra, W.; Mathur, S.: Molecular Co(II) and Co(III) heteroarylalkenolates as efficient precursors for chemical vapor deposition of Co_3O_4 nanowires, *Inorganic Chemistry*, 2014, 53: 10928–10936

[28] Jürgensen, L.; et al.: Controlled growth of Cu and CuO_x thin films from subvalent copper precursors, *Dalton Transactions*, 2020, 49: 13317–13325

[29] Czympiel, L.; Pfrommer, J.; Tyrra, W.; Schäfer, M.; Mathur, S.: Ligand-modulated chemical and structural implications in four-, five-, and six-fold coordinated aluminum heteroaryl alkenolates, *Inorganic Chemistry*, 2015, 54: 25–37

[30] Czympiel, L.; Frank, M.; Mettenbörger, A.; Hühne, S. M.; Mathur, S.: High activity heterogeneous catalysts by plasma-enhanced chemical vapor deposition of volatile palladium complexes on biomorphic carbon, *Comptes Rendus Chimie*, 2018, 21: 943–951

[31] Czympiel, L.: Heteroarylsubstituierte Alkenolate von Palladium, Platin und Gold sowie deren Anwendung in der Synthese neuer wiederverwendbarer heterogener Katalysatoren mittels CVD-Verfahren [Heteroaryl-substituted alkenolates of palladium, platinum and gold and their application in the synthesis of new reusable heterogeneous catalysts using CVD processes] [Thesis], University of Cologne, 2014

[32] Nandi, M.; Rhubright, D.; Sen, A.: Pyrolytic transformation of metal alkoxides to oxides: Mechanistic studies. Pyrolysis of homoleptic titanium alkoxides, *Inorganic Chemistry Communications*, 1990, 29: 3066–3068

[33] Schläfer, J.: Neuartige homo- und heterometallische Cer-Alkoxide [Dissertation], University of Cologne, 2014

[34] Graf, D.; Schläfer, J.; Garbe, S.; Klein, A.; Mathur, S.: Interdependence of structure, morphology, and phase transitions in CVD grown VO_2 and V_2O_3 nanostructures, *Chemistry of Materials*, 2017, 29: 5877–5885

[35] Brückmann, L.; Tyrraw; Stucky, S.; Mathur, S.: Novel air-stable and volatile bis (pyridylalkenolato) palladium(II) and- platinum(II) derivatives, *Inorganic Chemistry*, 2012, 51: 536–542

[36] Brune, V.; et al.: A novel molecular synthesis route to Li_2S loaded carbon fibers for lithium–sulfur batteries, *Journal of Materials Chemistry A*, 2022, 10: 9902–9910

[37] Brune, V.; et al.: Influence of the choice of precursors on the synthesis of two-dimensional transition metal dichalcogenides, *Dalton Transactions*, 2021, 50: 12365–12385

[38] Czympiel, L.; Lekeu, J. M.; Hegemann, C.; Mathur, S.: Oxidative halogenation of Sn(II) heteroaryl alkenolates: Formation of unusual trans-dihalo Sn(IV) complexes, *Inorganica Chimica Acta*, 2017, 455: 197–203

[39] Mathur, S.; Shen, H.; Barth, S.; Cavelius, C.: Inorganic nanomaterials through chemical design, *SPIE Newsroom*, 2006, DOI: 10.1117/2.1200602.0080

[40] Fornalczyk, G.; Valldor, M.; Mathur, S.: Monomeric iron heteroarylalkenolates: Structural design concepts and investigations on their application in chemical vapor deposition, *Crystal Growth & Design*, 2014, 14: 1811–1818

[41] Mathur, S.; Veith, M.; Haas M.; Shen H.; Huch V.; et al.: Single-source sol-gel synthesis of nanocrystalline ZnAl2O4: Structural and optical properties, *Journal of the American Ceramic Society*, 2001, 84: 1921–1928

[42] Davazoglou, D.: Optical properties of SnO_2 thin films grown by atmospheric pressure chemical vapour deposition oxiding $SnCl_4$, *Thin Solid Films*, 1997, 302: 204–213

[43] Rosental, A.; Tarre, A.; Gerst, A.; Uustare, T.; Sammelselg, V.: Atomic-layer chemical vapor deposition of SnO_2 for gas-sensing applications, *Sensors and Actuators B: Chemical*, 2001, 77: 297–300

[44] Nagano, M.: Chemical vapor deposition of SnO_2 thin films on (100) surfaces of rutile single crystals, *Journal of Crystal Growth*, 1984, 69: 465–468

[45] Nagano, M.: Chemical vapor deposition of SnO_2 thin films on rutile single crystals, *Journal of Crystal Growth*, 1984, 67: 639–644

[46] Vetrone, J.; Chung, Y.: Organometallic chemical vapor deposition of SnO_2 single crystal and polycrystalline films, *Journal of Vacuum Science & Technology A: Vacuum, Surfaces, and Films*, 1991, 9: 3041–3047

[47] Kwoka, M.; Ottaviano, L.; Passacantando, M.; Santucci, S.; Czempik, G.; Szuber, J.: XPS study of the surface chemistry of L-CVD SnO_2 thin films after oxidation, *Thin Solid Films*, 2005, 490: 36–42

[48] Kwoka, M.; Ottaviano, L.; Szuber, J.: AFM study of the surface morphology of L-CVD SnO_2 thin films, *Thin Solid Films*, 2007, 515: 8328–8331

[49] Abokifa, A. A.; Haddad, K.; Raman, B.; Fortner, J.; Biswas, P.: Room temperature gas sensing mechanism of SnO_2 towards chloroform: Comparing first principles calculations with sensing experiments, *Applied Surface Science*, 2021, 554: 149603

[50] Maximov, M. Y.; et al.: Characterization and electrochemical performance at high discharge rates of tin dioxide thin films synthesized by atomic layer deposition, *Journal of Electronic Materials*, 2017, 46: 6571–6577

[51] Lim, S.-S.; et al.: Atomic layer deposition of SnO_2 thin films using tetraethyltin and H_2O_2, *Ceramics International*, 2019, 45: 20600–20605

[52] Müller, R.; Hernandez-Ramirez, F.; Shen, H.; Du, H.; Mader, W.; Mathur, S.: Influence of precursor chemistry on morphology and composition of CVD-grown SnO_2 nanowires, *Chemistry of Materials*, 2012, 24: 4028–4035

[53] Mathur, S.; Barth S.: Molecule-based chemical vapor growth of aligned SnO_2 nanowires and branched SnO_2/V_2O_5 heterostructures, *Small*, 2007, 3: 2070–2075

[54] Samà, J.; et al.: Locally grown SnO_2 NWs as low power ammonia sensor, *Procedia Engineering*, 2015, 120: 215–219

[55] Pan, J.; Song, X.; Zhang, J.; Shen, H.; Xiong, Q.: Switchable wettability in SnO_2 nanowires and SnO_2@ SnO_2 heterostructures, *Journal of Physical Chemistry C*, 2011, 115: 22225–22231

[56] Singh, T.; Fischer, T.; Singh, J.; Gurram, S. K.; Mathur, S.; Plasma enhanced CVD of transparent and conductive tin oxide thin films, in *Nanostructured Materials and Nanotechnology VII*, Wiley, 2013, pp. 99–105

[57] Chen, H.; Liu, W.; Zhu, A.; Xiong, X.; Pan, J.: Visible-light excited strong red emission of SnO_2 nanostructures, *Materials Research Express*, 2017, 4: 045019

[58] Maruyama, T.; Ikuta, Y.: Tin dioxide thin films prepared by chemical vapor deposition from tin(II) acetylacetonate, *Solar Energy Materials and Solar Cells*, 1992, 28: 209–215

[59] Maruyama, T.; Morishita, T.: Tin dioxide thin films prepared by photochemical vapour deposition from tin(II) acetate, *Thin Solid Films*, 1994, 251: 19–22

[60] Dimitrov, M.; Tsoncheva, T.; Shao, S.; Köhn, R.: Novel preparation of nanosized mesoporous SnO_2 powders: Physicochemical and catalytic properties, *Applied Catalysis B: Environmental*, 2010, 94: 158–165

[61] Xiao, L.; Shen, H.; Von Hagen, R.; Pan, J.; Belkoura, L.; Mathur, S.: Microwave assisted fast and facile synthesis of SnO_2 quantum dots and their printing applications, *Chemical Communications*, 2010, 46: 6509–6511

[62] Saravanakumar, K.; Prasath, J.; Rajesh, R.: Synthesis of SnO_2 and MoS_2 nano composites for supercapacitor applications, *Materials Today: Proceedings*, 2021, 46: 8189–8195

[63] Patel, G. H.; et al.: Sol-gel synthesis and thermal characterization of SnO_2 nanoparticles, *Physica B: Condensed Matter*, 2021, 613: 412987

[64] Fisenko, N. A.; et al.: Atmospheric pressure solvothermal synthesis of nanoscale SnO_2 and its application in microextrusion printing of a thick-film chemosensor material for effective ethanol detection, *Sensors*, 2022, 22: 9800

[65] Parveen, N.; Ansari, M. O.; Ansari, S. A.; Kumar, P.: Nanostructured titanium nitride and its composites as high-performance supercapacitor electrode material, *Nanomaterials*, 2022, 13: 105

[66] Tricker, A. W.; et al.: Mechanocatalytic ammonia synthesis over TiN in transient microenvironments, *ACS Energy Letters*, 2020, 5: 3362–3367

[67] Li, C.; Lv X.; Chen J.; Liu, X.; Bai, C.: Kinetics of titanium nitride synthesized with Ti and N_2, *International Journal of Refractory Metals and Hard Materials*, 2015, 52: 165–170

[68] Shin, D. H.; Hong, Y. C.; Uhm, H. S.: Production of nanocrystalline titanium nitride powder by atmospheric microwave plasma torch in hydrogen/nitrogen gas, *Journal of American Ceramic Society*, 2005, 88: 2736–2739

[69] Tiznado, H.; Zaera, F.: Surface chemistry in the atomic layer deposition of TiN films from $TiCl_4$ and ammonia, *Journal of Physical Chemistry B*, 2006, 110: 13491–13498

[70] Ramanuja, N.; et al.: Synthesis and characterization of low pressure chemically vapor deposited titanium nitride films using $TiCl_4$ and NH_3, *Materials Letters*, 2002, 57: 261–269

[71] Elam, J. W.; Schuisky, M.; Ferguson, J. D.; George, S. M.: Surface chemistry and film growth during TiN atomic layer deposition using TDMAT and NH_3, *Thin Solid Films*, 2003, 436: 145–156

[72] Paranjpe, A.; Islamraja, M.: Chemical vapor deposition TiN process for contact/via barrier applications, *Journal of Vacuum Science & Technology B: Microelectronics and Nanometer Structures*, 1995, 13: 2105–2114

[73] Ishihara, S.; Hanabusa, M.: Laser-assisted chemical vapor deposition of titanium nitride films, *Journal of Applied Physics*, 1998, 84: 596–599

[74] Sun, S. C.; Tsai, M. H.: Comparison of chemical vapor deposition of TiN using tetrakis-diethylamino-titanium and tetrakis-dimethylamino-titanium, *Thin Solid Films*, 1994, 253: 440–444

[75] Caubet P.; et al.: Low-Temperature low-resistivity PEALD TiN using TDMAT under hydrogen reducing ambient, *Journal of the Electrochemical Society*, 2008, 155: H625

[76] Musschoot, J.; Xie Q.; Deduytsche, D.; Van den Berghe, S.; Van Meirhaeghe, R. L.; Detavernier, C.: Atomic layer deposition of titanium nitride from TDMAT precursor, *Microelectronic Engineering*, 2009, 86: 72–77

[77] Marcadal, C.; Eizenberg, M.; Yoon, A.; Chen L.: Metallorganic chemical vapor deposited TiN barrier enhancement with SiH_4 treatment, *Journal of the Electrochemical Society*, 2002, 149: C52

[78] Norton, E. T.; Amato-Wierda, C.: Kinetic and mechanistic studies of the thermal decomposition of $Ti(N(CH_3)_2)_4$ during chemical vapor deposition by in situ molecular beam mass spectrometry, *Chemistry of Materials*, 2001, 13: 4655–4660

[79] Siribbal, S. M.; et al.: Air-stable gadolinium precursors for the facile microwave-assisted synthesis of Gd_2O_3 nanocontrast agents for magnetic resonance imaging, *Crystal Growth & Design*, 2018, 18: 633–641

3.4 Simulation of Layer Deposition

Ameya Kulkarni

3.4.1 Introduction

Simulation is becoming increasingly important in the computer-aided development of products and processes. In principle, any question can be answered using a simulation that can be described by time- and location-dependent differential equations or an equivalent variation principle. Finite element analysis (FEA) is often used, and development of FEA is an ongoing process. Increasing performance of computer systems and the development of new models mean that increasingly complex system modeling can be handled. Field problems or multi-physical problems are becoming more and more important, so that increasingly broad fields of application can be opened up. Simulation makes it possible to model coating processes on a practical scale. Through-flow simulations using computational fluid dynamics (CFD) make it possible to significantly reduce cost-intensive tests, as the optimum process parameters can be determined in advance.

The basis of every simulation is correct modeling and the transfer to the corresponding simulation software. Simplifications made in advance must be taken into account when setting up the simulation model and the subsequent evaluation. The steps required for model implementation, such as building the geometry and assigning the characteristic properties, loads, and boundary conditions, are carried out in the pre-processor of the simulation software. Before the corresponding analysis can be carried out, the model must be meshed with suitable elements. A suitable attachment function and an adequate mesh refinement must be defined. The simulation software then uses the specifications made in the pre-processor to generate the FEM model and the resulting system of equations. This is then computed in the solver, and the variables of interest, such as flow profiles or temperature fields, are calculated. In the post-processor the results are prepared graphically. The simulation engineer is then responsible for evaluating the results [1, 2].

The simulation engineer can then use the results to iteratively adapt the model or compare it with real tests. It is possible to adjust all variables defined in pre-processing so that a new simulation can be calculated and the result can be further optimized. Figure 3.38 summarizes the basic procedure and the iterative potential of an analysis using the finite element analysis [3].

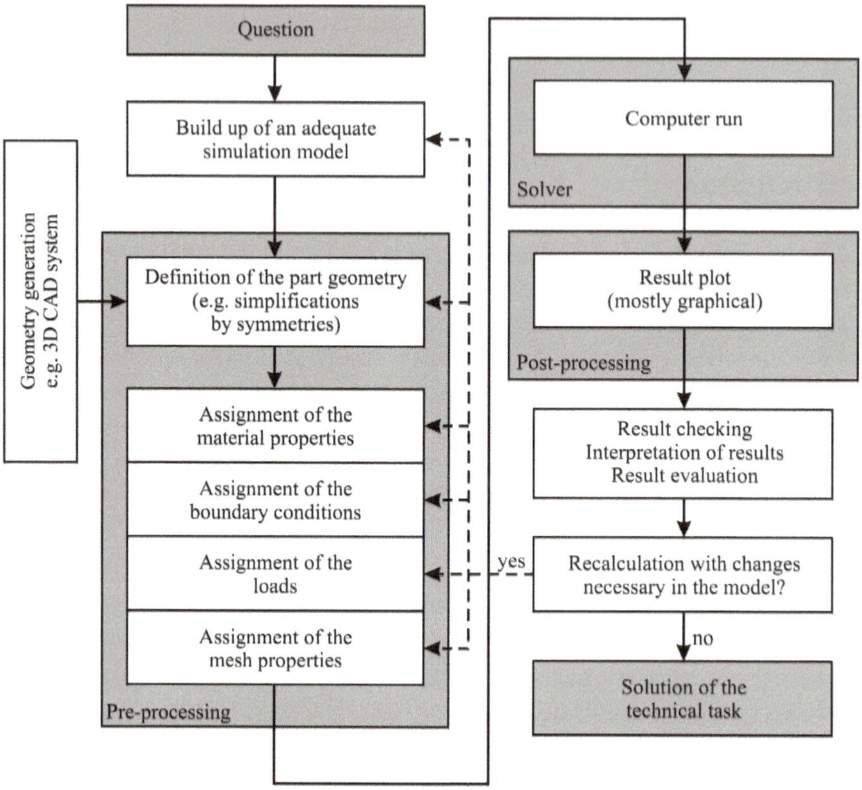

Figure 3.38 Sequence of a simulation using FEA (Image based on [3])

Steel tools used in industrial mass production processes such as injection molding are susceptible to wear and corrosion due to the rapid cyclical temperature and pressure fluctuations, and the use of abrasive polymers. To protect the mold surfaces, high-quality ceramic thin films can be applied by metalorganic chemical vapor deposition (MOCVD) [4]. In addition to the protective properties, ceramic materials such as yttrium-stabilized zirconia (YSZ) can thermally insulate tool surfaces, enabling more precise temperature control to avoid the formation of surface defects, such as the visibility of weld lines, on the manufactured plastic molded parts. At the same time, cycle times can be shortened and energy consumption during the molding process can be reduced [5].

Compared to other coating processes, CVD enables a homogeneous coating thickness distribution even on highly complex 3D surface geometries [6]. The main objective of the simulations is to find the optimum position of the test specimens in the reactor so that the coating on the test specimen is homogeneous. The influence of gas flow, temperature, and pressure on the flow behavior within the reactor, which contains a tool insert as a demonstrator, has been simulated with COMSOL Multiphysics®.

3.4.2 Theoretical Principles and Experimental Setup

A hot-wall CVD reactor is used to coat the samples and the mold inserts for injection molds, as shown in Figure 3.39. The coating process takes place at temperatures below 500 °C, in order to coat common steels without them losing their mechanical strength.

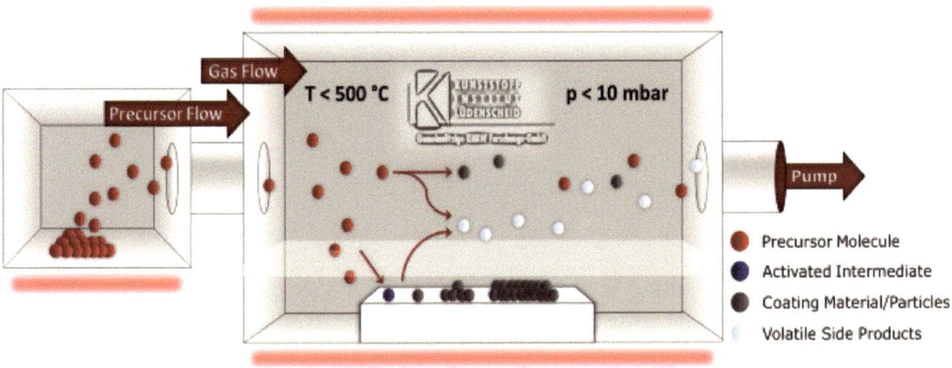

Figure 3.39 Schematic representation of the CVD reactor and the CVD process (Image source: Non-profit KIMW Forschungs-GmbH)

The MOCVD process is shown in Figure 3.39. In MOCVD, organometallic precursors are used, which decompose in situ at higher temperatures, in order to achieve the relatively low coating temperatures compared to conventional CVD. The precursor, a volatile molecular compound, is evaporated and fed into the reaction chamber, in which a vacuum with a pressure of a few mbar prevails. A thermally induced decomposition reaction takes place there, in which the precursor is first adsorbed on the surface, where it reacts with the desired material and thus completes the coating process.

The coating material itself is produced by specific chemical reactions, so an understanding of the physical and chemical mechanisms in the reaction chamber is necessary in order to develop a process. The properties of the coating to be applied depend primarily on the process parameters of temperature, pressure, gas flow, and precursor flow, which can be varied. The precursor material is evaporated in an evaporator and fed into the reactor chamber with a carrier gas through the main inlet. In order to homogenize the gas flow, the same amount of gas is also fed into the reactor through several smaller nozzles (ring dies) arranged in a circle around the main inlet. The basic structure and the defined boundary conditions are summarized in Figure 3.40.

SCCM=450(cm³/sec), T=385(°C), P= 500(Pa)

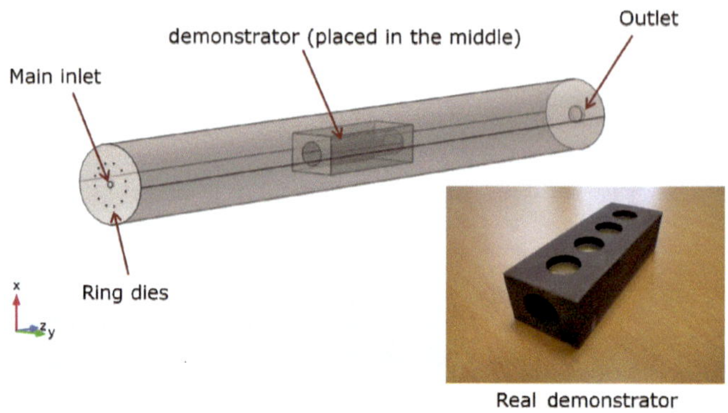

Figure 3.40 Initial model for the simulation (Image source: Non-profit KIMW Forschungs-GmbH)

To analyze the 3D conformity and the groove penetration properties of the coating, various types of demonstrator tools can be used. A rectangular tool with a cylindrical bore and four openings on the top, into which small metallic round bodies (coins) can be inserted, was used in the investigations. For reasons of simplification, the cut-outs in the demonstrator and the mounting of the demonstrator in the flow channel were not taken into account in the simulations, although these have an influence on the resulting flow profile. With this demonstrator, the coating thickness inside the demonstrator can be measured by measuring the coating thickness on the coins. The demonstrator can be surrounded by metal plates so that almost all of the gas flows through the demonstrator and not around it. Both modifications of the demonstrator are shown in Figure 3.41.

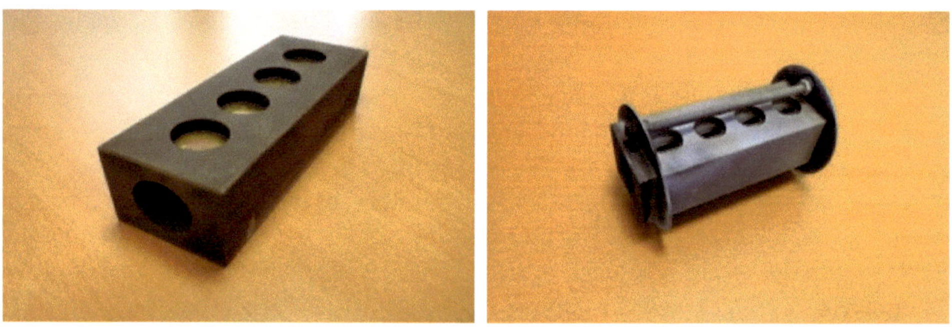

Figure 3.41 Demonstrator without metal plates (left) and with metal plates (right) (Image source: Non-profit KIMW Forschungs-GmbH)

The coating should be homogeneous on the entire inner surface of the demonstrator in order to achieve an optimal functionalized tool surface with the same properties regardless of the position on the tool. The coating thickness on the metallic coins was analyzed by calotte grinding.

A temperature of 385 °C and a pressure of 5 mbar (≈ 500 Pa) were used as process parameters for the coating. The tool geometry (with and without plates around the demonstration tool) and the amount of gas flow were varied during the process.

The CVD reactor itself (Figure 3.42) is made of ceramic and is heated by an external furnace. The actual coating takes place in a steel tube that is inserted into the CVD reactor. The furnace is insulated from the outer surface so that the ceramic tube also gets hot outside the furnace, but is cooled by the ambient air at the same time so that it does not reach such high temperatures. In order to achieve a perfect coating, the furnace is heated with a certain temperature gradient.

Figure 3.42 CVD reactor (white tube) and furnace (blue) (Image source: Non-profit KIMW Forschungs-GmbH)

3.4.3 The Equations of State

With the help of COMSOL Multiphysics®, it is relatively easy to simulate the gas flow within the reactor. The CFD module is used to perform all simulations. The type of flow considered for the simulations is based on the Reynolds number, which defines whether a flow is laminar or turbulent. From the experimental data on pressure and gas flow rate, it is clear that a laminar flow can be considered for this problem. Laminar flow is based on the Navier–Stokes equations (Equation 3.8–Equation 3.10) [7].

$$\rho \cdot \left(\frac{\partial \boldsymbol{u}}{\partial t} + (\boldsymbol{u} \cdot \nabla) \cdot \boldsymbol{u} \right) = -\nabla p + \nabla \cdot \left[\mu \cdot \left(\nabla \boldsymbol{u} + (\nabla \boldsymbol{u})^T - \frac{2}{3} \cdot (\nabla \boldsymbol{u}) \cdot \boldsymbol{I} \right) \right] + \boldsymbol{F} \tag{3.8}$$

$$\frac{\partial p}{\partial t} + \nabla \cdot (\rho \cdot \boldsymbol{u}) = 0 \tag{3.9}$$

$$\rho = \rho\,(p; T) \tag{3.10}$$

where

$\boldsymbol{u}$ = Velocity of the fluid (m/s)

ρ = Density (kg/m^3)

p = Pressure (Pa)

μ = Dynamic viscosity of the fluid (Pa s)

T = Temperature of the fluid (K)

$\boldsymbol{F}$ = External forces acting on the body (N)

$\boldsymbol{I}$ = Identity tensor (—)

Equation 3.10 shows that the density is dependent on the pressure and temperature of the fluid. In COMSOL Multiphysics® it is possible to define your own variables to take this dependency into account in the equations. For the processes considered for the experiments, the pressure is varied between 100 Pa and 500 Pa. Each time the pressure is varied, it is therefore necessary to calculate the corresponding density. This was simplified by defining an equation for the density with the variable “rho”, and later including this in a parametric sweep that varies the pressure and temperature for this case.

3.4.4 Experimental Procedure and Results

In the demonstrator setup without the plates, the gas can flow both through and around the demonstrator. The flow pattern on the coated coins shows very good gap penetration, as the coating is visible not only on the surface that is in direct contact with the inner surface of the demonstrator tool, but also on the areas covered by the metal of the demonstrator tool that holds the coins in position (Figure 3.43).

Figure 3.43
Coin coated in the demonstrator without plates at 385 °C and 500 Pa, and with gas flows of 450 sccm (main inlet) and 450 sccm (ring dies) (Image source: Non-profit KIMW Forschungs-GmbH)

In the case of the demonstration tool without the plates, the average coating growth rate across all four coins is approx. 0.27 µm/h. However, the gradient from the first to the last coin in the demonstrator is 0.08 µm (process 1, Table 3.11).

Table 3.11 Overview of the Layer Growth Rates on the Sample as a Function of the Gas Flow and the Geometry of the Demonstrator[a]

Process	Demonstrator geometry	Gas flow in main inlet/ring dies (sccm)	Layer growth rates on samples 1-4 (µm/h)				Average growth rate (µm/h)
1	Without plates	450/450	0.31	0.29	0.26	0.23	0.27
2	With plates	450/450	0.23	0.20	0.19	0.18	0.20
3	With plates	125/125	0.86	0.68	0.62	0.48	0.66

[a] The temperature for all coating experiments was 385 °C, the pressure in the reactor was kept at 5 mbar (≈ 500 Pa); sample 1 is closest to the inlet.

The coins coated with plates in the demonstrator tool (process 2, Table 3.11) showed a different flow pattern. The gap penetration capacity is poorer and the growth rate on the coin is slightly lower than in process 1. However, the layer growth gradient from the first to the last sample is lower than in process 1 at 0.05 µm, so that the homogeneity of the layer is higher (Figure 3.44).

Due to the higher gas velocity, the molecules may not have enough time to react with the metal surface of the coin, reducing the layer growth rate.

Figure 3.44
Coin coated with plates in the demonstrator at 385 °C and 500 Pa, and with gas flows of 450 sccm (main inlet) and 450 sccm (ring dies) (Image source: Non-profit KIMW Forschungs-GmbH)

As can be seen from the simulations, the gas velocity inside the demonstration tool with plates is seven times higher than without plates. Reducing the gas flow could therefore lead to an even better coating growth rate. Therefore, a third process was carried out with a lower gas flow rate, but maintaining the temperature and pressure of the first two processes.

One of the coins coated in this way is shown in Figure 3.45. The higher coating thickness is evident from the significantly darker appearance of the coating. The growth rate for this process (process 3, Table 3.11) is three times as high as for the previously mentioned sample (Figure 3.44), but the gap penetration capacity is poorer, as the metal coins do not exhibit a flow pattern on the side surfaces supported by the demonstrator material.

Figure 3.45
Coin coated with plates in the demonstrator at 385 °C and 500 Pa, and with gas flows of 125 sccm (main inlet) and 125 sccm (ring dies) (Image source: Non-profit KIMW Forschungs-GmbH)

The difference in the growth rate between the first and last samples (0.38 µm/h) is also much higher than in the other two methods, which means that the homogeneity of the coating within the demonstration tool is worse. It could be that most of the precursor already reacts in the first part of the reactor, as the gas flow is low.

3.4.5 Results of the Simulations

The simulation results show the flow path of the gas in the reactor. The velocity distribution in the reactor can be read from the velocity cross-sections. Vortices arise in the inlet sections of the reactor, as a result of which the velocity is higher in the front section of the reactor and the flow of the gas only becomes laminar later in the reactor. In preliminary tests, it was shown that the samples placed in the front section of the reactor had a very low coating, which is due to the turbulent flow behind the inlet. This effect was later confirmed by the simulation results, as the flow of the gas is so fast that the gas and precursor molecules have less time to react with the coins, so that they do not have a homogeneous coating.

Two cases were analyzed within the investigations, the first without plates and the second with plates (Table 3.12, process 1 and process 2). The difference in the velocity streamlines is shown in Figure 3.46 and Figure 3.48, and the differences in velocity due to the reduction in volume of the demonstrator with plates can also be clearly seen. The velocities obtained in the first case without plates are lower, which results in the gas and precursor molecules flowing slowly through the reactor. Therefore, we can observe a 3D capability and a very large groove penetration of the coating inside the demonstrator (Figure 3.47 and Figure 3.49).

Table 3.12 Overview of the Simulation Parameters Used

Process	Demonstrator geometry	Temperature (°C)	Pressure (Pa)	Gas flow in main inlet/ ring dies (sccm)
1	Without plates	385	500	450/450
2	With plates	385	500	450/450
3	With plates	385	500	125/125

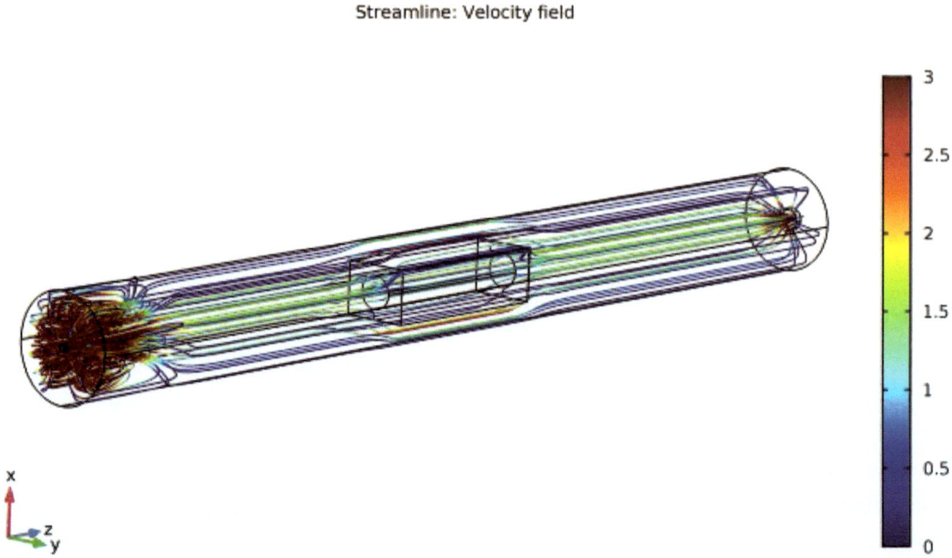

Figure 3.46 Velocity streamlines in the reactor volume with the demonstration tool without plates (Image source: Non-profit KIMW Forschungs-GmbH)

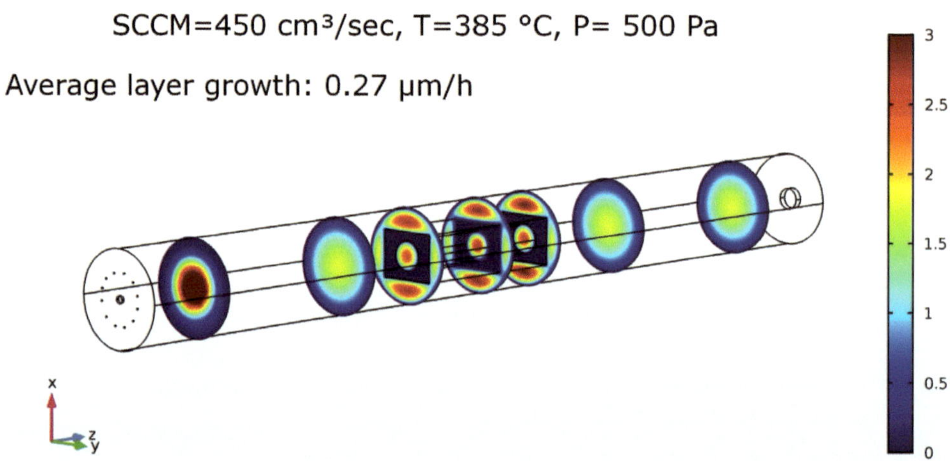

Figure 3.47 Velocity cross-sections in the reactor volume with the demonstration tool without plates (Image source: Non-profit KIMW Forschungs-GmbH)

In the second case with plates (Table 3.12, process 2), higher gas flow velocities are observed, as the gas and the precursor molecules flow through a reduced channel cross-section, so that the coating on the samples is lower (Figure 3.48 and Figure 3.49). By reducing the amount of gas, the velocity within the demonstrator can be reduced, potentially leading to a more uniform coating with higher 3D capability while conserving resources. The velocities achieved would therefore be much lower, allowing sufficient time for the gas and precursor molecules to react with the samples. Due to the lower volume, a larger amount of gas could be passed through the demonstrator, resulting in a more homogeneous coating of the samples.

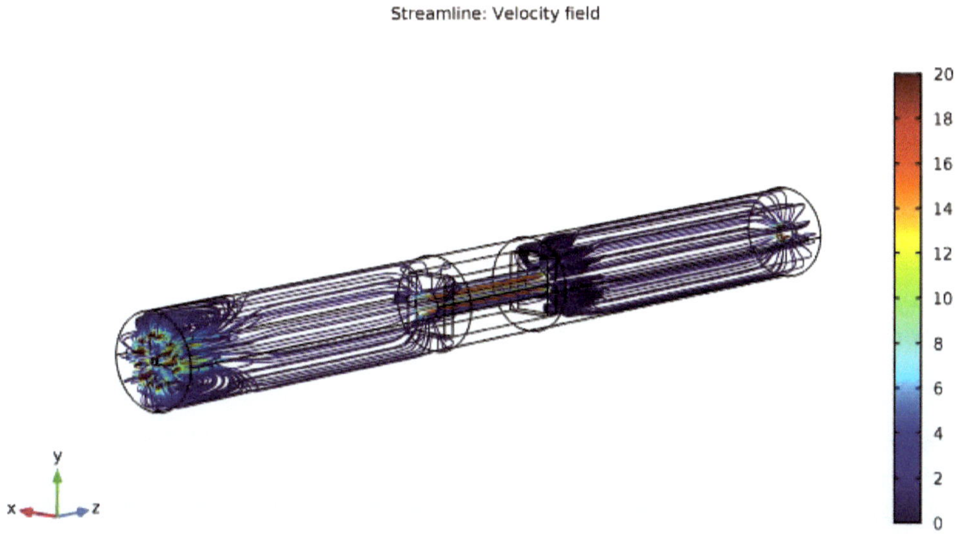

Figure 3.48 Velocity streamlines in the reactor volume with the demonstration tool with plates (Image source: Non-profit KIMW Forschungs-GmbH)

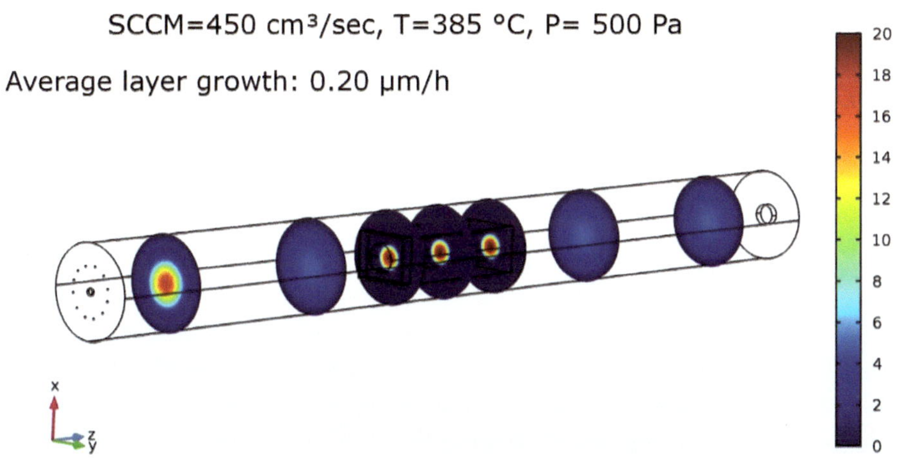

Figure 3.49 Velocity cross-sections in the reactor volume with the demonstration tool with plates (Image source: Non-profit KIMW Forschungs-GmbH)

To reduce the gas velocity in the demonstrator tool, the amount of gas was reduced while the temperature and pressure were kept constant. The experiment was carried out on the real CVD reactor (see process 3, Table 3.12) with a volume flow of 125 sccm of gas, which was introduced through both the ring dies and the main inlet.

The reduced gas velocities lead to higher layer growth rates (Figure 3.50 and Figure 3.51). As the precursor molecules have more time to react with the inner surface of the demonstrator, the growth rate of the coating is increased.

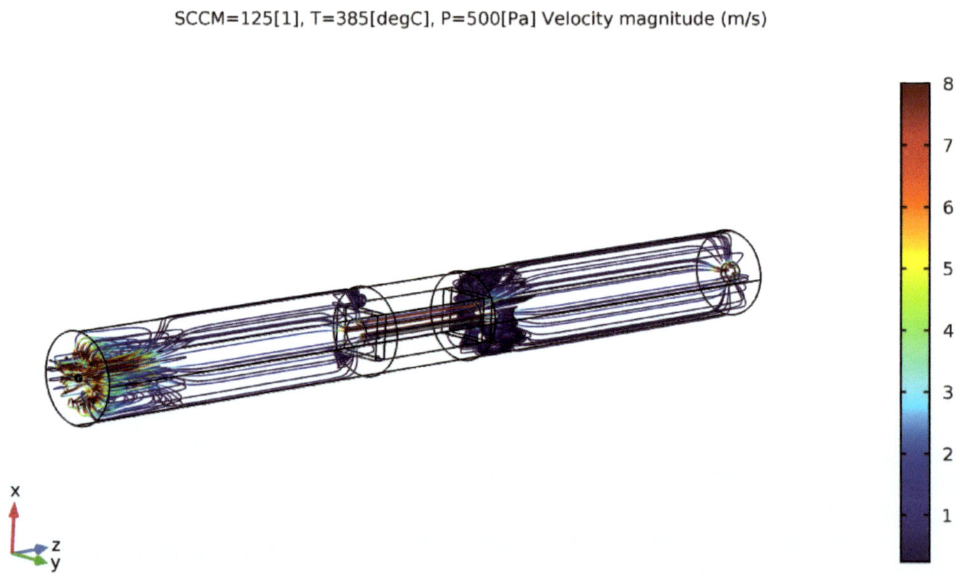

Figure 3.50 Velocity flow lines within the reactor volume with the demonstration tool with plates (reduced gas flow) (Image source: Non-profit KIMW Forschungs-GmbH)

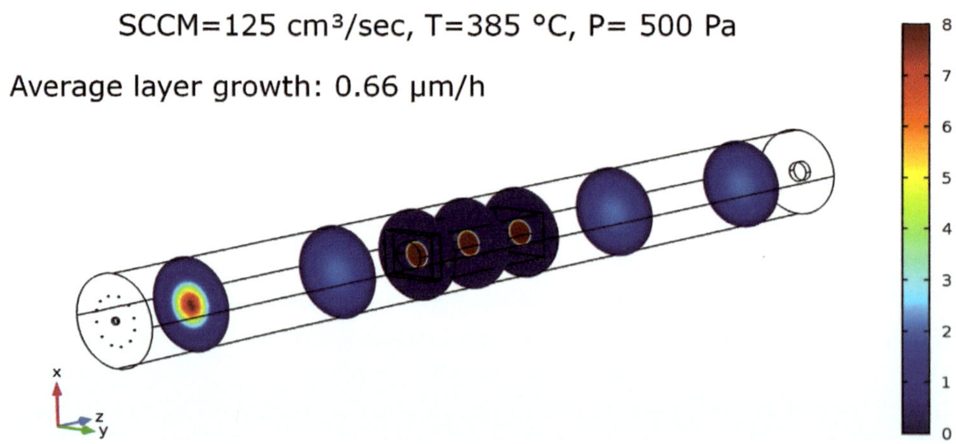

Figure 3.51 Velocity cross-sections in the reactor volume with the demonstration tool with plates (reduced gas flow) (Image source: Non-profit KIMW Forschungs-GmbH)

A heat transfer analysis was also carried out to investigate the temperature distributions within the CVD reactor. The 3D model shown in Figure 3.52 of the reactor is used. The simulations show how the heat is transferred from the copper (heating elements) to the ceramic and steel cylinders. The experiments show a better coating rate on the samples when a temperature gradient in the CVD reactor is taken into account. A high-temperature element is also considered in the outlet section, as it helps to decompose the precursor in a suitable manner. The simulation model was also created considering a temperature gradient.

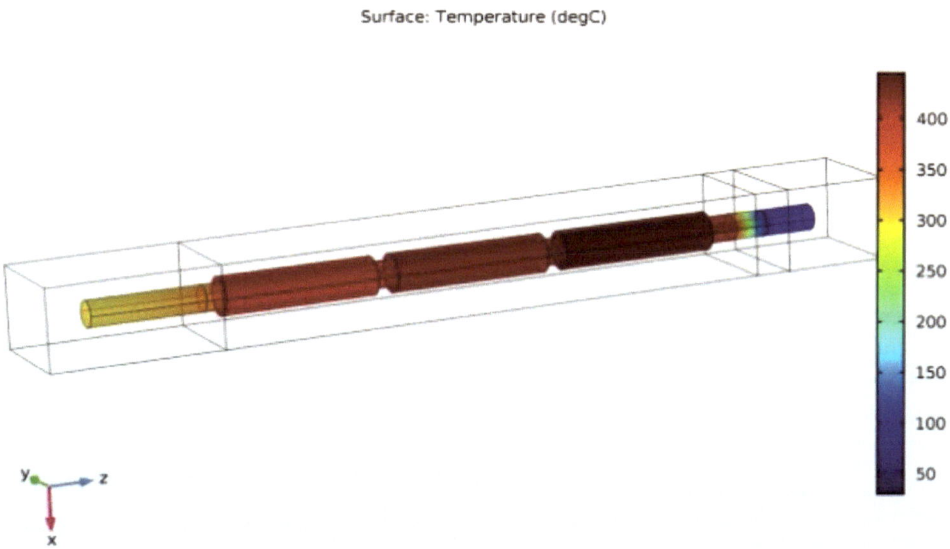

Figure 3.52 Temperatures in different sections of the reactor (Image source: Non-profit KIMW Forschungs-GmbH)

The temperatures were measured in various sections of the CVD reactor on the steel cylinder during the tests, and also checked in the simulations. By optimizing the simulation model, e. g. by defining a thin layer to achieve better insulation than in the real reactor, the temperatures in three sections show optimum values (Table 3.13 and Figure 3.53).

Table 3.13 Comparison of the Temperature Values Measured on the Real CVD Reactor and Those Taken From the Simulation at Intersection Points Corresponding to the Position of the Thermocouples in the CVD Reactor

Version	T1 (°C)	T2 (°C)	T3 (°C)
Experiment	394	406	412
Simulation	392	408	421

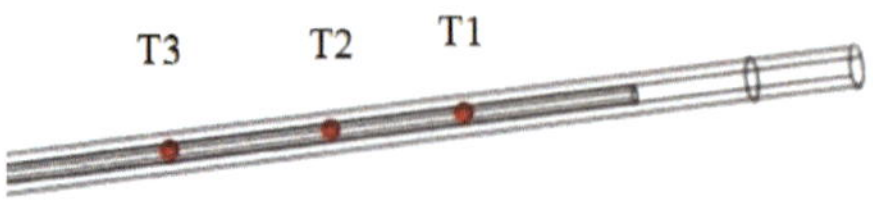

Figure 3.53
Positions of the segments within the reactor steel tube at which the resulting temperature was taken into account (Image source: Non-profit KIMW Forschungs-GmbH)

3.4.6 Conclusion

As part of the investigations, the simulation results were compared with the experimental results of the CVD coating tests. With the help of the simulations, it is possible to make predictions about the effects of the reduced gas flow during the coating time. In order to be able to predict the coating thickness, further background information about the reaction processes during the coating process is required. In addition, a deposition rate for the coating material on the surface must be defined in order to be able to take the chemical reactions into account. Of course, the chemical reaction module can also be helpful in addressing these questions.

For future simulations that reflect the real coating processes in the reactor to a greater extent, a multi-physics simulation should also be developed in which the temperature gradient present in the CVD reactor is used and influences the gas flow through the reactor.

References for Section 3.4

[1] Bathe, K. J.: *Finite Element Method* (2nd edn), Springer, 2002

[2] Knothe, K.; Wessels, H.: *Finite Elemente, Eine Einführung für Ingenieure*, Springer, 1999

[3] Vajna, S.; Weber, C.; Zemann, K.; et al.: *CAx für Ingenieure* (3rd edn), Springer Vieweg Verlag, 2018

[4] Heiroth, S.; Ghisleni, R.; Lippert, T.; Michler, J.; Wokaun, A.: Optical and mechanical properties of amorphous and crystalline yttria-stabilized zirconia thin films prepared by pulsed laser deposition, *Acta Materialia*, 2011, 59: 2330–2340

[5] Atakan, B.; Khlopyanova, V.; Mausberg, S.; Mumme, F.; Kandzia, A.; Pflitsch, C.: CVD and analysis of thermally insulating ZrO_2 layers on injection molds, *Physica Status Solidi C*, 2015, 12: 878–885

[6] Fornalczyk, G.; et al.: Yttria-Stabilized zirconia thin films via MOCVD for thermal barrier and protective applications in injection molding, *Key Engineering Materials*, 2017, 742: 427–433

[7] Navier–Stokes Equations, *https://www.comsol.de/multiphysics/navier-stokes-equations*, accessed on 01 December 2022

4 Measurement Technology for Coating Characterization

4.1 Crater Grinding Method

Dr. Ruben Schlutter

Crater grinding is often used to determine coating thicknesses in accordance with ISO 26424-11 [1] and to evaluate the wear resistance of coatings and materials in accordance with ISO 26423 [2]. The same test setup is used for both questions, whereby the coating is completely penetrated in tests to determine the coating thickness, while it is only abraded in a wear measurement [3].

The experimental setup (Figure 4.1, left) consists of a drive shaft with traction rings. These prevent the test ball from slipping so that the grinding path covered by the ball during the test can be determined. The ball usually has a diameter of 30 mm and is made of hardened steel. The shaft is set in rotation. An abrasive applied to the ball is used to grind a crater, the so-called calotte, into the sample. The grinding time and the grinding path are specified as test parameters. Depending on the grinding depth, the radius of the calotte in the layer or even individual layers up to the substrate can be evaluated microscopically (Figure 4.1, right). As the radius of the sphere is several orders of magnitude larger than the layer thicknesses to be examined, the section is made at a very shallow angle, which widens the layer. The contact force of the ball should remain constant during the tests. The angle of attack of the ball is 60°. The distance between the sample and the shaft is standardized using a setting gauge so that a constant contact force of the ball of 0.54 N is ensured. The contact force has a significant influence on the wear rate [4, 5].

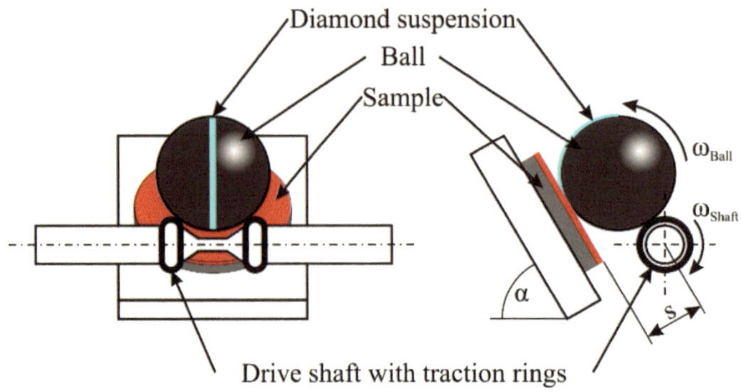

Figure 4.1 Testing device for spherical ground joints (Image based on [1, 4])

Before starting the test, the sample must be cleaned of dirt, grease, and oil using a suitable cleaning agent. The ball must also be thoroughly cleaned before the test. The area to be measured is adjusted. As the calotte impressions are very small, several measurements can be carried out on one sample. A minimum distance of 1–2 mm between the calottes is recommended [3, 5].

As the surface roughness also has an influence that should not be neglected, several measurements should always be carried out on a sample in order to be able to make a valid statement about the coating thickness. As a rule of thumb, it can be assumed that the thickness of the coating should be a factor of five greater than the R_z value of the coating. The repeatability can also be determined in this way. When evaluating the calottes, care must be taken to ensure that there is a sharp calotte edge without scoring or chipping. If this is the case, the abrasive may be unsuitable for the coating or coating system, and should be adapted by changing the grain size and shape of the abrasive particles or the carrier medium (liquid or paste). The abrasive must be adapted empirically [3].

4.1.1 Determination of the Coating Thickness

If the aim of the experiment is to determine the thickness of the coating, then the coating must be completely ground through so that the substrate is visible. Figure 4.2 shows the basic concept of the test with the relevant parameters used to determine the coating thickness [4].

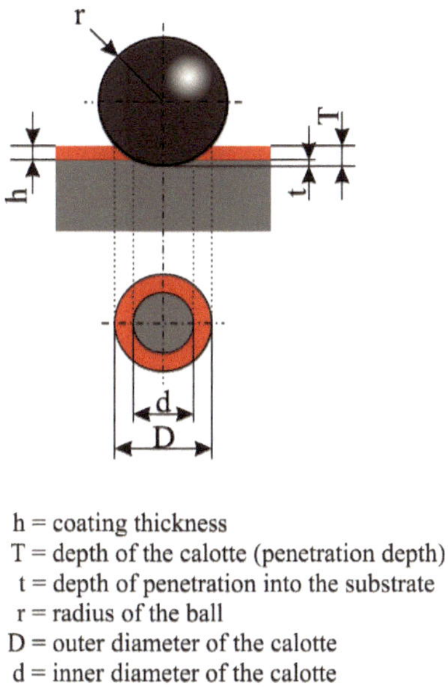

Figure 4.2
Coating thickness measurement
(Image based on [4])

Figure 4.3 shows a coating thickness measurement on a multi-layer coating. During the coating process, alternating layers of yttrium and phosphorus-doped zirconium oxide were deposited. This procedure is necessary in order to achieve a high coating thickness without cracks running through the entire coating thickness or the coating flaking off due to its brittle properties. The yttrium-doped zirconium oxide layer forms crystalline layers. These offer good thermal insulation properties, but are very porous. This layer is therefore combined with a phosphorus-doped layer, which forms an amorphous layer. This layer is very dense, but also very brittle, which is why high layer thicknesses cannot be deposited. By alternating the deposition of the layer systems, the advantages can be combined and the disadvantages eliminated [5].

The penetration depth of the ball depends on the radius of the ball and the outer calotte diameter and can be calculated according to Equation 4.1.

$$T = r - \sqrt{r^2 - \frac{D^2}{4}} \tag{4.1}$$

T = Depth of the calotte (penetration depth) (mm)

r = Radius of the ball (mm)

D = Outer calotte diameter (mm)

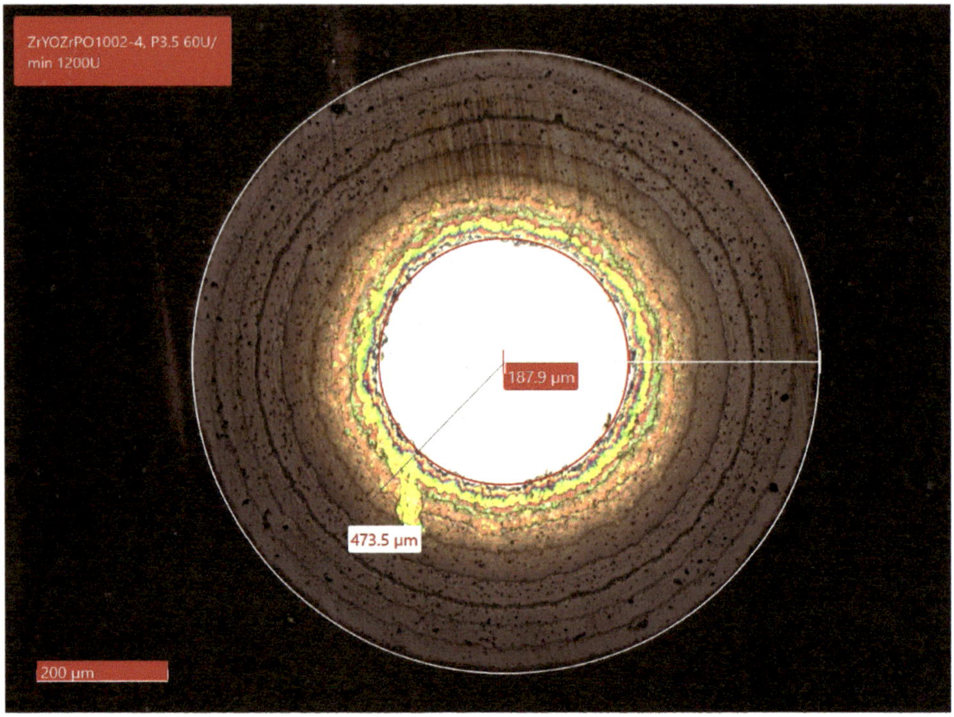

Figure 4.3 Coating thickness measurement on a multi-layer yttrium and phosphorus-doped zirconium oxide coating (Image source: Non-profit KIMW Forschungs-GmbH)

As the ball has completely penetrated the layer, the penetration depth of the ball into the substrate can be determined according to Equation 4.2.

$$t = r - \sqrt{r^2 - \frac{d^2}{4}} \tag{4.2}$$

T = Penetration depth into the substrate (mm)

R = Radius of the ball (mm)

D = Inner calotte diameter (mm)

The coating thickness h is then the difference between the penetration depth of the ball and the penetration depth of the ball into the substrate, so that this is calculated according to Equation 4.3.

$$h = T - t$$

$$h = \sqrt{r^2 - \frac{D^2}{4}} - \sqrt{r^2 - \frac{d^2}{4}} \tag{4.3}$$

As a rule, only a small amount is ground into the substrate. In addition, the coatings are very thin compared to the diameter of the ball, so that Equation 4.3 can be simplified and Equation 4.4 is used to calculate the layer thickness [4].

$$h = \frac{D^2 - d^2}{8 \cdot r} \tag{4.4}$$

The accuracy of determining the coating thickness depends largely on the exact determination of the calotte diameter. The grinding depth should also be selected so that the outer calotte diameter is approximately twice as large as the inner calotte diameter (Equation 4.5).

$$D \approx 2 \cdot d \tag{4.5}$$

4.1.2 Determination of Wear Resistance

In contrast to the determination of the coating thickness, the coating is not completely penetrated by the ball when examining the wear resistance. Figure 4.4 shows the basic end of the test with the relevant parameters used to determine the coating thickness. The ground-in calotte is considered to be ideally spherical. The wear volume can then be calculated using Equation 4.6 [6].

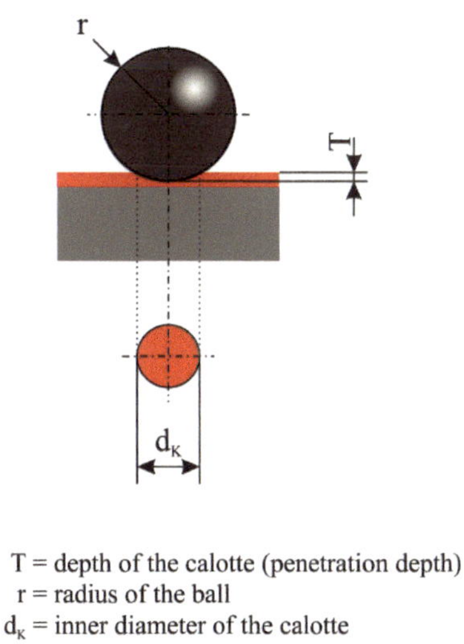

Figure 4.4
Determination of the wear rate (Image based on [4])

$$V_W = \frac{\pi}{3} \cdot T^2 \cdot (3 \cdot r - T) \tag{4.6}$$

V_W = Wear volume (mm^3)

T = Depth of the calotte (mm)

r = Radius of the sphere (mm)

Figure 4.5 shows the determination of the wear rate on a tungsten carbide coating. The coating was not completely ground off during the test. The radius of the resulting calotte is 230.7 µm. The wear rate according to Equation 4.8 is approximately 8×10^{-15} $m^3/(m\ N)$.

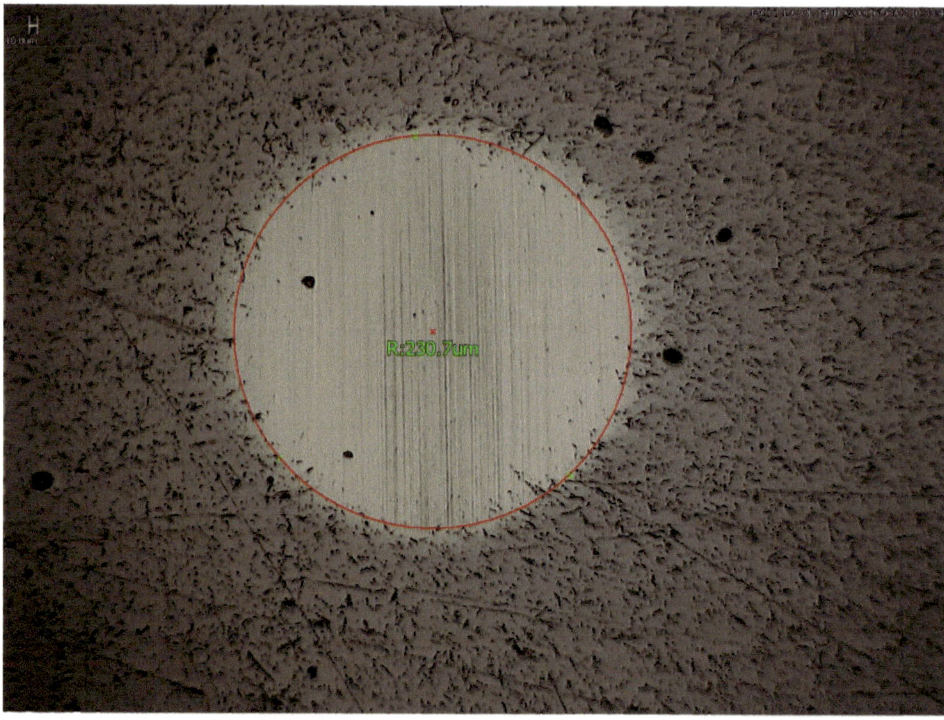

Figure 4.5 Determination of the wear rate on a tungsten carbide coating (Image source: Non-profit KIMW Forschungs-GmbH)

The depth of the calotte can be determined using the microscopically determined calotte diameter according to Equation 4.7.

$$T = r - \sqrt{r^2 - r_K^2} \tag{4.7}$$

r_K = Radius of the calotte (mm)

The wear rate is calculated according to Equation 4.8, and describes the abraded volume per grinding path and ball force.

$$V_R = \frac{V_W}{s \cdot F_S} \tag{4.8}$$

V_R = Wear rate ($mm^3/(mm\ N)$)

S = Grinding path (mm)

F_S = Spherical force (N)

Since the ball diameter (or the ball radius) and the number of ball rotations are known, the grinding path can be calculated according to Equation 4.9.

$$s = 2 \cdot \pi \cdot r \cdot n \tag{4.9}$$

n = Number of ball rotations (—)

References for Section 4.1

[1] ISO 26424-11:2008: Fine ceramics (advanced ceramics, advanced technical ceramics) – Determination of the abrasion resistance of coatings by a micro-scale abrasion test, 2008

[2] ISO 26423:2009: Fine ceramics (advanced ceramics, advanced technical ceramics) – Determination of coating thickness by crater-grinding method, 2009

[3] Bethke, R.; Näcker, N.: Schichtdicke und Verschleiß präzise bestimmen, *Journal für Oberflächentechnik*, 2020, 60: 72–77

[4] kaloMAX II Manual [Version 1.02], BAQ GmbH

[5] Fornalczyk, G.; Sommer, M.; Mumme, F.: Yttria-stabilized zirconia thin films via MOCVD for thermal barrier and protective applications in injection molding, *Key Engineering Materials*, 2017, 742: 427–433

4.2 Scanning Electron Microscopy

Dr. Ruben Schlutter

4.2.1 Introduction

Microscopic methods are often the most direct and best way to record and examine structural features of surfaces. Suitable preparation creates a contrast due to the different reflectivity and absorptivity of the material structure in the microscope image.

It is important that adjacent structures appear separate from each other so that they can be viewed individually and hence resolved. The resolution capacity of a microscope is limited by the numerical aperture of the objective and the wavelength of the light used, and can be calculated using Equation 4.10. Raleigh defined this relationship for optical devices in general. Abbe proved that it is also valid in microscopy [1, 2].

$$d = \frac{\lambda}{A} \tag{4.10}$$

A = Numerical aperture of the lens (—)

d = Resolution capacity of the microscope (nm)

λ = Wavelength of the light used (nm)

The numerical aperture of the lens can be calculated according to Equation 4.11 and depends on the refractive index of the medium surrounding the lenses and the aperture angle of the front lens.

$$A = n \cdot \sin\alpha \tag{4.11}$$

n = Refractive index of the surrounding medium (—)

α = Half aperture angle of the front lens (—)

In a vacuum, the refractive index is 1, as light propagates exactly at the speed of light in a vacuum. An increase in resolution can thus be achieved by using shorter wavelength light or lenses with an enlarged aperture. In practice, a numerical aperture of 0.95 can be achieved, whereby the maximum resolution up to which two points can be distinguished from each other is limited to approx. 0.5 µm [3].

The wavelength at which the electrons are accelerated can be determined using the de Broglie wavelength according to Equation 4.13, and depends on the speed of the electrons and therefore on the acceleration voltage. The higher the accelerating voltage, the higher the speed of the electrons and the shorter their wavelength. As the speed of the electrons increases, relativistic effects also come into play, so that Equation 4.15 must be used for the calculation [4, 5].

$$\lambda_{\mathrm{DB}} = \frac{h}{m_{\mathrm{e}} \cdot v} \cdot 10^9 \tag{4.12}$$

$$\lambda_{\mathrm{DB}} = \frac{h}{\sqrt{2 \cdot m_{\mathrm{e}} \cdot e \cdot U_{\mathrm{B}}}} \cdot 10^9 \tag{4.13}$$

λ_{DB} = de Broglie wavelength (nm)

h = Planck's quantum of action = 6.626×10^{-34} (J·s)

m_{e} = Mass of the electron = 9.109×10^{-31} (kg)

v = Velocity of the electron (m/s)

e = Charge of the electron = 1.602×10^{-19} (A·s)

U_{B} = Acceleration voltage (V)

$$\lambda_{\mathrm{DB}} = \frac{h \cdot c}{m_{\mathrm{e}}} \cdot \sqrt{\frac{1}{v^2} - \frac{1}{c^2}} \cdot 10^9 \tag{4.14}$$

$$\lambda_{\mathrm{DB}} = \frac{h \cdot c}{\sqrt{2 \cdot e \cdot U_{\mathrm{B}} \cdot m_{\mathrm{e}} \cdot c^2 + (e \cdot U_{\mathrm{B}})^2}} \cdot 10^9 \tag{4.15}$$

c = Speed of light = 2.998×10^8 (m/s)

The resulting velocities and wavelengths of the electrons are shown in Table 4.1. Magnetic fields are used to bundle and focus electron beams. These act as electron lenses [6]. With modern scanning electron microscopes, the electron beam can be focused to a diameter of 5–10 nm, with a resolution of 8 nm being achieved in favorable cases [3].

Table 4.1 Velocities and Wavelengths of an Electron Beam as a Function of the Acceleration Voltage

Acceleration voltage (V)	v/c (%)	Wavelength of the electrons (nm)
1000	6.26	0.0388
2000	8.85	0.0274
5000	13.89	0.0173
10,000	19.50	0.0122
20,000	27.19	0.0086
50,000	41.27	0.0054

4.2.2 Instrument Setup

Scanning electron microscopy (SEM) is used to detect the signals generated when an electron beam is scanned over a sample surface to be imaged. Figure 4.6 shows the basic structure of a scanning electron microscope and the different detectors. The signals that can be obtained by irradiation with an electron beam include secondary electrons (SE), backscattered electrons (BSE), and characteristic X-rays. As these enable a detailed surface analysis of the sample, the scanning electron microscope is used in a variety of areas, from analyzing surface structures to material development.

Scanning electron microscopes use electrons instead of light waves to transport information and visualize surfaces. The electrons are shot from the cathode to the anode by an accelerating voltage of 1–50 kV from the cathode to the anode. The cathode is heated to make it easier for the electrons to be released from the cathode. Between the cathode and the anode there are condenser lenses, deflection coils, and the objective lens, which focus the electrons into a circular beam with a diameter of 5–10 nm. The electron beam scans the surface of the sample in a grid pattern [3].

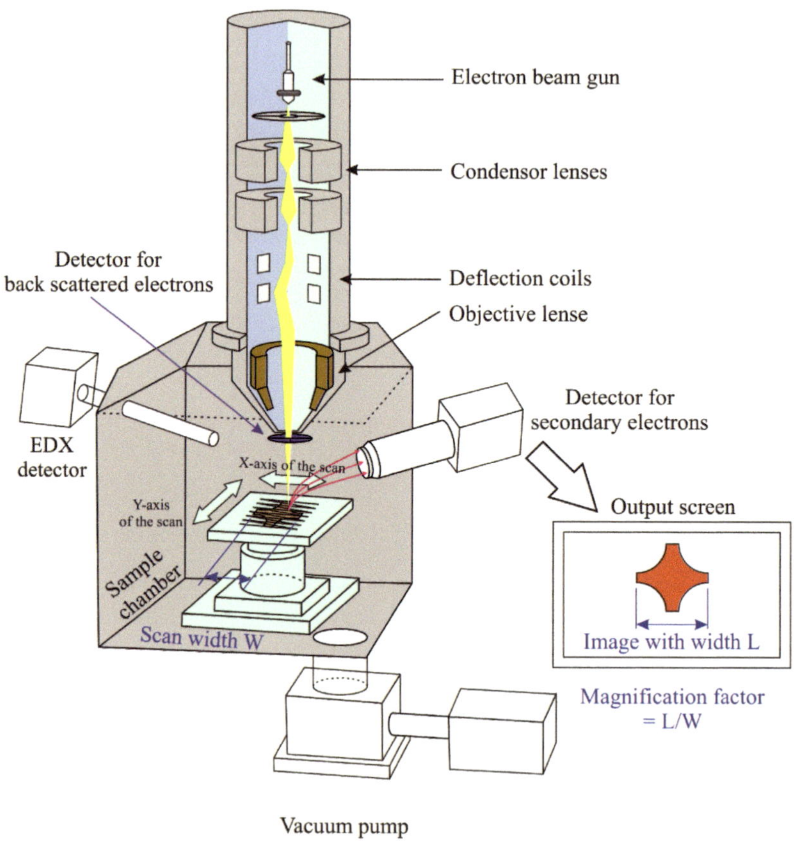

Figure 4.6 Basic structure of a scanning electron microscope (Image based on [7])

When a non-conducting sample is examined, the sample surface is charged by the incident electrons and imaging artifacts such as abnormal contrast and image aging occur. Low-vacuum SEM observation was developed as a method to avoid or reduce these problems. In low-vacuum SEM, positive charges generated by collisions between the incident electrons and the residual gas molecules in the chamber are used to neutralize the negative charge on the sample surface, thus reducing the effects of charge build-up [8].

The image to be output is then compiled on the computer from the individual measured values, point by point and line by line. The electron beam retains its cylindrical shape as far as possible regardless of the sample height and surface inaccuracy, so that higher and lower lying areas can be easily captured. The depth of field is very good. The sharpness is limited by the scattering during reflection and the receiving surface of the sensors for collecting the electrons [3].

When the electron beam hits the sample, the electrons interact with the electrons in the sample. The electrons influenced by the electron beam can be detected by various detectors and used to generate images. Secondary and backscattered electrons are mainly analyzed (Figure 4.7).

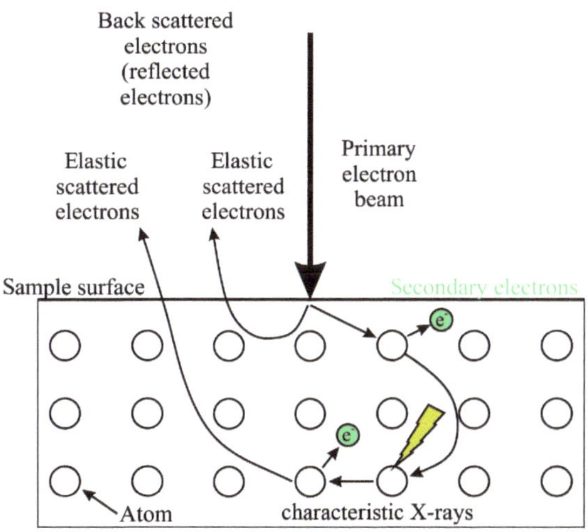

Figure 4.7
Scattering of the primary electrons in the sample (Image based on [7])

The mode for detecting secondary electrons is the most important, since these electrons, thanks to their low exit energy of a few eV, can be easily collected by means of a positively biased collector lattice placed on one side of the sample. Behind the collector lattice, the SEs are accelerated onto a scintillator biased at +10 kV, and the light quanta generated are recorded by a photomultiplier [9].

Secondary electrons are produced by inelastic collisions of the primary electrons with an electron bound in an atom or the atom itself. In some cases, a bound electron, the secondary electron, is released from the atom. Energy is lost due to the inelastic collision, so that the secondary electrons have a low energy, as shown in Figure 4.8. They must therefore originate from the area near the surface (see Figure 4.9). Due to the low energy, the best resolution can be achieved with the secondary electrons. The secondary electrons can also be easily deflected, which is why the SE detectors are often arranged laterally in the sample chamber [3].

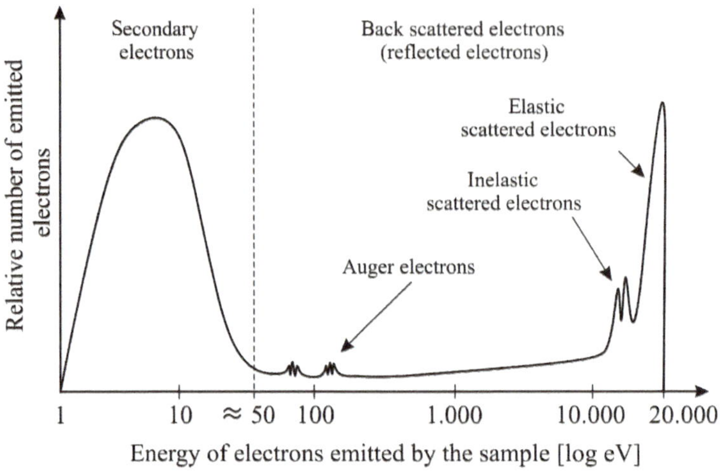

Figure 4.8 Energy distribution of the emitted electrons (Image based on [7])

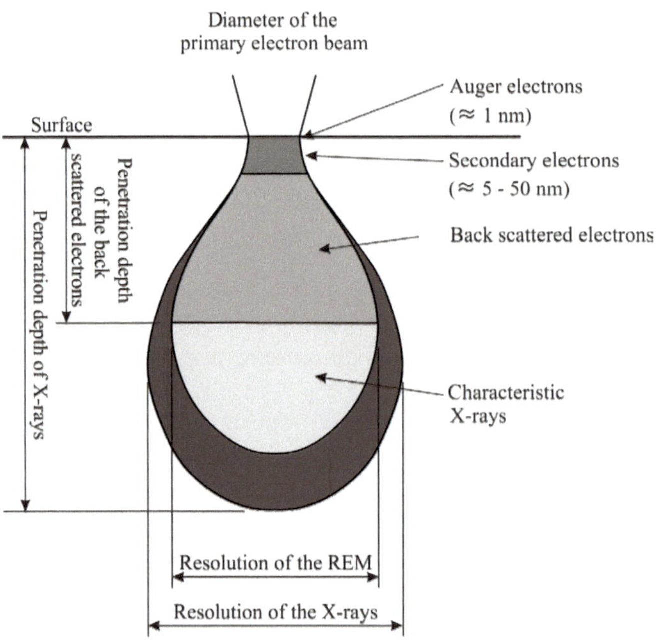

Figure 4.9 Areas where electrons and X-rays are formed in the sample (Image based on [3])

In contrast to the secondary electrons, backscattered electrons have a significantly higher energy (see Figure 4.8). They move along straight paths and are not influenced by electrostatic fields. They originate from deeper areas of the sample, as shown in Figure 4.9. Usually, the detector for detecting the backscattered electrons is positioned directly below the objective lens, as the backscattered electrons are reflected directly in the direction of the electron beam. The sample is viewed directly from above. The information content of the backscattered electrons is strongly dependent on the atomic number, because heavy elements emit more electrons than light elements. Individual areas of the detector can be evaluated separately so that the structure of a three-dimensional image can be calculated from the shadows cast by the individual areas [3, 7].

Figure 4.10 shows the dependence of the proportion of scattered electrons on the atomic number of the elements. This dependence produces a contrast in the backscattered electron image. Elements with a lower atomic number appear darker and elements with a higher atomic number appear brighter. A rough conclusion about the element distribution is possible with the help of the backscattered electron image.

It is also known that the electron reflectivity changes depending on the angle of incidence of the electron beam, and increases as the tilt angle of the sample increases. This means that the topographical information of the sample surface is also contained in the signal of the backscattered electrons.

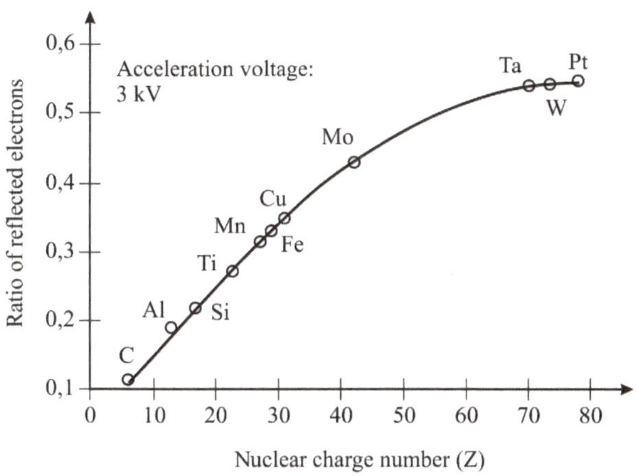

Figure 4.10 Proportion of scattered electrons (Image based on [7])

Below the region in which the backscattered electrons are generated, X-rays are produced (Figure 4.7 and Figure 4.9). These can be examined by energy-dispersive X-ray (EDX) spectroscopy.

An electron is knocked out of a lower shell by an electron impact, as shown in Figure 4.11. The free space is immediately occupied by an electron from an energetically higher shell. During the transition to the lower shell, the electron emits an X-ray photon with a characteristic energy. Since the energy levels between the shells are specific to the element, the energy of the photon can enable identification of the shell jump and element involved. The corresponding line in the spectrum characterizes the element, and the height of the peak is a measure of the fraction of the corresponding element present [7, 10].

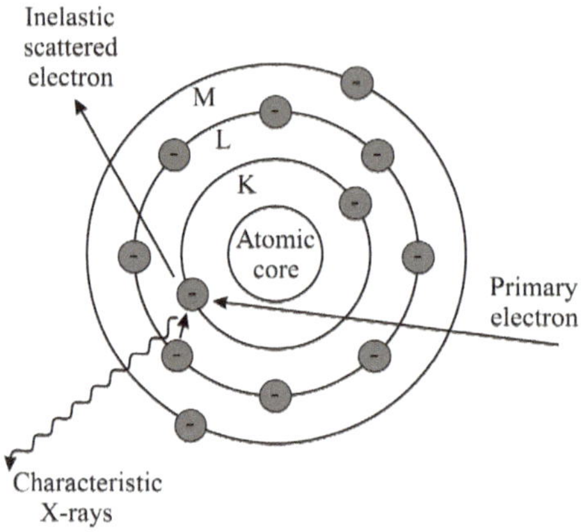

Figure 4.11
Generation of the characteristic X-rays in a magnesium atom (Image based on [7])

Figure 4.12 shows the qualitative regions from which the individual shell jumps originate. As the EDX spectroscopy is a microscopic analytical method, the origin of the photons must be known. The acceleration voltage of the primary electrons must be greater than the energy of the X-ray photons in order to knock the electrons out of their shells. The size of the range depends on the acceleration voltage and the type of electron jump. As electron jumps to a K-shell are lower in energy, the radius at which these jumps can occur is also smaller. In principle, higher-energy electron jumps and higher acceleration voltages increase the range in which the electron jumps can take place [7].

The photons generated during the electron jump are detected using the EDX sensor. Each incident photon generates a flash of light in the sensor, which is transmitted via an optical fiber and converted into an electrical signal in a multiplier. The pulses are processed in such a way that the entire X-ray spectrum of the scanned area can be recorded and evaluated [3].

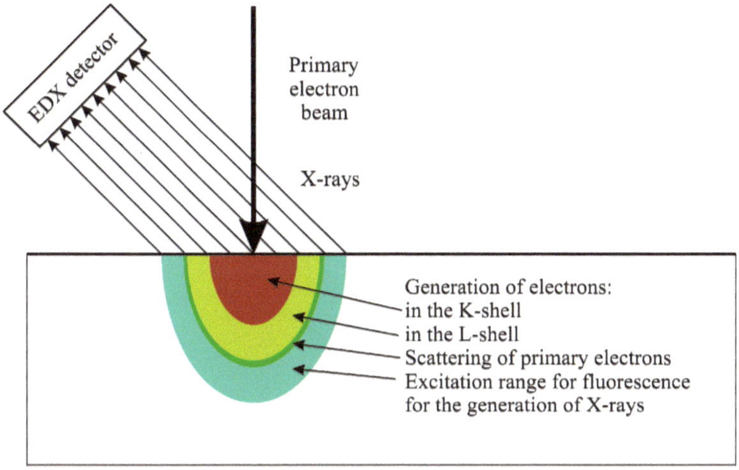

Figure 4.12 Areas for generating X-rays (Image based on [7])

4.2.3 Preparation of the Sample

The preparation of the sample depends on the type of analysis to be carried out. In the area of damage analysis, samples are generally not cleaned, so that any existing coatings can also be examined. When examining fracture patterns or coatings, cleaning is necessary, as the coatings impair the examinations. Depending on the type and degree of contamination, cleaning with water or ethanol may be sufficient, or cleaning in an ultrasonic bath may be necessary.

In addition, the samples must not change under electron bombardment. They must also not become electrically charged, as this would lead to overexposure of the image. A poorly conducting sample can either be examined with a UVD, as there is a certain electrical conductivity due to the remaining gas molecules in the low vacuum. On the other hand, sputtering, usually with gold or palladium, is possible, involving coating the sample with a thin electrically conductive layer, so that the electrons can flow off.

4.2.4 Sensors in a Scanning Electron Microscope

4.2.4.1 SE Sensor

The secondary electrons are produced by inelastic collisions of the primary electrons directly on and under the examined component surface. Due to the low energy of the electrons as they leave the sample, they contain information from the area close to the surface. The evaluation of images generated from secondary electrons is also close to the surface.

Figure 4.13 shows the secondary electron-image of a nozzle coated with zirconium oxide. The coating is damaged at the corner, and deposits of the original coating and the steel substrate can be seen.

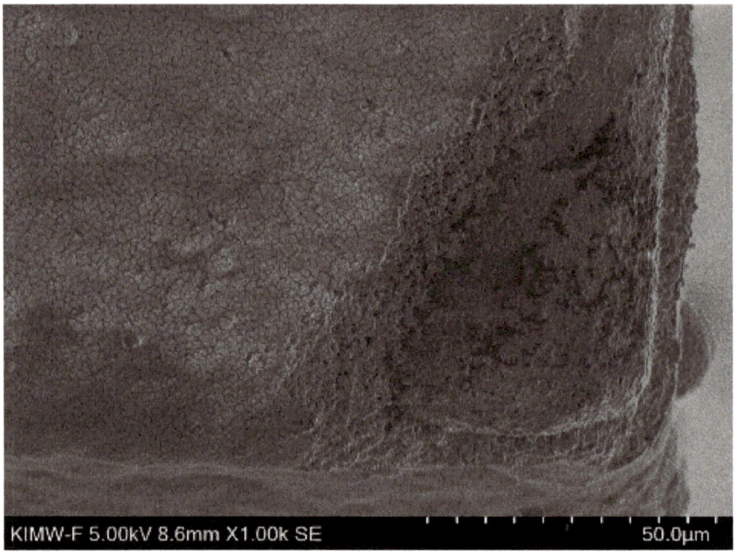

Figure 4.13 Secondary electron image of a zirconium oxide layer that has been broken off at one corner (Image source: Non-profit KIMW Forschungs-GmbH)

4.2.4.2 UVD Sensor

As it is difficult to detect low-energy secondary electrons by observation in the low-vacuum range, the ultra-variable pressure detector (UVD) was developed for investigations under low vacuum. Figure 4.14 shows the operating principle of the UVD. In order to efficiently detect secondary electrons generated by the sample when it is irradiated by an electron beam in a low vacuum, a bias voltage is applied to the front edge of the detector. This creates an electric field between the detector and the sample, which accelerates the secondary electrons generated on the sample surface. The secondary electrons then collide with the gas molecules in the low vacuum environment and ionize the molecules into positive ions and electrons. At the same time, light is generated. By capturing this light with the UVD during beam scanning, images can be taken that reflect topological information about the sample surface [8].

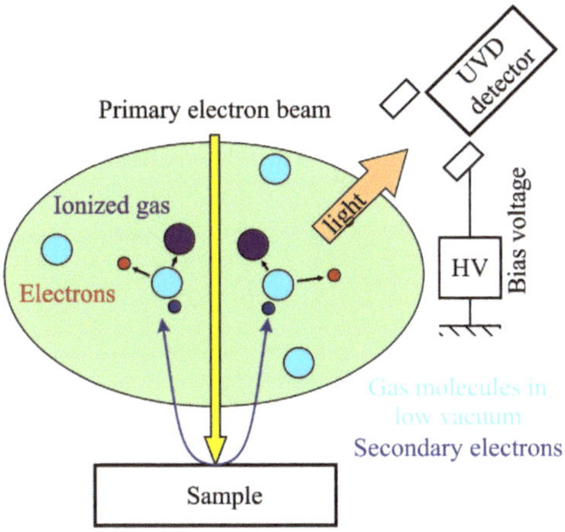

Figure 4.14
Schematic operating principle of the ultra-variable pressure detector (UVD) (Image based on [8])

The UVD can be used to analyze the surface in detail. The gas molecules present under low vacuum conditions can prevent charge build-up of a poorly conductive surface, so it is not necessary to sputter the sample to create an electrically conductive layer [8].

Figure 4.15 shows the fracture edge of a zirconium oxide layer on a steel substrate. The fracture edges of the individual layers are clearly visible.

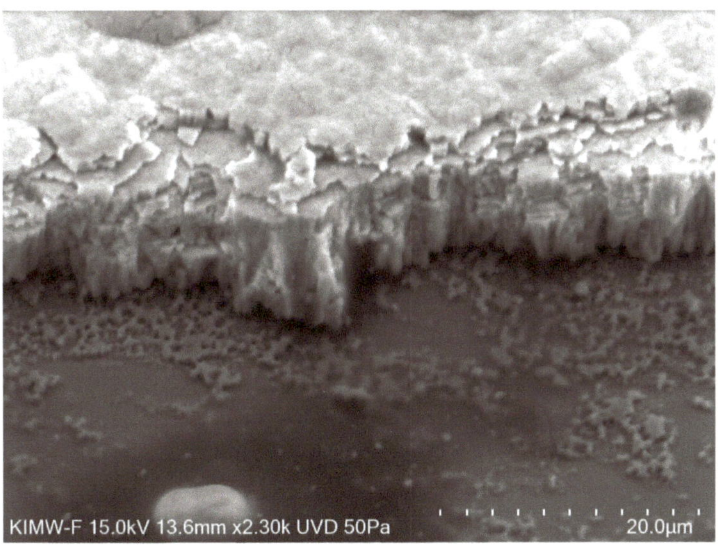

Figure 4.15 UVD image of a broken zirconium oxide layer (Image source: Non-profit KIMW Forschungs-GmbH)

4.2.4.3 BSE Sensor

When analyzing the backscattered electrons, the distribution of the sample composition and the topography of the sample can be examined. Figure 4.16 shows the backscattered electron image of a nozzle coated with zirconium oxide. In contrast to Figure 4.13, no surface details of the coating can be seen. In Figure 4.16, the zirconium oxide coating appears brighter, and the steel substrate can be seen in the dark areas at the bottom right.

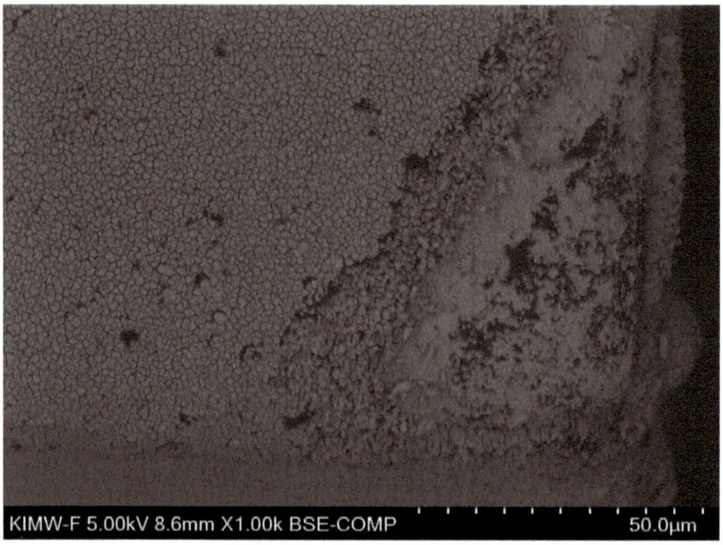

Figure 4.16 Backscattered electron-image of a zirconium oxide layer fractured off at a corner (Image source: Non-profit KIMW Forschungs-GmbH)

Figure 4.17 shows the evaluation of the individual sensor components of the BSE sensor. Depending on the direction, different shadows can be seen. The topology of the sample can be deduced from this shadow cast by calibrating the BSE sensor. The calibration depends on the acceleration voltage and the magnification of the sample.

In Figure 4.18 shows the topological profile of the sample after the calibration has been performed. All further investigations, such as the analysis of the 3D profile or an analysis of the surface roughness, can be carried out with this data.

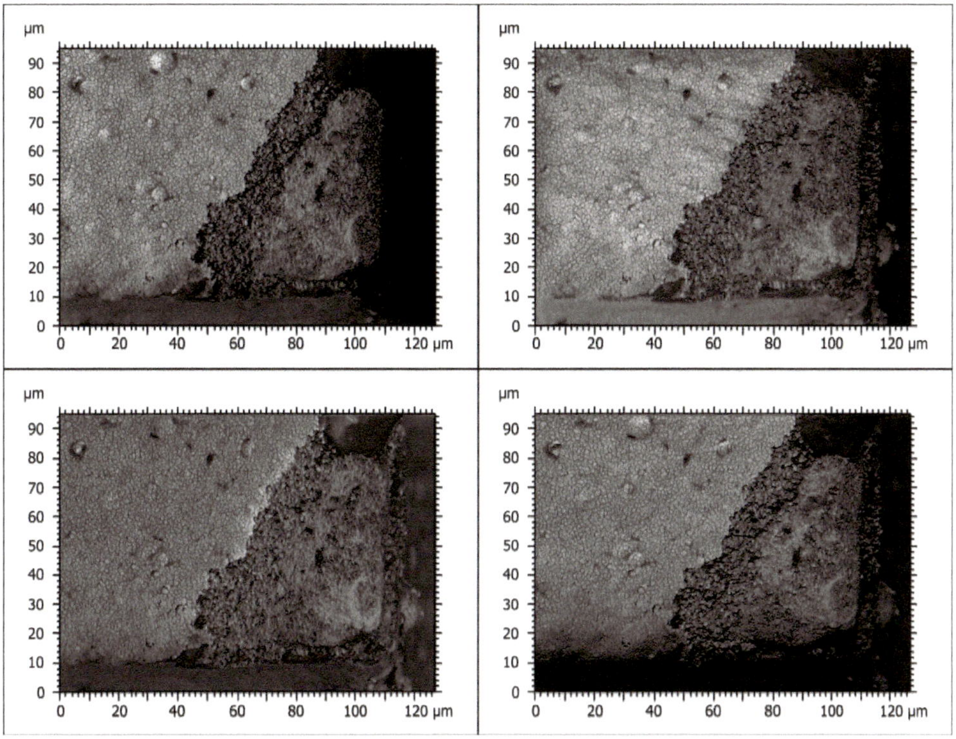

Figure 4.17 Evaluation of the individual sensor areas of the backscattered electron sensor to generate a 3D image (Image source: Non-profit KIMW Forschungs-GmbH)

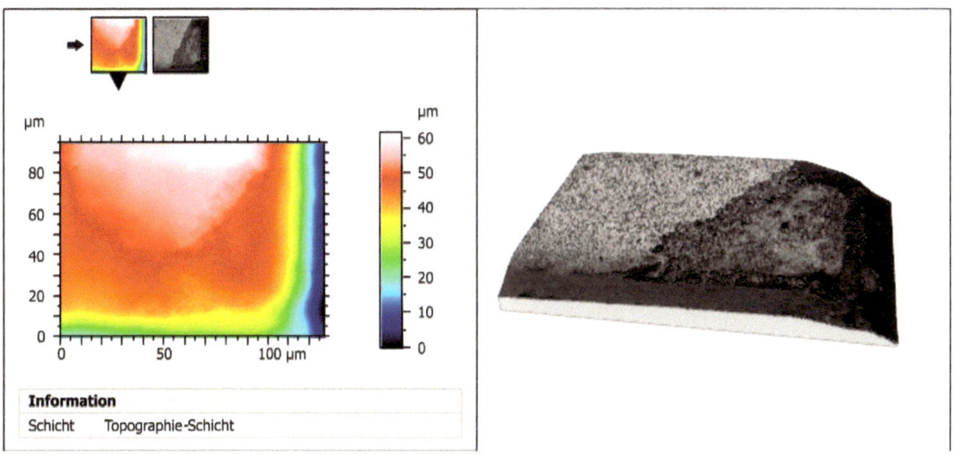

Figure 4.18 Calculation and output of the 3D image (Image source: Non-profit KIMW Forschungs-GmbH)

Table 4.2 summarizes the results of the investigation of the surface roughness of the coating. It is also possible to determine a surface roughness over individual lines. The height parameters were determined over the definition area (the red and white area in Figure 4.18) and calculated according to ISO 25178-2. The definitions of the parameters are briefly explained below [11]:

- S_q describes the mean square value of the height.
- S_{sk} is a measure of the inclination of the sample surface, and is calculated as the quotient of the mean third power of the ordinate values and the third power of the mean square height S_q.
- S_{ku} describes the curvature (kurtosis) of the sample surface and is calculated as the quotient of the mean fourth power of the ordinate values and the fourth power of the mean square height S_q.
- The largest value of the tip height describes the maximum tip height S_p.
- S_v is the largest value of the sinking depth.
- S_z is the sum of the largest value of the peak height and the largest value of the sink depth.
- S_a is the arithmetic mean of the absolute ordinate values.

Table 4.2 Analysis of the Height Profile According to ISO 25178

Parameter	Value	Unit
S_q	0.0985	µm
S_{sk}	-0.875	—
S_{ku}	22.5	—
S_p	0.725	µm
S_v	1.24	µm
S_z	1.97	µm

Figure 4.19 and Figure 4.20 show an analysis of the height profile of the fractured coating. The jump in the height characterizes the end of the coating and the lower plateau the steel substrate. The thickness of the deposited layer is approx. 7 µm.

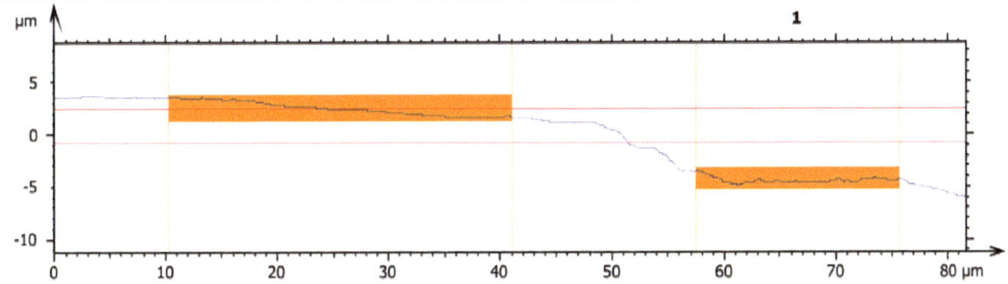

Parameter	Stufe 1	Einheit
Breite	18.3	µm
Maximale Tiefe	7.34	µm
Durchschnittstiefe	6.92	µm

Figure 4.19 Height profile of the fractured coating (Image source: Non-profit KIMW Forschungs-GmbH)

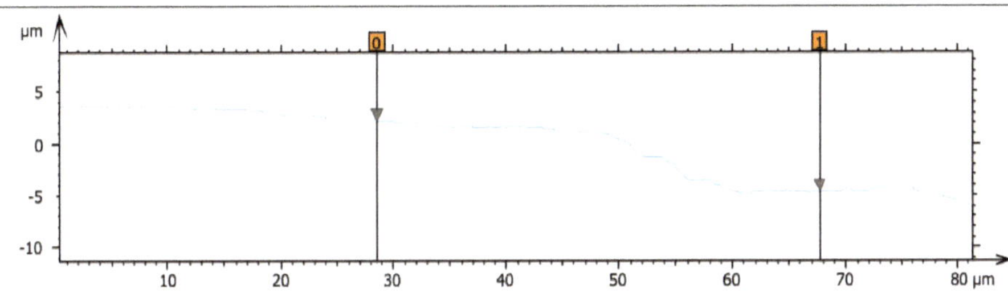

Parameter	0-1	Einheit
Horizontaler Abstand	39.2	µm
Höhenunterschied der Punkte	-6.85	µm

Figure 4.20 Spot height of the fractured coating (Image source: Non-profit KIMW Forschungs-GmbH)

4.2.4.4 EDX Sensor

Figure 4.21 shows the backscattered electron image of the sample surface to be examined. It is mainly used for orientation in order to be able to evaluate interesting areas. The sample surface was not cleaned before the examination.

Figure 4.22 shows the distribution of the most common elements in the EDX spectrum elements. It can be seen that the coating consists of zirconium. The fracture of the coating at the corner is almost complete, as the iron of the base substrate can be seen here. In the center of Figure 4.22 an accumulation of carbon can be seen. This is a deposit of polypropylene that adhered to the fracture edge during use of the coated nozzle. Residues of carbon are also present in the fracture edge. It is unclear whether this is also residue from the processed plastic or contamination from touching the nozzle after use.

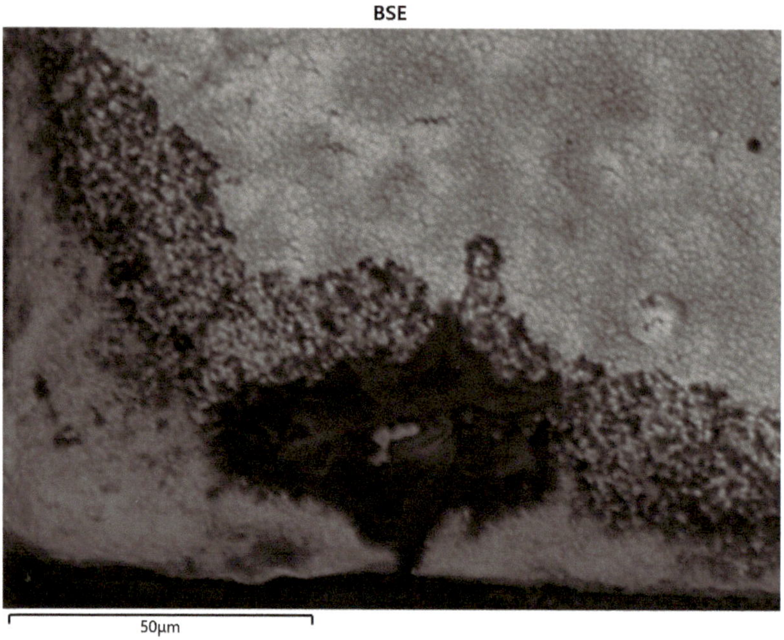

Figure 4.21 Backscattered electron image of the zirconium oxide layer fractured off at the corner (Image source: Non-profit KIMW Forschungs-GmbH)

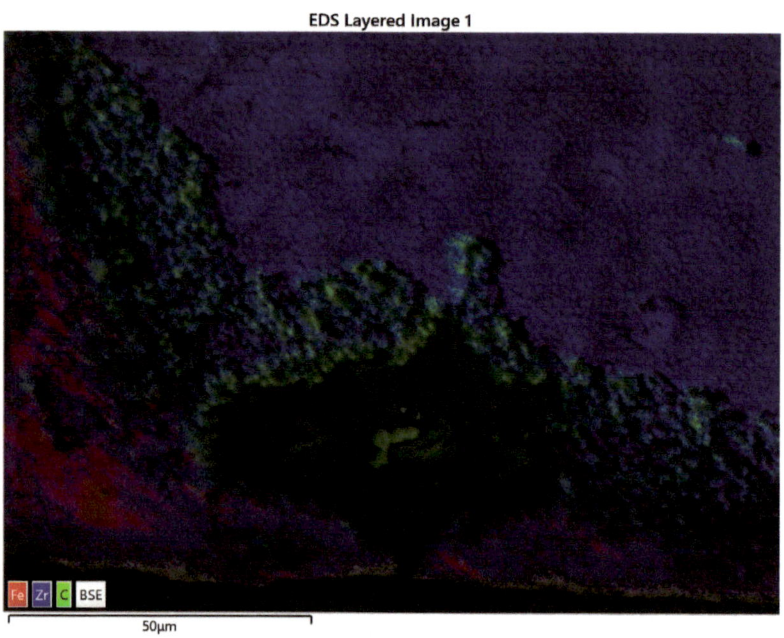

Figure 4.22 Color plot of the main elements at the fractured corner (Image source: Non-profit KIMW Forschungs-GmbH)

Figure 4.23 shows an element-by-element breakdown of the distribution of the most frequently occurring elements in the EDX spectrum, which was generated and analyzed together with Figure 4.22. It can be seen that the coated areas in particular, as well as the areas in which the substrate is partially visible, also contain oxygen. Since the layer does not consist of pure zirconium, but also of zirconium oxide, the presence of oxygen is to be expected.

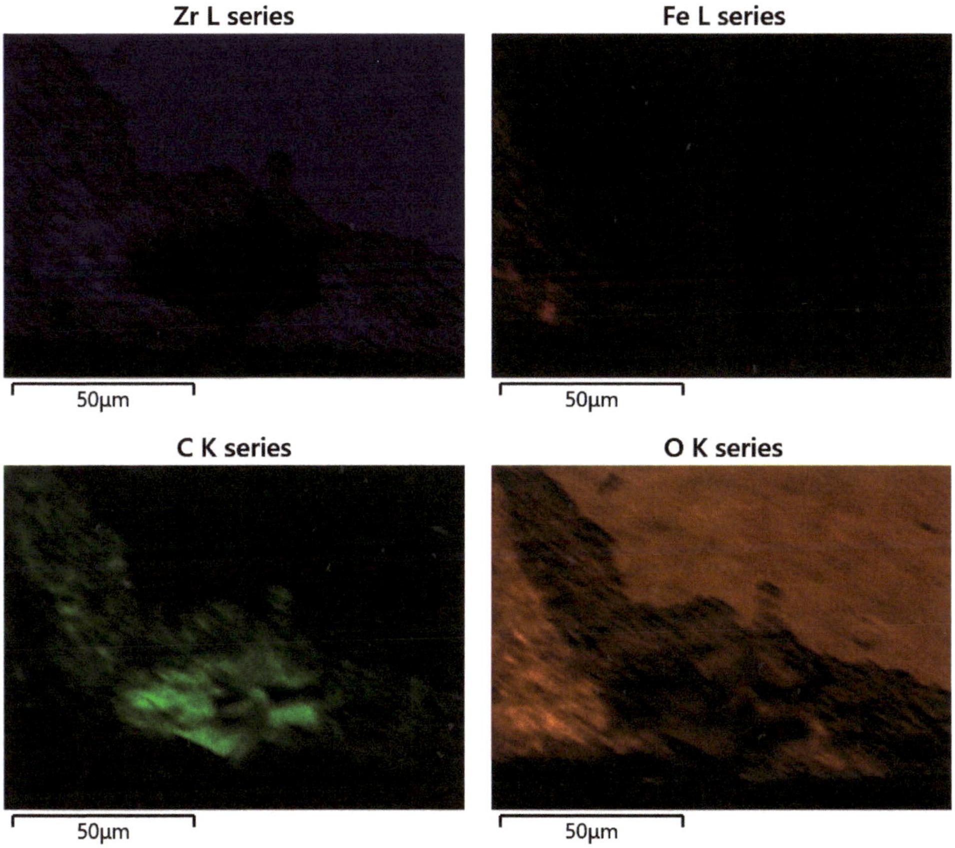

Figure 4.23 Distribution of the elements zirconium, iron, carbon, and oxygen at the fractured corner (Image source: Non-profit KIMW Forschungs-GmbH)

For a more precise evaluation of the elements in the sample, different areas were defined in which the entire EDX spectrum of the respective surface is evaluated. These areas are shown in Figure 4.24. The recorded spectra are shown in Figure 4.25–Figure 4.27. A quantitative summary of the detected elements can be found in Table 4.3.

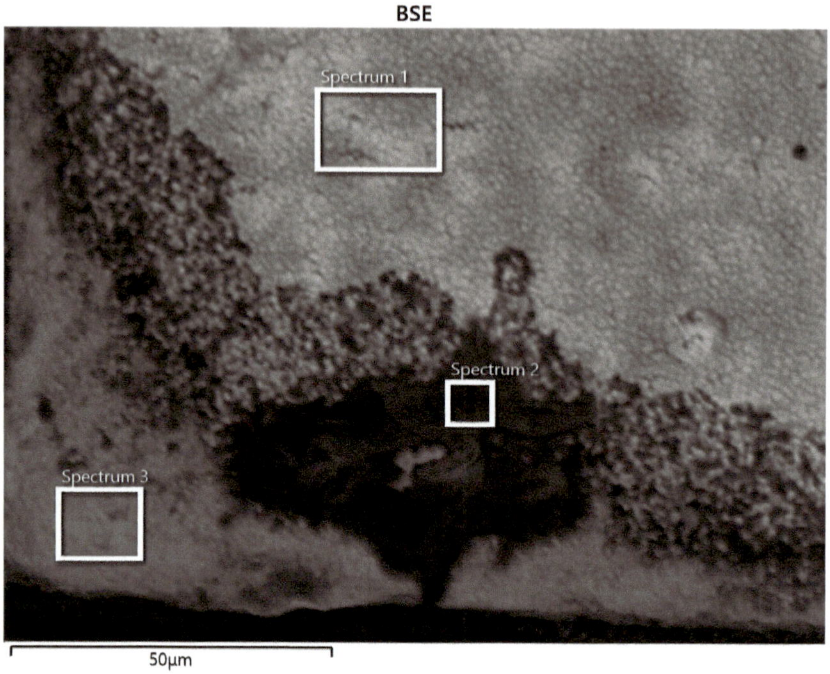

Figure 4.24 Position of the spectra for the EDX analysis (Image source: Non-profit KIMW Forschungs-GmbH)

Figure 4.25 shows the EDX spectrum for spectrum 1, which is located on the coated surface of the sample. Accordingly, the main elements detected are zirconium and oxygen. The proportion of carbon also present indicates impurities within the coating, which result from the precursor used in the coating. This is an organometallic molecule that has bound to the zirconium oxide and is dissolved by the acetylacetonate. During the coating process, the precursor decomposes so that the zirconium oxide is deposited on the surface to be coated. However, decomposition products of the precursor and the solvent can also be deposited in the coating.

Figure 4.26 shows the EDX spectrum of the attached plaque. It can be seen that it is mainly carbon and oxygen. The carbon results from the polymer used in the nozzle, while the sodium, chlorine, potassium, and oxygen are probably impurities, such as skin sweat, which have adhered to the nozzle before and after it came into contact with the polymer.

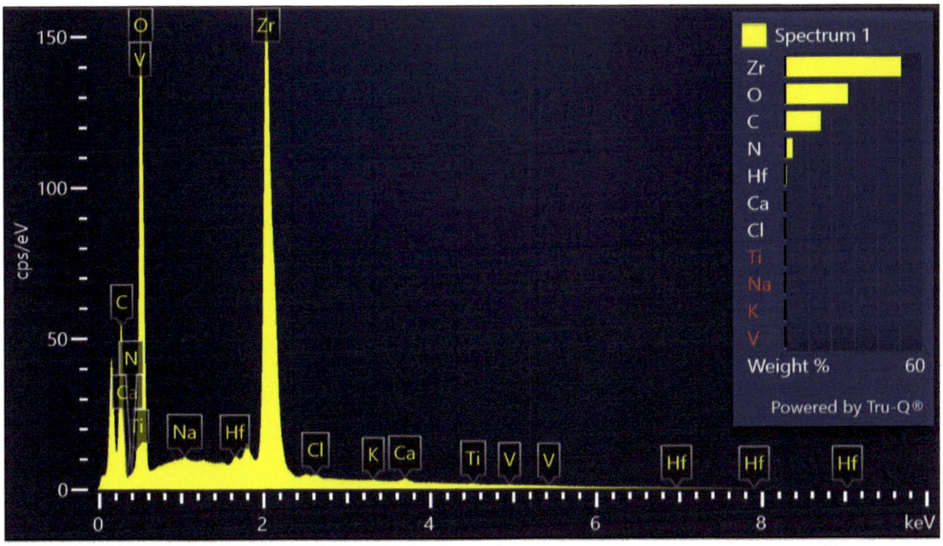

Figure 4.25 Elemental analysis for spectrum 1 (Image source: Non-profit KIMW Forschungs-GmbH)

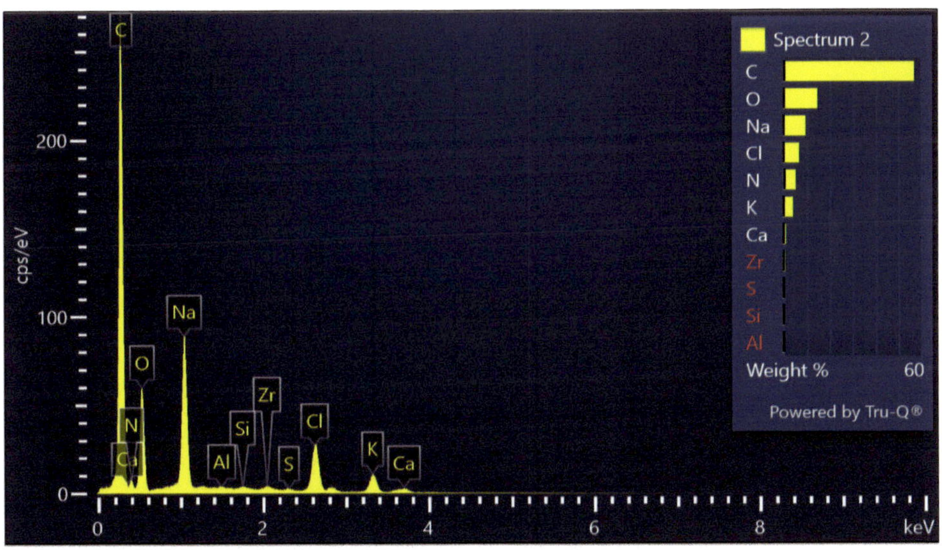

Figure 4.26 Elemental analysis for spectrum 2 (Image source: Non-profit KIMW Forschungs-GmbH)

Figure 4.27 shows the EDX spectrum in the area of the fracture. Areas of the zirconium oxide layer are still visible here. The steel substrate is also visible, which explains the proportions of boron and chromium.

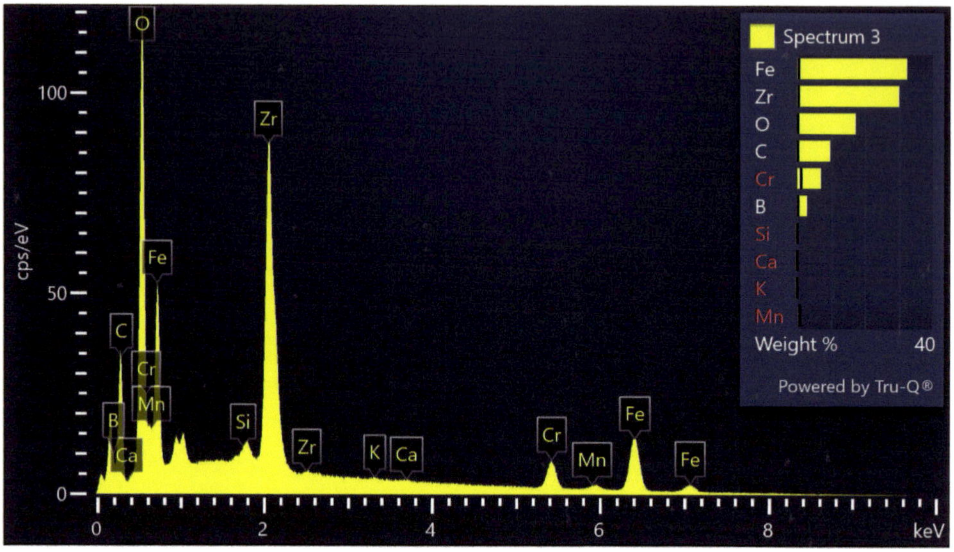

Figure 4.27 Elemental analysis for spectrum 3 (Image source: Non-profit KIMW Forschungs-GmbH)

Table 4.3 Elemental Analyses of the EDX Spectra

Element	Spectrum 1		Spectrum 2		Spectrum 3	
	(wt%)	(atom%)	(wt%)	(atom%)	(wt%)	(atom%)
B	—	—	—	—	2.89	8.29
C	15.87	33.99	57.10	69.72	9.76	25.18
N	3.50	6.43	5.27	5.52	—	—
O	27.58	44.36	14.76	13.53	17.42	33.75
Na	—	—	9.59	6.12	—	—
Cl	—	—	6.77	2.80	—	—
K	—	—	4.09	1.53	—	—
Cr	—	—	—	—	7.02	4.19
Fe	—	—	—	—	32.39	17.98
Zr	51.09	14.41	—	—	30.09	10.23
Other	1.96	0.81	2.42	0.78	0.43	0.38

References for Section 4.2

[1] Hering, E.; Martin, R.; Stohrer, M.: *Physik für Ingenieure* (8th edn), Springer, 2002

[2] Hecht, E.: *Optics (Global Edition)* (5th edn), Pearson Higher Education, 2017

[3] Ehrenstein, G.: *Mikroskopie* (1st edn), Hanser, 2020

[4] Grundwissen de-BROGLIE-Wellenlänge, *https://www.leifiphysik.de/quantenphysik/quantenobjekt-elektron/grundwissen/de-broglie-wellenlaenge*, accessed on 11 December 2022

[5] De-Broglie-Wellenlänge von schnellen Elektronen, *https://virtuelle-experimente.de/elektronenbeugung/wellenlaenge/de-broglie-relativistisch.php*, accessed on 11 December 2022

[6] von Ardenne, M.: *Elektronen-Übermikroskopie*, Justus von Springer, 1940

[7] Let's familiarize ourselves with the SEM, Hitachi High-Technologies Corporation, 2016

[8] Nagaoka, Y.; Hirato, T.: New material observation method by SEM using ultra variable-pressure detector (UVD) – Overview and example observations, *Technical Magazine of Electron Microscope and Analytical Instruments*, 2021, 16: 03

[9] Reimer, L.: *Scanning Electron Microscopy* (2nd edn), Springer, 1998

[10] Energy dispersive X-ray spectroscopy, *https://www.chemie.de/lexikon/Energiedispersive_R%C3%B6ntgenspektroskopie.html*, accessed on 11 December 2022

[11] ISO 25178-2:2021: Geometrical product specifications (GPS) – Surface texture: Areal – Part 2: Terms, definitions and surface texture parameters, 2021

4.3 Laser Microscopy

Dr. Stefan Svoboda

4.3.1 Basic Principle

A laser scanning microscope is a light microscope in which, instead of a normal light source, e.g. a halogen lamp, a focused laser beam is guided over the sample to be examined in a grid pattern. This means that a complete image is never generated at the same time, but the image is subsequently constructed on the basis of the measured light intensities.

The design of the microscope is based on the confocal measuring principle (confocal means "with focus", or focused on the same point), in which a pinhole diaphragm is integrated in the beam path. This ensures that light is only transmitted from a very small, sharply focused volume and all out-of-focus Z planes are blocked. This is due to the fact that the light from the lens is focused on the pinhole before it reaches the detector. The point of light in the center of the pinhole and the illumination point on the sample are simultaneously in focus, i.e. confocal to each other (Figure 4.28) [1].

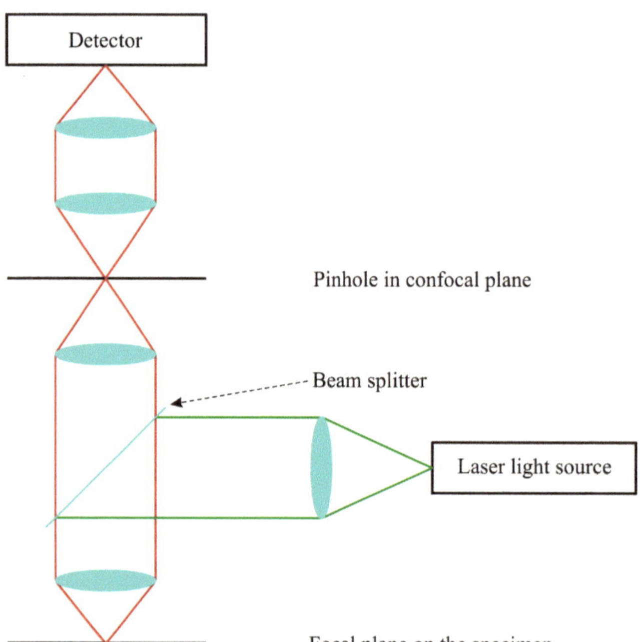

Figure 4.28 Schematic drawing of the confocal principle (Image source: University of Applied Sciences Schmalkalden)

To generate a three-dimensional image, individual images are taken in different focal planes one after the other. To do this, the distance in the Z direction is varied, e.g. by moving the table height in a corresponding interval. The closer the focal plane is to the sample surface, the higher the intensity. The spatial image is then assembled from the recorded individual images. The resulting image can be displayed as a two-dimensional image (in which all different height ranges are displayed in focus, i.e. an "all-in-focus image") or as a three-dimensional topographical image [2].

4.3.2 Taking a Picture

An LSM 700 laser scanning microscope from Carl Zeiss was used for the following illustrations. The centerpiece here is a solid-state laser with a wavelength of 405 nm and an output from the fiber of 5 mW. For image acquisition, the laser is put into operation and all presets (shown in Figure 4.29) are activated. After the sample has been moved to the correct focus and the desired sample area has been set with the reflected light microscope, the live image is started in the laser microscope.

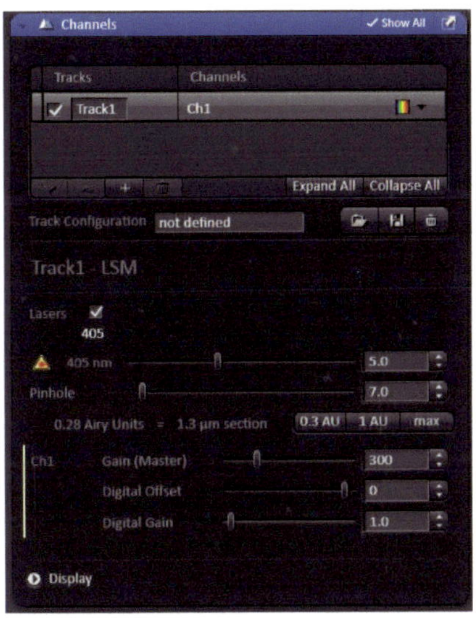

Figure 4.29
Basic settings of the laser microscope
(Image source: University of Applied Sciences Schmalkalden)

Due to the wavelength of the laser light (405 nm in the example) and the wide aperture of the lenses, lateral details can be resolved down to 120 nm and height differences of a few tens of nanometers can be determined.

When taking images, it is important that no areas of the sample surface to be examined reflect the light too strongly. Accordingly, the laser parameters must be optimized for all Z stacking planes (the areas which are shown in red in Figure 4.30 would result in reflections that are too strong).

Figure 4.30
Overexposed laser image of a cracked sol–gel coating (Image source: University of Applied Sciences Schmalkalden)

In the next step, the Z range must be selected and the increment for the Z stack must be defined. The number of levels should be within a reasonable range (e. g. 30–200). In Figure 4.31, for example, a Z-stack height of 10.5 µm with a step size of 0.1 µm is set, which corresponds to 106 individual planes.

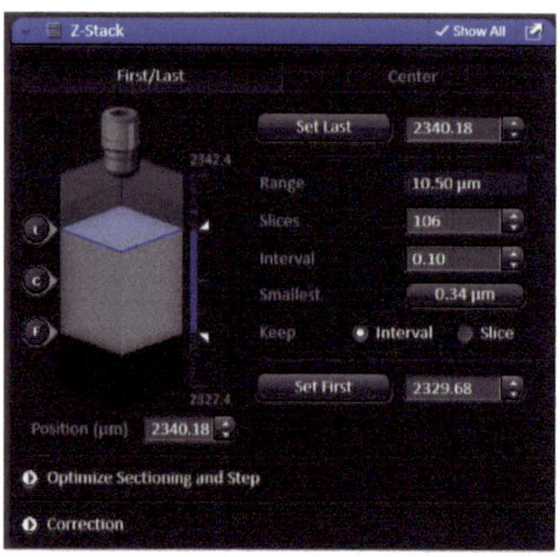

Figure 4.31
Z-stack parameters (Image source: University of Applied Sciences Schmalkalden)

The individual Z levels are now recorded (see Figure 4.32) and then combined to form an overall image.

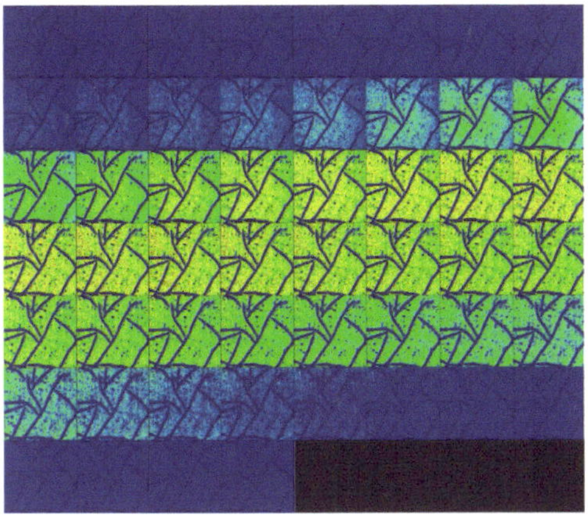

Figure 4.32
Z-stack gallery (Image source: University of Applied Sciences Schmalkalden)

Figure 4.33 shows the intensity distribution over the surface. The red areas in the base of the central scratch mark (scratch test) and the scratches in the coating (coating surface shown in blue) of a tool steel, which have the lowest light intensities, can be clearly seen.

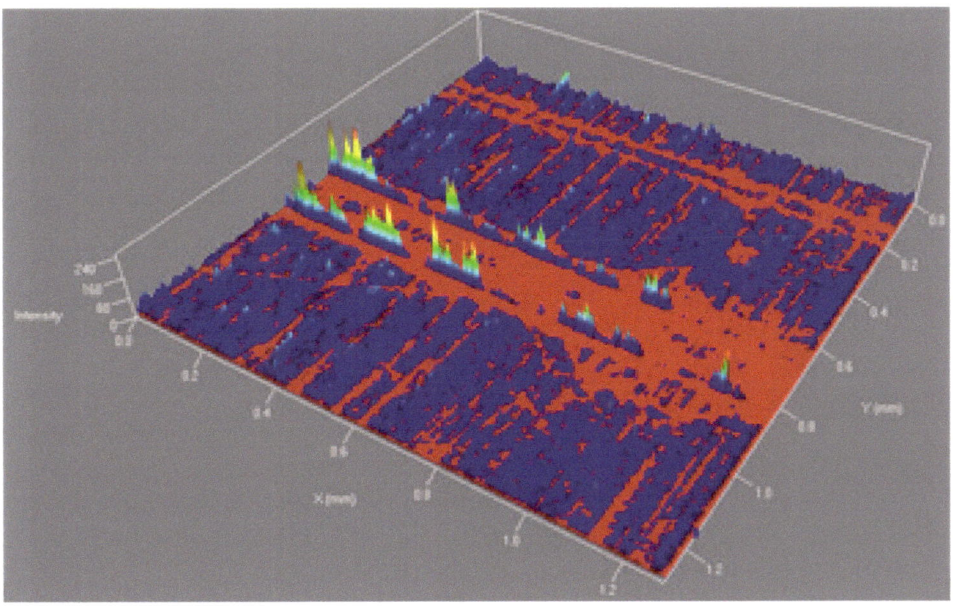

Figure 4.33 Intensity distribution of a scratch track to determine the adhesive strength of a sol-gel coating (Image source: University of Applied Sciences Schmalkalden)

The topographical representation enables a color assignment to the different height ranges, as shown in Figure 4.34. The volume above or below can be determined by shifting the Z-plane.

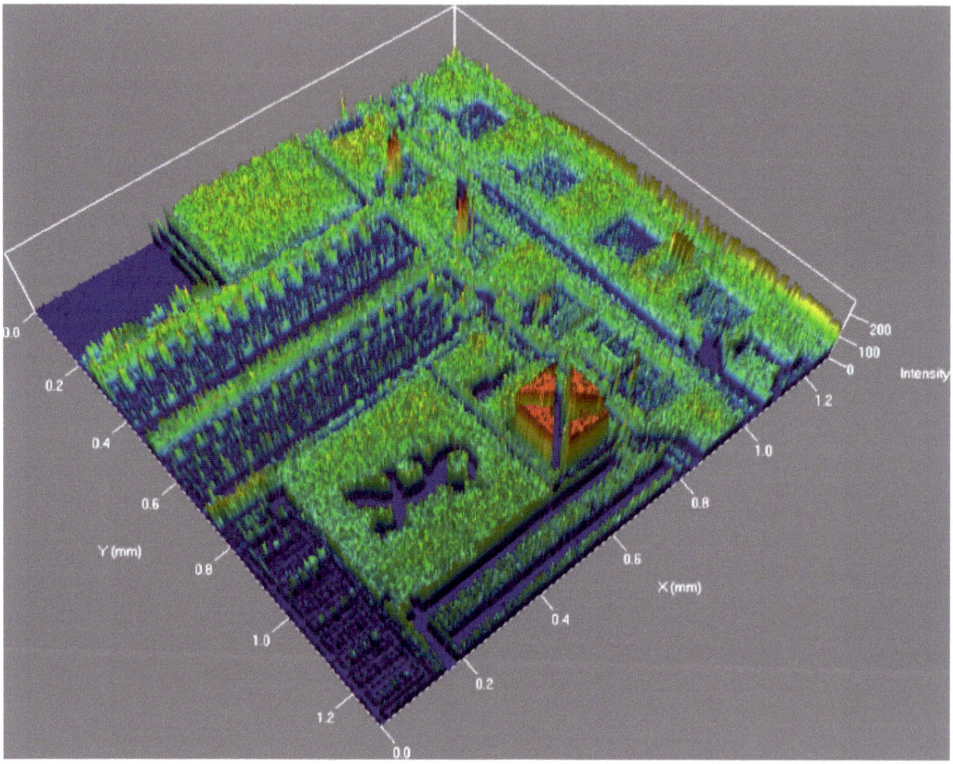

Figure 4.34 Topographical representation of a wafer chip section (Image source: University of Applied Sciences Schmalkalden)

If larger sample areas need to be examined or even smaller areas with high resolution (i. e. when using larger lenses), this is not possible with a single image section. In this case, images can be combined from several individual images (tile scan). An overlap of 5–10% should be set in order to assemble the image with pixel accuracy. This very quickly produces large amounts of data that need to be processed. Figure 4.35 shows the Z-gallery of a composite image (3×1).

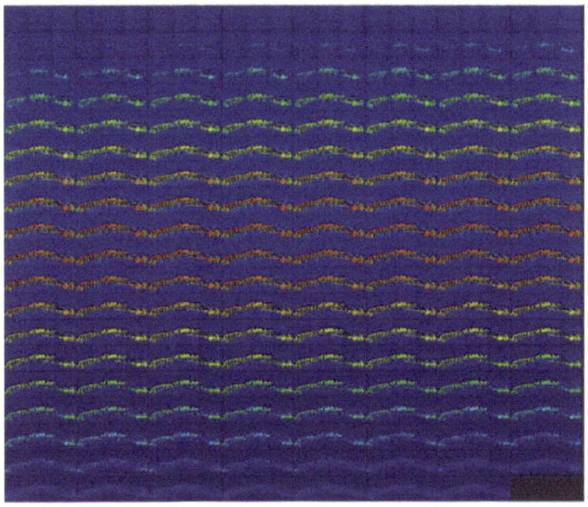

Figure 4.35
Z-stack gallery (3×1) section of a pin-on-disc wear track (Image source: University of Applied Sciences Schmalkalden)

This allows larger areas of a sample to be visualized in high resolution. Figure 4.36 shows a larger section of a wear track (pin-on-disc) with the corresponding light intensity distribution.

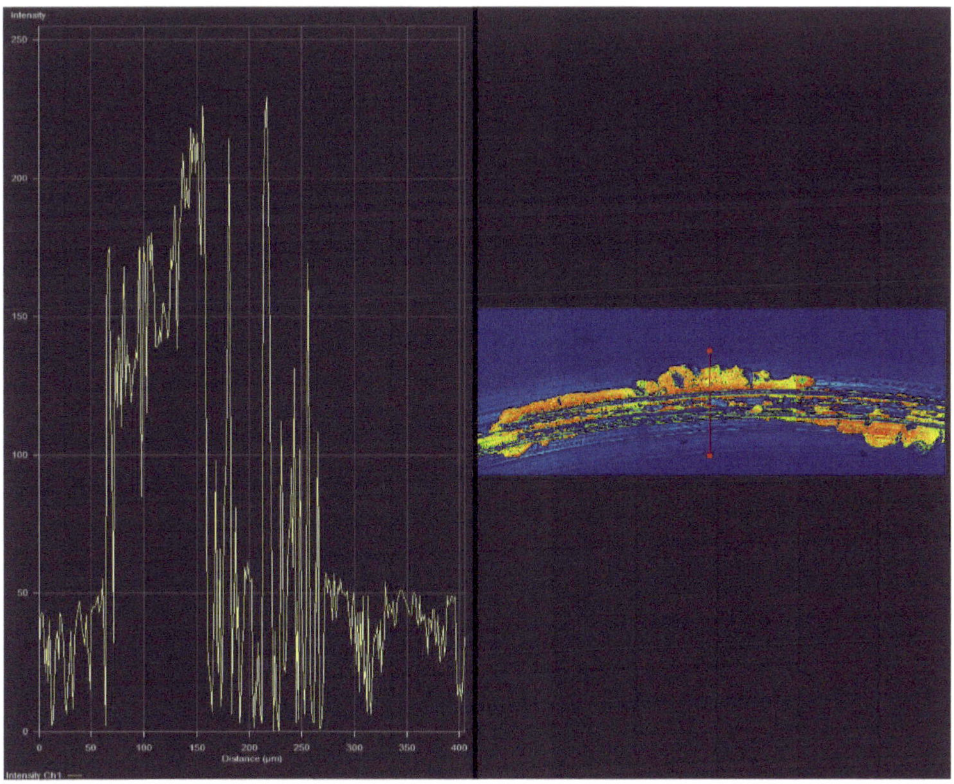

Figure 4.36 Section of a wear track (from 3×1 tile scan) showing the light intensity over the cross-section (Image source: University of Applied Sciences Schmalkalden)

As determination of the height of the surface structure is carried out with pixel-precision, this can be used to determine a surface roughness (Figure 4.37). When measuring a surface profile, a line roughness can also be specified for this purpose. The results correlate quite well with classic roughness measurement methods. Especially with soft surfaces, e.g. plastics, this type of roughness measurement is more suitable than conventional mechanical methods, where the results can be rendered invalid by surface deformations.

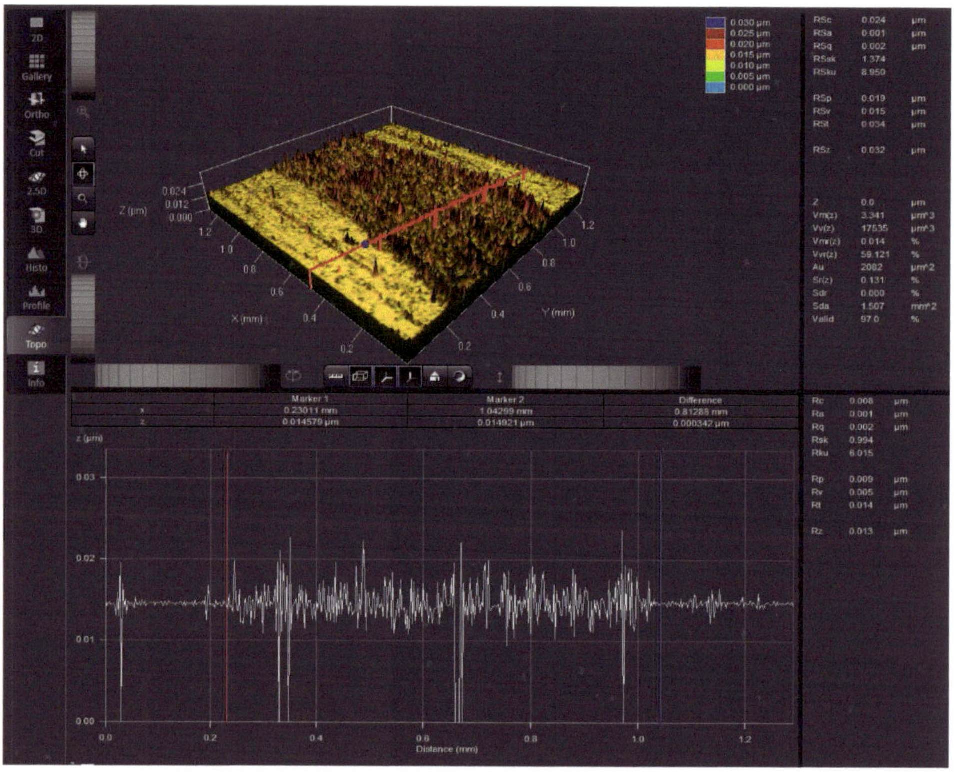

Figure 4.37 Roughened surface of a wear track (pin-on-disc) with surface roughness (R_{Sa} etc.) and line roughness data (R_a etc.) (Image source: University of Applied Sciences Schmalkalden)

4.3.3 Application Examples

4.3.3.1 Crack Network in Sol–Gel Coating

In a cracked wear-resistant sol–gel coating of Al_2O_3 on a tool steel, the crack network and the coating thickness can be visualized (Figure 4.38).

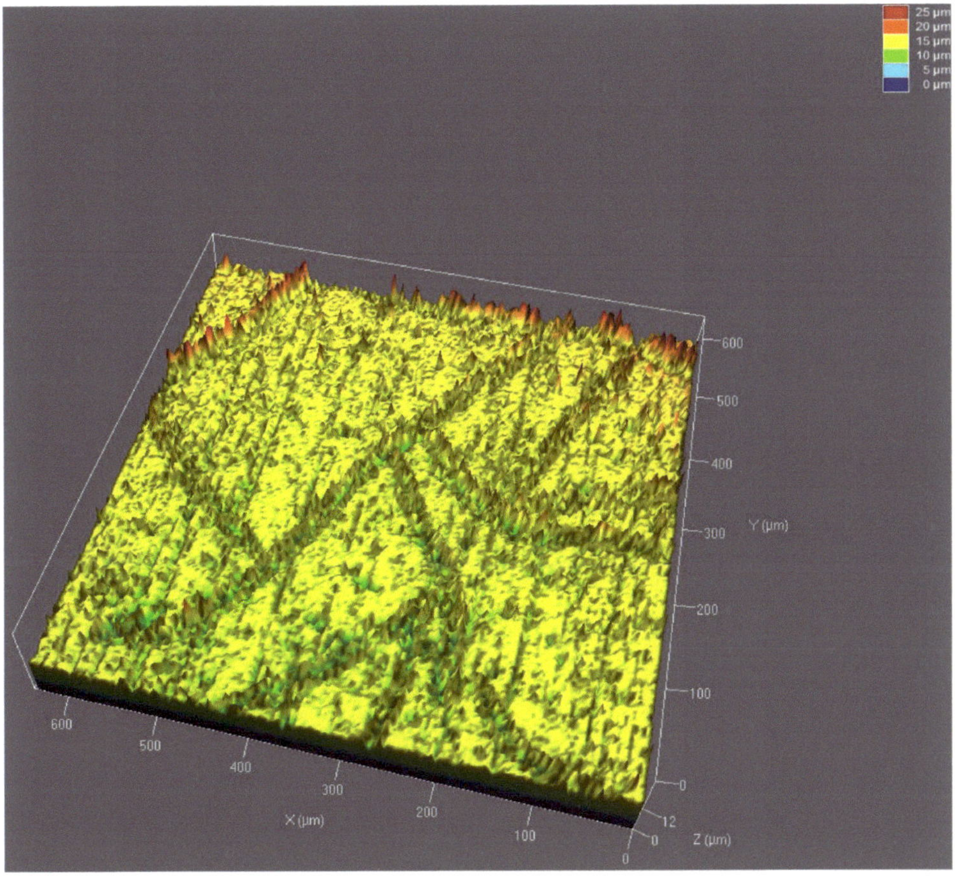

Figure 4.38 Topography of a cracked sol–gel layer (Image source: University of Applied Sciences Schmalkalden)

The intensity distribution over the surface is more suitable for visualizing the crack progression (Figure 4.39). The layer thickness can be indicated by the topographic color planes with a maximum of about 5 µm.

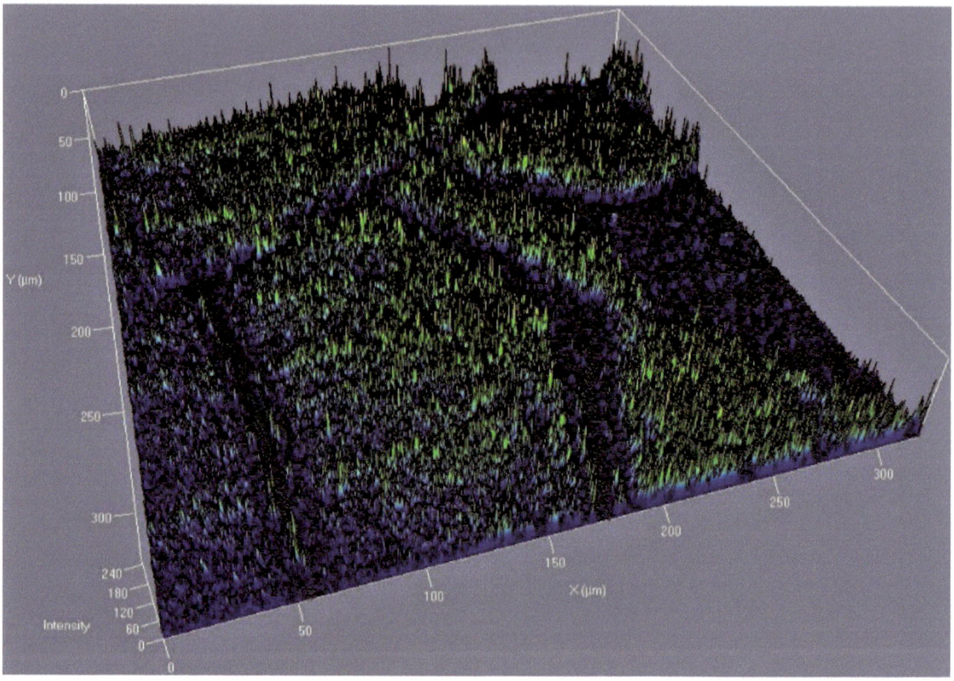

Figure 4.39 Intensity distribution of the cracked sol–gel layer (Image source: University of Applied Sciences Schmalkalden)

4.3.3.2 Representation and Evaluation of Crater Grinding

The crater grinding method is very suitable for determining the thickness of thin coatings. This method can also be used to determine the wear resistance of surfaces. The sample is mounted at an angle in a holder and positioned in front of a drive shaft. A steel ball is placed between the shaft and the sample surface, wetted with an abrasive (e. g. diamond suspension), and the shaft is moved at a defined speed. At the end of the test, the produced calotte is analyzed using a topography and profile analysis, as shown in Figure 4.40. The generated penetration depth, the area, and the wear volume can be determined from this. The deposits of wear particles around the generated calotte are also clearly visible.

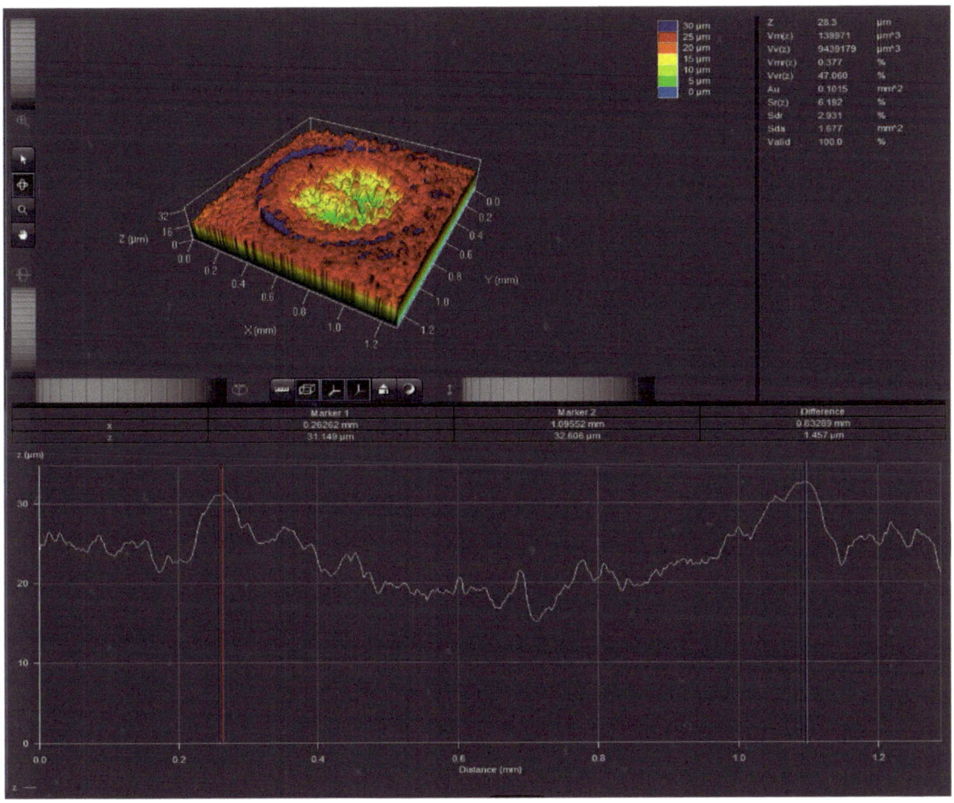

Figure 4.40 Topography and profile analysis of a calotte section (Image source: University of Applied Sciences Schmalkalden)

4.3.3.3 Roughness Measurement on a Plastic Sample

The usual roughness measurement using a stylus dragged across the surface is not well suited for soft surfaces, as surface deformation can occur. Optical surface roughness measurement in accordance with ISO 25178 is a real alternative here. By defining suitable profile lines, a line roughness can also be displayed (Figure 4.41).

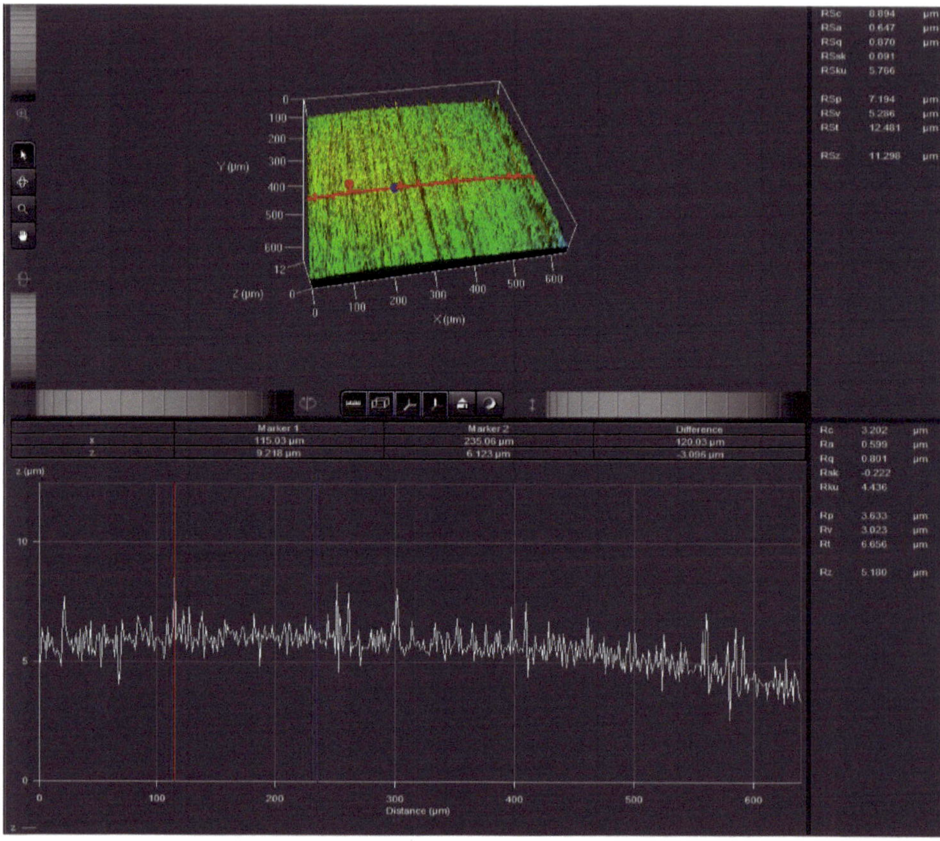

Figure 4.41 Surface roughness measurement with topographical surface representation and roughness profile along a selected line (Image source: University of Applied Sciences Schmalkalden)

4.3.3.4 Evaluation of a Wear Test

A wide variety of coatings are used to improve tool surfaces subject to wear. An essential component of coating selection is their wear behavior under the corresponding application conditions. There are a number of suitable wear test methods for this purpose. In the example, a test was carried out with oscillating movement of a ceramic ball, which was subjected to a corresponding normal force. In addition to the wear progression over the test time, the total wear volume at the end of the test is an important parameter. Figure 4.42 shows the topographical 3D image of the wear track. The evaluation of the test is shown in Figure 4.43.

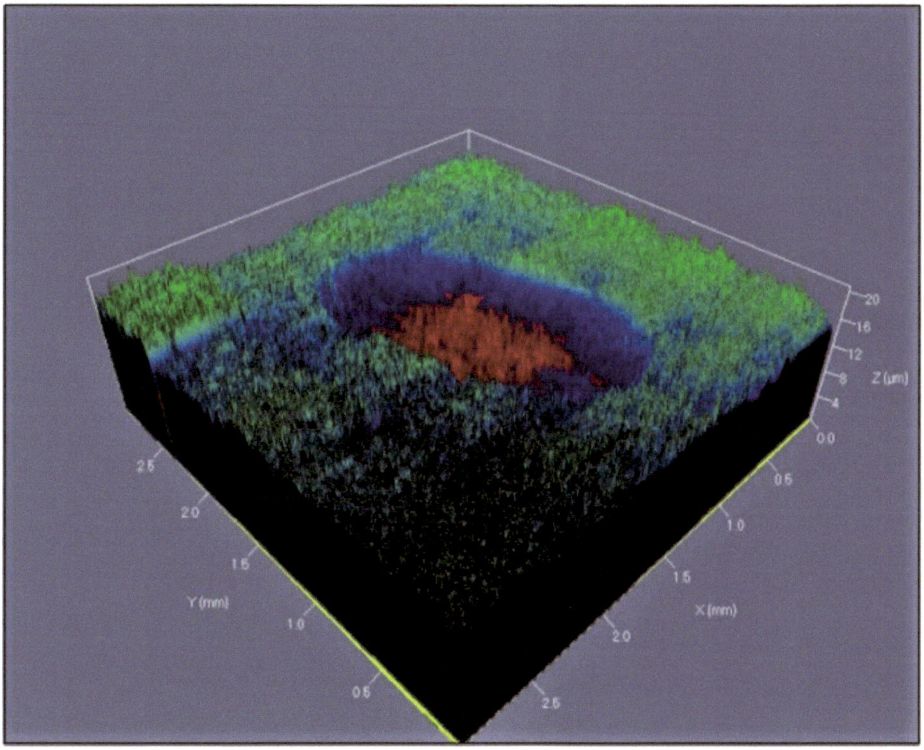

Figure 4.42 Topography wear track (Image source: University of Applied Sciences Schmalkalden)

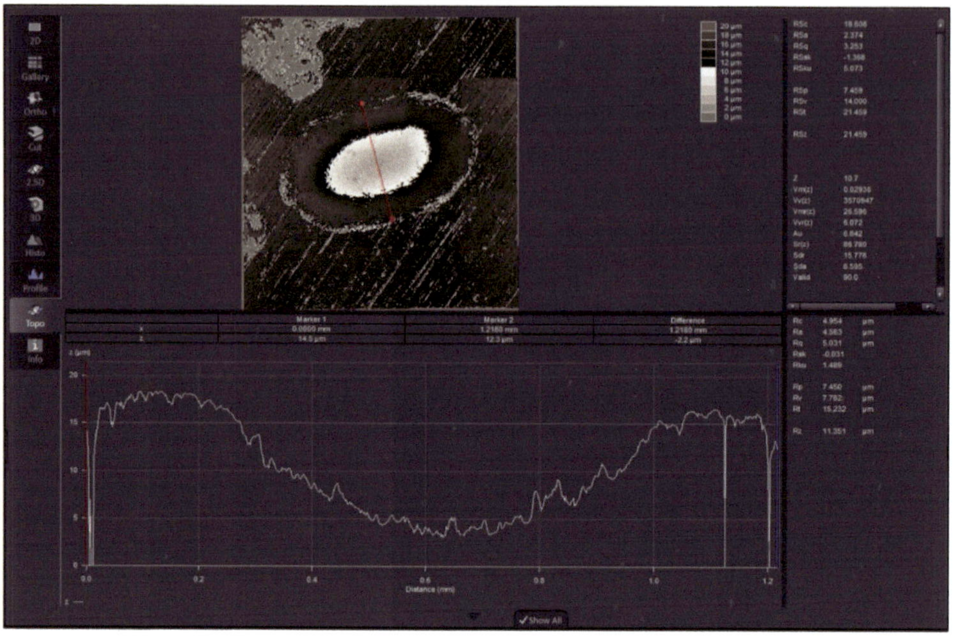

Figure 4.43 Topography with indication of the wear volume and profile analysis (Image source: University of Applied Sciences Schmalkalden)

References for Section 4.3

[1] LSM 700 flexible and non-contact [Brochure], Carl Zeiss Microscopy

[2] Schmeidel, J.: Konfokale Laser-Scanning-Mikroskopie in der Materialforschung [Brochure], Carl Zeiss Microscopy

4.4 White Light Interferometry

Dr. Andreas Balster

4.4.1 Introduction

The mold wall limits the flow path of the plastic melt and thus ensures the overall shape of the part. However, the contact of the molding compound with the tool also shapes the surface of the molded part on a microscopic level and thus forms the surface impression that is presented to the user both visually and haptically.

The appearance of a surface is decisively influenced by its microstructure. Depending on the roughness of the interface, the proportion of directional and diffuse reflection of irradiated light (gloss) changes. We perceive this interplay as high-gloss in one case and matt in another. In an industrial context, a quantitative description of the degree of roughness and the resulting surface properties is therefore of great importance (Figure 4.44) [1, 2].

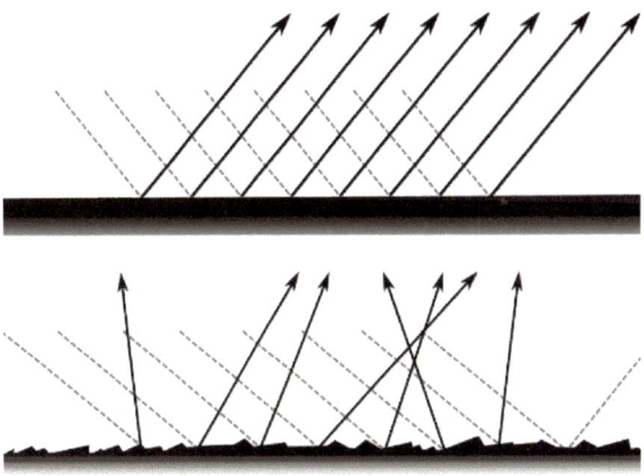

Figure 4.44 Directed (top) and diffuse (bottom) reflection of light on surfaces of different roughness (Image source: own figure, Andreas Balster)

As greater roughness also changes the size of the surface, other properties are also affected, for example tribological behavior, wetting with liquids, tendency to soiling, and scratch-sensitivity.

4.4.2 Roughness as a Measured Variable

Although practically everyone has a certain amount of first-hand haptic experience and can compare two surfaces in terms of their roughness, measuring and quantifying this property is anything but trivial. If you want to describe the roughness of a surface objectively, this is done by specifying height differences on the respective surface. For technical parts, the order of magnitude of height differences is in the micrometer (µm) range. The lateral resolution capacity of the human sense of touch (the so-called spatial threshold) is between 1.5 mm and 4 mm, depending on age and practice, which is around three orders of magnitude higher than the typical roughness of technical molded parts. By contrast, roughness is measured with lateral resolutions that are also in the micrometer range [1].

Unevenness of a surface can in turn occur on completely different scales and also overlap each other (Table 4.4). Height differences that only become apparent and can only be detected when a surface is viewed over a large area are summarized under the term "waviness". Waviness and roughness can overlap on one and the same surface, as can acoustic waves in the form of beats, for example. According to DIN 4760, shape deviations of technical surfaces are generally divided into six different degrees, with 1st- to 4th-order shape deviations overlapping and thus determining the overall perceived surface quality of a body. The visually noticeable roughness, as shown in Figure 4.44, belongs to the 4th and especially the 5th order of shape deviations [1].

Table 4.4 Shape Deviations According to DIN 4760

Shape deviation	Designation
1st order	Shape deviations
2nd order	Waviness
3rd order	Roughness in the form of grooves
4th order	Roughness in the form of grooves, scales, and crests
5th order	Roughness of the microstructure
6th order	Lattice structure of the material

DIN 4760 defines terms and measurement conditions for determining roughness values of technical surfaces. First, a representative measuring section is selected. The sensing distance of the (in this case tactile) measuring device is divided into seven equal sections; the first and last sections are not taken into account when describing the surface (Figure 4.45). The measuring distance l_m results from the sensing distance l_t minus a leading and trailing distance, each of which corresponds to the length of a partial measuring distance l_e. This in turn also represents the numerical limit wavelength λ_c, which separates the roughness from the waviness.

The average roughness value R_a is the arithmetic mean of the amounts of all profile values of the roughness profile. R_a corresponds to the size of a rectangle of the length of the measuring section l_m and the height R_a (see Figure 4.45).

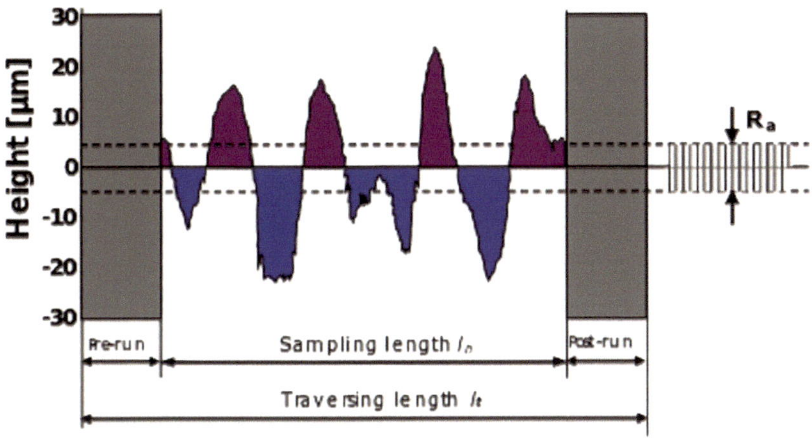

Figure 4.45 Average roughness value R_a

The roughness depth R_z is the mean value of the five individual roughness depths of the section of the measuring section l_m. The individual roughness depth z occurs within a partial measuring section l_e and is the greatest difference between the highest peak and the deepest valley. The maximum roughness depth R_{max} is the greatest individual roughness depth within the measurement section (Figure 4.46).

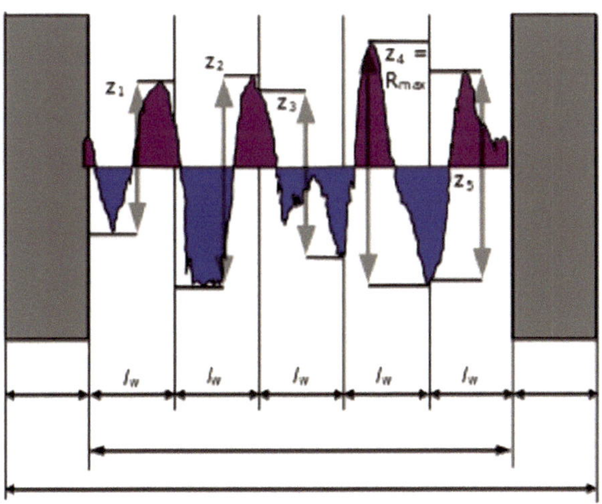

Figure 4.46 Roughness depth R_z and maximum roughness depth R_{max}

The measuring distance according to ISO 4768 for 0.5 µm < R_z < 10 µm is 4 mm, with an additional 0.8 mm each for lead and lag. This is the most common roughness range for plastic molded part surfaces. Larger roughness depth values are determined with longer measuring distances, smaller ones with correspondingly shorter ones.

This determination of roughness is based on a line measurement which in its beginnings was exclusively mechanical. However, the topography of a surface is only revealed by adding the second dimension, i.e. by measuring the area. Averaging two orthogonal measurement lines is not always meaningful. Roughness values alone therefore do not allow any conclusions to be drawn about the actual appearance of the surface, for example whether it is an eroded or etched tool structure.

Today, there are various methods and procedures for measuring the roughness or topography of surfaces, which are essentially classified by whether it is a contact or non-contact method and by the physical technique. Mechanical scanning of a surface is an invasive method. This means that the contact of the measuring probe with the surface to be measured influences the shape of the surface itself to a greater or lesser extent depending on the material – either the probe tip or the surface can undergo irreversible deformation. Repeated measurement in close spatial proximity in order to generate a two-dimensional, realistic image is therefore hardly possible. In addition, both the lateral resolution and the ability to detect indentations are limited by the shape and size of the probe tip. If a non-contact and therefore non-destructive method is used, these obstacles are removed.[1)] However, in view of the different types of interaction and the jump in resolution capacity, there remains a question of what exactly the "surface" represents, because the interaction with light takes place at the level of the electron shells of the atoms or molecules located at the interface and therefore has its own physical dependencies. For this reason, among others, the standards ISO 14406 and ISO 25178 introduce terms such as the "electromagnetic surface" and the "primary surface", and define procedures (filters and operators) whereby a data set describing the surface is transformed in such a way that it can be compared with others, so enabling a distinction to be made between "surface parameters" and "topographic parameters". The analogs corresponding to these roughness parameters in two dimensions are defined using infinitesimals and are less descriptive than these. To illustrate this point, Part 2 of ISO 25178, which deals with the definition of parameters for describing surfaces, contains well over 100 pages [3].

---

[1] However, the fundamental problem of not being able to detect undercuts by measuring from the z-direction remains. For obvious reasons, this point does not play a role for tool surfaces.

4.4.3 White Light Interferometry method

White light interferometry is a powerful method for measuring surface profiles, thickness, and thickness variations in materials. It is based on the interference of light waves that are reflected by the sample and generates interference patterns that are recorded on a detector.

4.4.3.1 Measuring Principle of White Light Interferometry

White light interferometry utilizes the principle of interference of light waves. When light waves collide, they can either be in phase or out of phase, which leads to amplification or extinction of the light. In a so-called Michelson interferometer, this interference is converted into a light signal with a time-fluctuating amplitude by an arrangement of mirrors.

Such an interferometer consists of the arrangement of a semi-transparent mirror, which splits an incident beam of light into two beams of equal intensity but propagating perpendicular to each other, and two arms with totally reflecting mirrors at their respective ends, which superimpose the light previously split into two portions by reflection. The two recombined light beams emerge from the interferometer perpendicular to the original direction. One of the two mirrors is movable along the optical axis and can therefore change the distance to the beam splitter. The second arm is fixed at a defined distance. If the path difference of the split light beams is an integer multiple of the wavelength λ, the recombined light beam has the maximum amplitude (constructive interference, Figure 4.47, left). If the movable mirror is moved to a position that differs by $\lambda/2$ from that of the fixed mirror, the two beams *annihilate* each other completely during recombination (destructive interference, Figure 4.47, right).

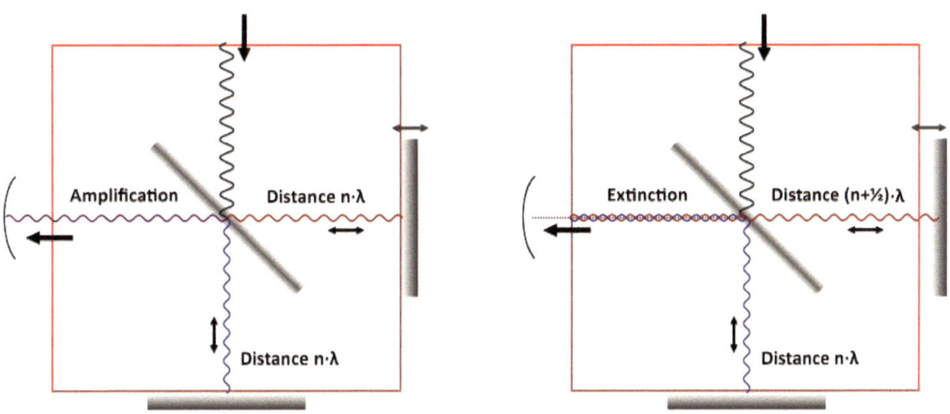

Figure 4.47 Functional principle of a Michelson interferometer. The incident light beam (black) is split, reflected by two mirrors, and recombined again. Depending on the mirror position, the intensity is increased or decreased

An observer who is in the position of the detector sees a regular back-and-forth movement of the moving mirror, and when a monochromatic, coherent light source such as a laser is used, the light signal regularly increases and decreases.

This "flashing signal" is generated for each irradiated wavelength λ, but the interference condition $n{\cdot}\lambda$ for constructive interference is fulfilled for each wavelength at a separate position of the moving mirror. At the zero crossing of the mirror, however, the superimposed waves are amplified without exception: at equidistance (i. e. identical distance of the two arms to the beam splitter), positive interference is generated for all existing wavelengths.

Figure 4.48 illustrates the signal generated in the case of interference of a few wavelengths. It can be seen that the superposition for mirror positions outside the equidistance generally leads to an attenuation of the overall signal. When the mirror passes through the central position, a flash of light of maximum intensity is generated. In the so-called interferogram, this burst is called a center burst (or center peak) [2].

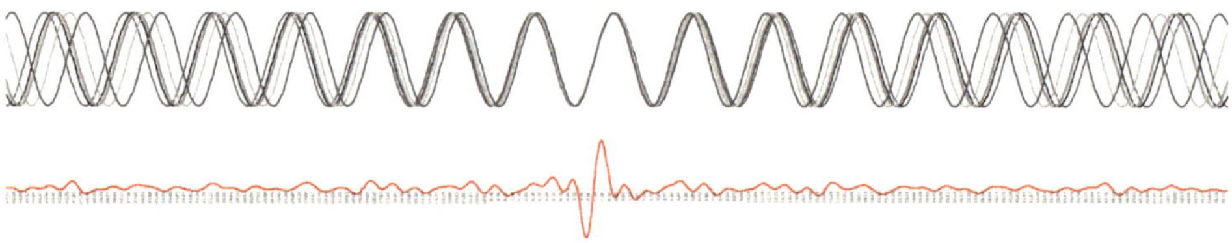

Figure 4.48 Interference pattern of waves of different frequencies superimposed on each other in an interferometer after recombination of the partial paths. Constructive interference for all wavelengths involved is only present for the middle mirror position, where both arms are the same length

In white light interferometry, white light (which consists of a continuum of different wavelengths) is used that is as coherent as possible. The intensity pattern that develops over time when the mirror is moved is correspondingly complex, but always produces the described center burst.[2] This means that the measuring arrangement can be used to determine distances. The arrangement of the interferometer is selected so that the sample surface corresponds to the moving arm of the interferometer. The light reflected by the static mirror on the one hand and the sample surface on the other is recombined, resulting in the interference described above. If the distance between the interferometer and the sample is now periodically changed (which corresponds to the back and forth of the moving mirror), the collective amplification of all irradiated wavelengths – the center burst – occurs simultaneously for all points in a certain plane perpendicular to the light beam at any given time. The mapping of

2 The temporal development of this signal contains all the information about the frequency dependency of the signal. This information can be recovered by means of a Fourier transformation. This principle is also used in FTIR spectroscopy (see Section 4.5).

such interference maxima over the measuring range therefore provides information on which points lie on a common height level. In Figure 4.49 this is illustrated as an example for three different positions of the measuring sample. It should be noted that the sections shown on the right-hand side cannot be used completely for mapping, as parts of them lie inside the sample volume or in the shadow and cannot be measured.

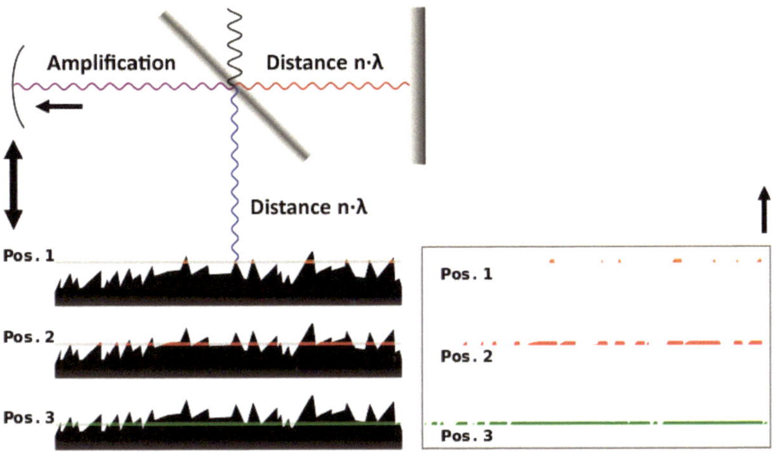

Figure 4.49 Planes of constructive interference at three example positions of a sample surface. Only the points on the surface where the reflection can reach the detector can be mapped. In position 3, the majority of the green plane is inaccessible

The interference patterns are recorded using a suitable detector and then processed to calculate the surface profile of the sample. The accuracy of the measurements depends on various factors, such as the quality and design of the interferometer, the reflective properties of the surface, and the sensitivity of the detector. However, the z-resolution is not limited in principle by the wavelength of the selected light, as is the case in classical imaging microscopy due to the Abbe limit. Typical measuring devices have magnifications in the range of 2.5× to 230× (image/object size). The measurement of transparent layers can usually be achieved by technical additions. Using white light interferometry, resolutions of 40 nm to 16 µm in the lateral direction and 0.8 nm in the z-direction can be achieved.

4.4.3.2 Applications of White Light Interferometry

White light interferometry has many applications in various fields (see Figure 4.50 and Figure 4.51). In the semiconductor industry, for example, it is used to measure the surface profiles of microchips and other semiconductor components. In biomedicine, it is used to examine and characterize tissue and cells. White light interferometry is also used in nanotechnology to measure the thickness variations of thin layers and layer systems (Figure 4.51) [4, 5].

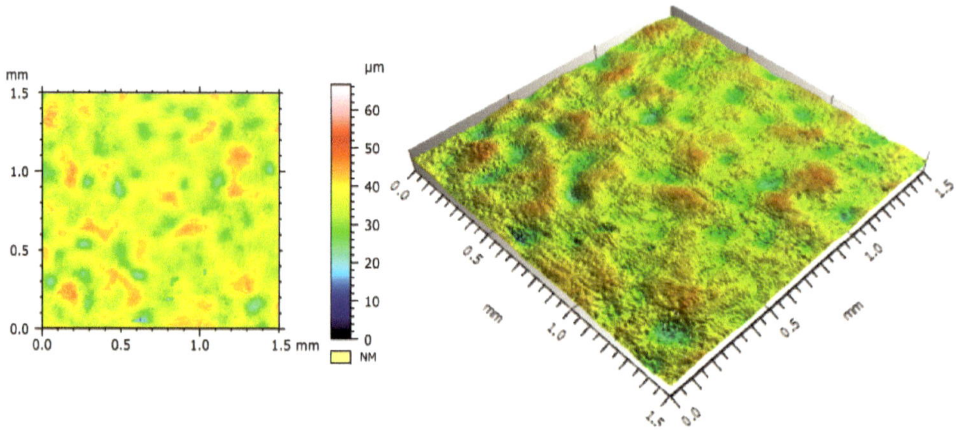

Figure 4.50 Height mapping of a surface in false colors (left) and in a 3D view (right). In the underlying case, the condition of a tool surface coated with tungsten carbide was considered. The roughness is in the scale of micrometers (Image source: Non-profit KIMW Forschungs-GmbH)

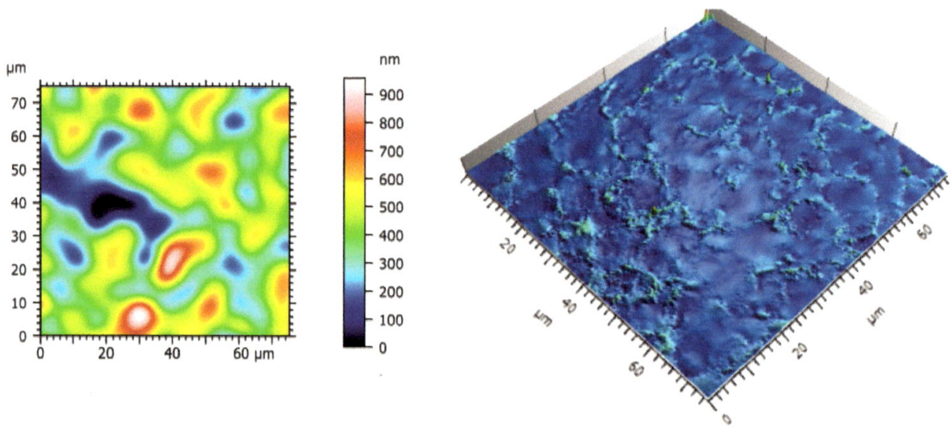

Figure 4.51 Height mapping of a surface in false colors (left) and in a 3D view (right). In the underlying case, the condition of a tool surface coated with titanium nitride was considered. The roughness is in the sub-micrometer scale (Image source: Non-profit KIMW Forschungs-GmbH)

In optics and photonics, white light interferometry is used to characterize optical components such as lenses and mirrors. In addition, white light interferometry can also be used in the manufacture of optical components to check their quality. When inspecting the surfaces of tools and molded parts, statements can be made about their roughness and imaging fidelity, and signs of wear can be quantified. In conjunction with gloss and color measurements, precise supplier and quality agreements can be made.

4.4.3.3 Limitations of White Light Interferometry

Although white light interferometry is a powerful method for measuring surface profiles, it also has some limitations. One of these is the limited measurement speed, as the interference patterns on the detector need to be carefully aligned to achieve accurate measurements.

Another associated limitation is the sensitivity to vibrations and thermal fluctuations. Small movements or temperature changes can disturb the interference pattern and lead to inaccurate measurements. For this reason, experiments must be carried out in a stable and controlled environment to minimize interference.

Another limitation is the limited depth that can be measured with white light interferometry. The method is best suited for surface profile measurements, and is less suitable for measuring deep structures or volume measurements [5].

References for Section 4.4

[1] Sorg, H.: *Praxis der Rauheitsmessung und Oberflächenbeurteilung*, Fachbuchverlag Leipzig, 1995

[2] Balster, A.: Schadensanalyse an Kunststoffen [Lecture notes] (1st edn), FH Südwestfalen, 2011

[3] ISO 25178-2:2021: Geometrical product specifications (GPS) – Surface texture: Areal – Part 2: Terms, definitions and surface texture parameters, 2021

[4] Meschede, D.: *Optics, Light, and Lasers: The Practical Approach to Modern Aspects of Photonics and Laser Physics*, Wiley-VCH, 2017

[5] Todhunter, L. D.; Leach, R. K.; Lawes, S. D. A.; Blateyron, F.: Industrial survey of ISO surface texture parameters, *CIRP Journal of Manufacturing Science and Technology*, 2017, 19: 84–92

4.5 Infrared Spectroscopy

Dr. Andreas Balster

4.5.1 Introduction

Infrared (IR) spectroscopy is one of the most important analytical techniques in materials science, and is used in many areas due to its diverse applications. Especially in the characterization of polymers, the method is absolutely essential, because it provides unique information about the structure and composition of organic compounds. In this chapter, the basics of FTIR spectroscopy are explained, and applications in the field of synthesis, processing, and quality control of polymeric materials are shown.

The term "FTIR spectroscopy" stands for Fourier transform infrared spectroscopy, and it is a technique based on measuring the absorption of IR radiation by a sample. IR radiation is electromagnetic radiation in the spectral range between visible light and microwave radiation. In IR spectroscopy, a sample is exposed to this IR radiation, whereby certain frequencies are absorbed by the atoms and molecules of the sample. The absorption of these rays is characteristic of the chemical composition and structure of the sample, and can be measured using a detector.

4.5.2 Physical Basics

Figure 4.52 shows the electromagnetic spectrum. The IR spectral range of the electromagnetic spectrum is on the long-wave side of the light visible to humans, with the wavelength of IR light being generally between 800 nm and approx. 1 mm. When we talk about IR spectroscopy, we generally mean the mid-infrared range (MIR) with wavelengths between 2.5 µm and 25 µm. The shorter wavelength range, which is closer to the visible part of the light spectrum, is called the near-infrared (NIR), which requires special spectrometers. There are fewer compound-specific signals in this spectral range, but the available measurement technology allows spectra to be recorded very quickly, which is useful in automated quality control (e. g. for food), as well as in the sorting of plastic waste. In the medical sector, optical coherence tomography is an imaging method based on NIR radiation. The so-called far-infrared has less analytical significance [1, 2].

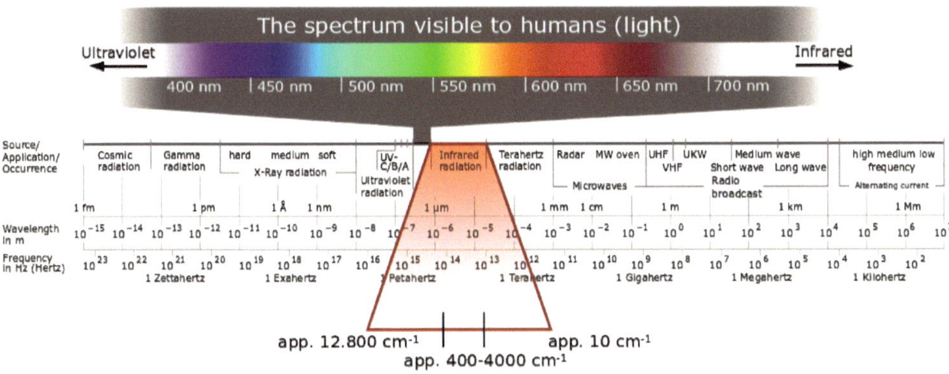

Figure 4.52 The electromagnetic spectrum and the range of IR radiation (Image based on [3])

For historical reasons, the wavelength λ (Equation 4.16) is not used in connection with IR spectra; instead, the number of wave trains per centimeter is used. The resulting unit of measurement is therefore the reciprocal of the wavelength (in cm), and is known as the wavenumber (with units of 1/cm or cm^{-1}) [1, 2].

$$\lambda \cdot \nu = c = 2.998 \times 10^8 \text{ m/s} \quad (4.16)$$

$$\nu = \frac{c}{\lambda}$$

c = Speed of light (m/s)

λ = Wavelength (m)

ν = Frequency (1/s)

IR light excites molecules to vibrate. IR spectroscopy therefore utilizes the energy absorption of molecules, which is associated with changes in the vibrational states of chemical compounds. As this does not involve any rearrangement of electrons, no chemical reactions are induced, and it is a non-destructive analysis method. An IR spectrum shows the absorption of the IR radiation as a function of frequency, which provides information about the chemical composition and structure of the sample.

IR spectra are recorded in the range of approx. 400–4000 cm^{-1} (see Figure 4.52). These limits vary slightly depending on the recording technique. Another special feature of IR spectra is the fact that they are plotted from left to right with decreasing energy. For example, the left-hand limit of an IR spectrum might therefore be 4000 cm^{-1}, and the right-hand limit 400 cm^{-1}.

The frequency of IR radiation is in the same order of magnitude as that of molecular vibrations. Interactions and energy transfer between IR radiation and a molecule are only possible if the frequency of the radiation is the same as the vibrational frequency within a molecule. This is called the resonance frequency. If the molecule absorbs this radiation, it continues to vibrate at this frequency, only with greater amplitude. So if IR light with a broad spectrum is sent through a sample, some frequencies are absorbed, while the rest is transmitted without absorption. The absorbed frequencies correspond to the natural frequencies (resonance frequencies) of the vibrational modes in the molecules or an integer multiple of these frequencies (these are called harmonics, as in acoustics). If IR radiation is absorbed by a molecule, its energy content increases and it vibrates more strongly. However, this excited state is only short-lived. The excited molecule very quickly releases its excess energy to neighboring molecules through collisions, which results in an increase in the temperature of the sample.

Each molecule can vibrate in several ways. The more atoms a molecule contains, the more vibrational possibilities are available. The vibrational modes are determined by the structure of the molecule and are molecule-specific. The so-called normal vibrations are derived from considerations from group theory. Vibrations of molecules are differentiated on the basis of the relative atomic motions [4].

- Valence vibrations (ν) along the bond plane:
  - Symmetrical
  - Asymmetrical
- Deformation vibrations (δ):
  - Spreading vibrations (“bending”)
  - Pendulum oscillations (“rocking”)
  - Torsional vibrations (“twisting”)
  - Tilting vibrations (“wagging”).

In Figure 4.53 these possible vibrations are illustrated for a carbon atom. For all normal vibrations carried out by a molecule, the center of gravity of the molecule is not changed. Since carbon atoms in all stable organic compounds have four bonds (these can be divided into single, double, and triple bonds), structural information can be obtained from the position and intensity of the bands in the spectrum [4].

The resulting normal vibrations are superimposed and lead to combined vibrations. While this can make the assignment of the vibrations very complicated, the advantage for the user is that the analyzed substances generate a characteristic “fingerprint”, which can be used to identify them easily with suitable evaluation algorithms.

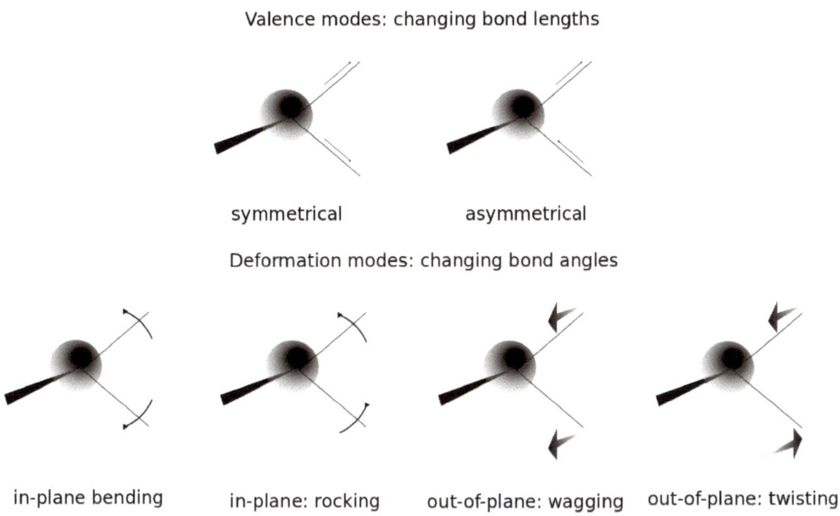

Figure 4.53 Possible oscillations of a tetrahedral symmetry center. Wagging and twisting oscillations leave the plane of the drawing

4.5.3 The Application of FTIR Spectroscopy to Polymers: Material Identification

FTIR spectroscopy is therefore a valuable technique for the characterization of polymers, as it provides information about the chemical composition and structure of the polymer. FTIR spectroscopy can be used to identify the bonds in polymers, to determine the molecular structure and conformation of the polymer, and to elucidate the polymer morphology and structure.

Tasks that can be performed with IR spectroscopy in connection with plastics are, for example:

- Identification of polymer samples
- Detection of impurities and material mix-ups
- Determining the content of polymer blends
- Determination of additives (qualitative/quantitative)
- "Mapping" of substance distributions in a matrix
- Detection of hydrolytic or oxidative damage
- Determination of end groups, and from this an estimation of the molecular weight.

The main advantages are the specificity even for structurally very similar substances, the very simple sample preparation for solids, the quick and easy execution, and the possible automation of the evaluation. Furthermore, it is an analytical method that can be used for quantification. Together with the affordable costs for acquisition and maintenance, even for small laboratories, IR spectroscopy can rightly be described as the "workhorse" in laboratories dealing with polymer materials. In some cases, e.g. with complex mixtures of materials or trace analysis, it may nevertheless be necessary to use additional analytical techniques such as GC/MS (see Section 4.11) in order to obtain clear information about the chemical composition of an engineering plastic [5, 6].

For any type of sample preparation, it is essential that the part of a sample of interest can be directly exposed to the IR beam. If one is interested in the nature of an inclusion in a MOCVD layer, it must first be prepared and isolated from the layer. A direct measurement of organic substances under or behind other materials – especially metals – is not possible due to the described interaction of the sample with the IR radiation. IR spectroscopy, on the other hand, is quite suitable for checking the identity and purity of organic precursors.

4.5.4 Identification and Structure Clarification

IR spectroscopy can be used to determine the molecular structure and, in some cases, the conformation of polymers.[3)] The identification of polymers by IR spectroscopy is based on the fact that each type of bond in the polymer has a characteristic IR absorption pattern. Polymers consist of repeating units that are linked together by covalent bonds. These bonds generate specific vibrations and deformations of the molecules, which can be recorded when irradiated with IR radiation. The different bonds in the polymer absorb IR radiation at different frequencies and intensities, resulting in a specific IR spectrum. The IR spectrum of a polymer can therefore be regarded as a "fingerprint" of the polymer, as it is characteristic of its chemical composition [7].

By automatically comparing the IR spectrum of the polymer with known spectra of other polymers, it can be identified in this way. However, it should be noted that some polymers may have similar IR spectra if they have similar structures or functional groups. In such cases, it may be helpful to use additional analytical techniques such as NMR spectroscopy [9] to characterize the polymer.

In addition to identifying the polymer, FTIR spectroscopy can also be used to obtain information about special features of the molecular structure and the conformation of the polymer. This can be helpful, for example, when investigating polymers with different physical and mechanical properties, in order to understand the causes of these differences.

---

[3] If the infrared light is linearly polarized, statements can also be made about crystallinity [7].

The absorption bands in the IR spectra of engineering plastics depend on their chemical composition and structure, as the resonance frequencies depend on the strength of the bonds and the masses of the atoms involved. Some typical absorption bands that can be observed in the IR spectra of engineering plastics are listed below:

1. Aliphatic C–H stretching: The stretching of C–H bonds in aliphatic groups such as methyl, methylene, and methine groups produces strong absorptions at about 2900 cm^{-1}.
2. Aromatic C–H stretching: The stretching of the C–H bonds in aromatic rings leads to a characteristic absorption at about 3050 cm^{-1}.
3. C=O stretching: The stretching of the C=O double bond in carbonyl groups produces an absorption at about 1750 cm^{-1}. This is an important absorption band for polycarbonate, polyketones, polyamides, and polyesters such as PET and PBT.
4. C–O valence vibration: The stretching of the C–O single bond in alcohols, ether groups, carboxylates, and esters leads to absorption bands in the range of 1100–1,300 cm^{-1}. Polyvinyl alcohol, polyethylene glycol, and polyacrylates, for example, show absorptions at 1150 cm^{-1} and 1260 cm^{-1}.
5. C=C stretching: The stretching of the C=C double bond in unsaturated hydrocarbons leads to an absorption at about 1650 cm^{-1}.
6. N–H stretching: The stretching of the N—H bonds in amine groups leads to absorption bands in the range of 3300–3,500 cm^{-1}. Polyamides, for example, show absorptions at around 3300 cm^{-1}.
7. O–H stretching: The stretching of the O–H bonds in alcohols and carboxylic acids leads to a characteristic absorption at about 3400 cm^{-1}.

However, these absorption bands can depend on the type of plastic, its chemical composition, and its structure. This results in an absorption pattern for a sample that is characteristic of its composition which can be stored in databases and compared with other spectra. In this way, foreign substances, fluctuating compositions, material changes due to ageing, and many other processes can be recognized in the spectra.

The depth of information clearly depends on the complexity and symmetry of a ligand system. For example, $W(CO)_6$ has a very simple IR spectrum in which even small quantities of other substances can be easily identified, provided they are present in sufficient concentration and have a homogeneous distribution. In contrast, acetylacetonates exhibit characteristic spectra with many intrinsic absorptions in which foreign compounds are more difficult to identify [10, 11]. If one has access to the spectra of the pure substances, identification and corresponding quality control is possible.

Figure 4.54 shows the IR spectrum of acetylacetone (2,5-pentanedione). The most intense absorption band can be assigned to the valence vibrations of the double bonds between the oxygen atoms and the corresponding carbon atoms, and is located at a rather low-frequency position for this carbonyl bond at approx. 1600 cm^{-1} [12].

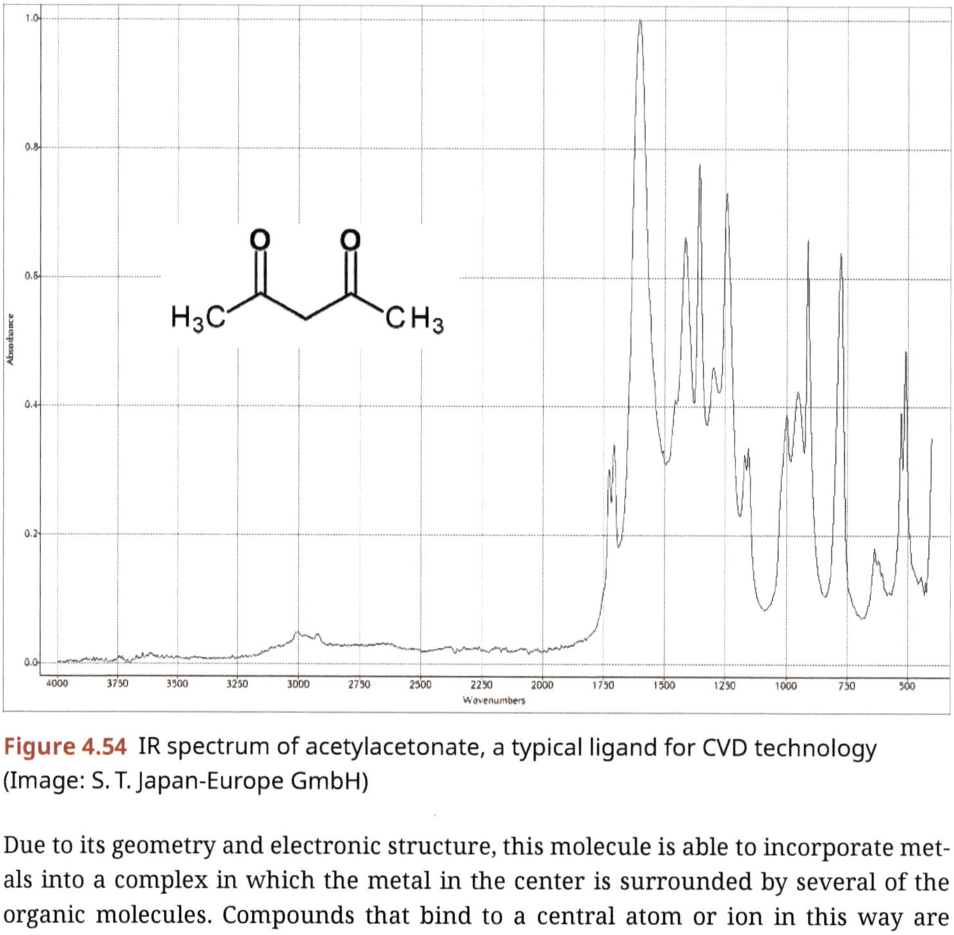

Figure 4.54 IR spectrum of acetylacetonate, a typical ligand for CVD technology (Image: S. T. Japan-Europe GmbH)

Due to its geometry and electronic structure, this molecule is able to incorporate metals into a complex in which the metal in the center is surrounded by several of the organic molecules. Compounds that bind to a central atom or ion in this way are called ligands. Bonding to a metal center changes the binding ratios and strengths, which is reflected in the shift of absorption bands. Figure 4.55 shows the vibrational spectrum of such a complex between nickel and two molecules of acetylacetone. In this case, it is actually nickel acetylacetonate dihydrate, which has two attached water molecules, which can be removed in a vacuum at approx. 50 °C [13]. A comparison of the two spectra reveals clear differences in band position and intensity, which can be used for structural clarification and the detection of chemical processes taking place.

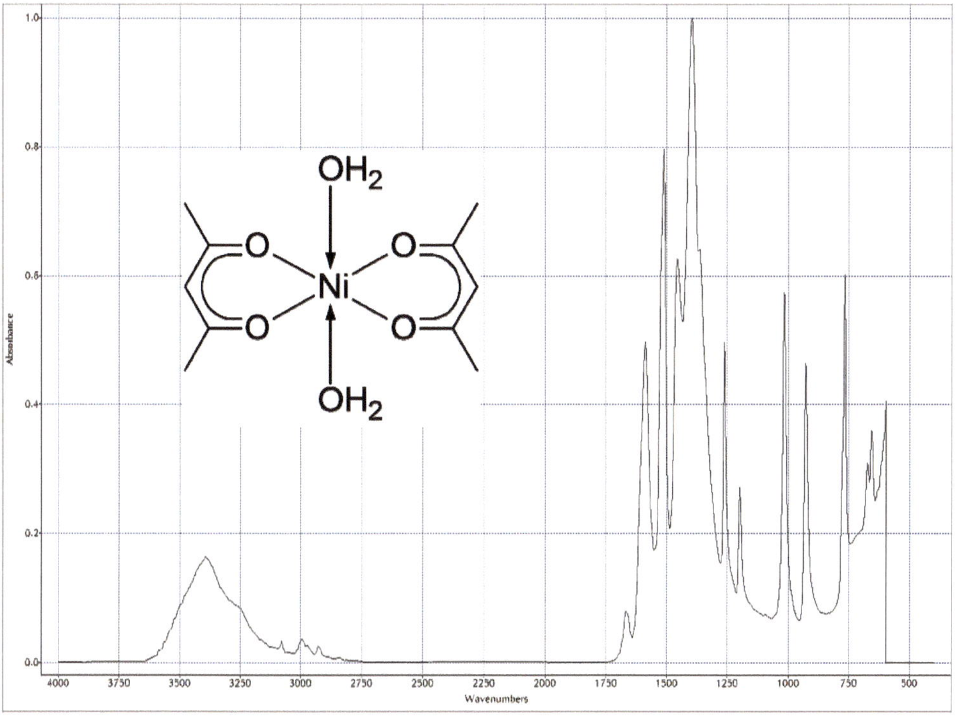

Figure 4.55 IR spectrum of nickel acetylacetonate dihydrate, a precursor for CVD technology (Image: S. T. Japan-Europe GmbH)

It may therefore be necessary to carry out additional analysis techniques such as mass spectrometry or chromatography to obtain clear information about the chemical composition of an engineering plastic.

4.5.5 Quantification of Components

FTIR spectroscopy can be used to identify and quantify the different bonds in polymers. The intensity of an absorption in a spectrum depends on only a few factors, of which the number of bonds is a decisive factor in addition to the change in the dipole moment of the molecule during the vibrational process. The relationship is linear: twice as many bonds results in double the so-called extinction, which in turn has a logarithmic relationship to the measured absorption at a certain wavenumber. The relationship is known as the Beer–Lambert law (Equation 4.17), which forms the basis for the quantitative application of IR spectroscopy [5, 6].

$$E_\lambda = \log\left(\frac{I_0}{I}\right) = \varepsilon_\lambda \cdot c \cdot d \tag{4.17}$$

E_λ = Absorbance at the considered wavelength λ (—)

I_0 = Input intensity of the light at the considered wavelength (W/m^2)

I = Residual intensity (transmitted light) (W/m^2)

ε_λ = Extinction coefficient (a wavelength- and substance-specific factor) (m^2/mol)

c = Concentration of the substance under consideration (more precisely of the vibration involved) (mol/m^3)

d = Layer thickness of the sample (in the case of the ATR technique, the cumulative indentation depth) (m)

As ε_λ and d are constant (or can be kept constant in an experiment), there is a direct linear relationship between the measured absorbance and the concentration. Investigations into the purity of substances and the control of the composition of preparations are therefore possible with comparatively little experimental effort. For ATR-FTIR spectroscopy, the detection limit and the precision of quantification are measured in the order of percent, at best in the per mille (parts per thousand) range. For trace analysis, IR spectroscopy is therefore inferior to many other methods.

4.5.6 Metrological Aspects of FTIR Spectroscopy

FTIR spectroscopy uses the Fourier transform technique to measure the absorption of IR radiation. While the first IR spectrometers in the 20th century sent the light from the source through the sample using suitable optical components and the spectral range was scanned successively by rotating an optical grating, today interferometry is used to expose the sample to all frequencies simultaneously. With the aid of a Michelson interferometer, a time-dependent signal reaches the detector, whose amplitude curve contains the required spectrometric information.[4] The signal is "translated" from the time domain to the frequency domain using Fourier transformation, which generates the FTIR spectrum (see Figure 4.56).

[4] For the structure of an interferometer and details on developing an interferogram, see Section 4.4. Note the analogy between Figure 4.56 and Figure 4.48.

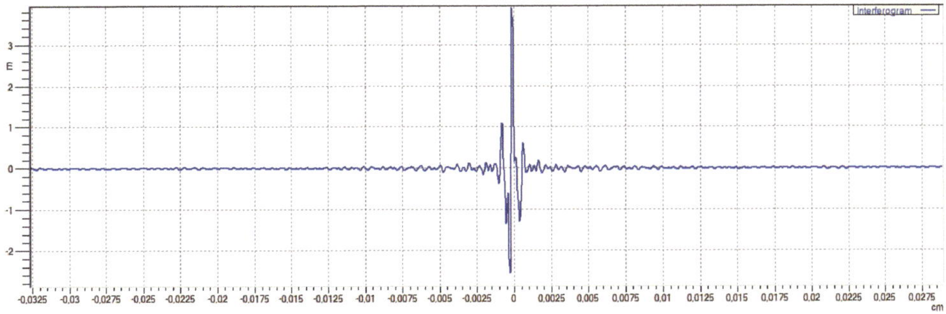

Figure 4.56 Interferogram generated by an FTIR spectrometer (Source: pro3dure medical GmbH)

Thanks to this setup, it is possible to record a spectrum in under a second, whereas the classic setup generated a spectrum in periods that could be measured in minutes. This increase in speed means that several spectra (scans) can be recorded from one and the same sample, which in turn significantly improves the signal-to-noise ratio. As the spectral resolution depends on the path length of the light beam (or, more precisely, on the path length of the moving part of an interferometer), the spectrometers cannot be built arbitrarily small. Over time, however, geometric designs have been developed that have made it possible to build devices with small footprints and high performance. As a result, FTIR spectrometers used for routine tasks are not much bigger than a shoebox, are robust, and can be operated easily after a short training period.

4.5.7 ATR-FTIR Spectroscopy

The attenuated total reflection (ATR) technique is a method in FTIR spectroscopy that makes it possible to obtain IR spectra of solids and liquids without prior sample preparation. The ATR technique is based on the use of a crystal that is brought into contact with the sample and generates a total reflection of the IR radiation at the surface of the sample. It is the most widely used preparation method today for routine examinations of solid or liquid samples. Figure 4.57 shows the schematic structure of an ATR-FTIR spectrometer.

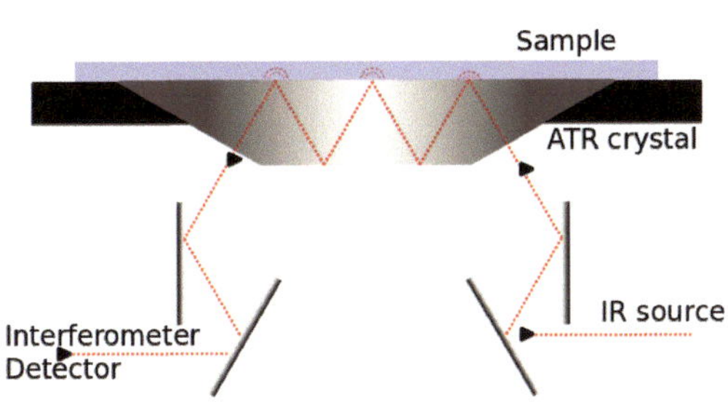

Figure 4.57 Schematic beam path of the IR beam through a sample that is in close contact with the ATR crystal

The ATR crystal usually consists of a highly refractive material such as diamond, zinc selenide (ZnSe), or germanium, and is pressed against the sample at a certain angle. When the IR radiation hits the crystal, it would be totally reflected at the interface between the crystal and the sample and would not penetrate the sample. Due to quantum-mechanical effects, however, there is a non-zero probability of the light wave in question being present beyond the interface. This is called an evanescent wave. If a sample is brought into this space, for example by pressing it closely against the crystal surface, part of the IR radiation is absorbed by the so-called evanescent field region at the interface between the crystal and the sample. The absorption of the IR radiation by the sample leads to an additional, frequency-dependent attenuation of the intensity. The attenuation of the reflected light, i. e. the difference between the irradiated light and the light of a given frequency measured after interaction with the sample, is measured as an ATR signal and interpreted as an IR spectrum. As the material of the crystal itself, like the laboratory atmosphere, causes absorption, ATR spectra are calculated with the so-called background spectrum, also known as the blind spectrum. The background spectrum (Figure 4.58) is freshly recorded at regular intervals, ideally before each sample measurement, and subtracted from the sample spectrum.

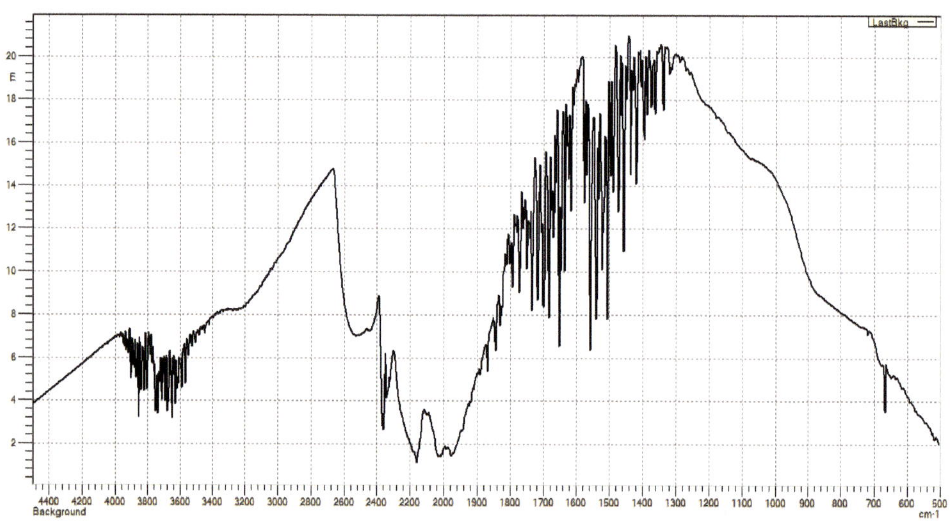

Figure 4.58 Typical background spectrum (blank measurement), caused by self-absorption of the ATR crystal used and IR-active components of the laboratory atmosphere (Image source: pro3dure medical GmbH)

The advantage of the ATR technique is that it enables fast and non-destructive analysis of solids and liquids without the need for extensive sample preparation. In addition, the ATR technique can also be used for samples with low volume or high viscosity. However, it should be noted that the ATR signal can be influenced by the surface layer of the sample and therefore does not always reflect the internal structure of the sample. This is particularly important in the case of contaminated component surfaces or in the presence of surface-active functional additives such as release agents or antistatic agents. The penetration depth of the light roughly corresponds to the wavelength of the radiation. This also means that the intensity distribution of ATR spectra is frequency-dependent, which is compensated for in the device software (ATR correction).

4.5.8 Application in Molding Tool Technology

IR spectroscopy is generally not suitable for analyzing metals, as metals do not have characteristic absorption bands in the IR range. This is due to the completely different type of bonding between the atoms. While covalent bonds in organic molecules are localized between two atoms and therefore lead to the absorption phenomena described, metal bonds are are virtually non-directional forces in which the atoms are held together by the joint "use" of electrons, whereby these electrons (in the so-called "conduction band") are more or less free to move. In general, metals reflect or absorb IR radiation very strongly, which can lead to high background signals that can interfere with the analysis of other components in a sample.

However, IR spectroscopy can be used in the investigation of organic molecules adsorbed on the surface of metals. This means that it can be used, for example, to detect the presence of tool coatings (plaque) due to degradation products of the processed materials, plastic additives, release agents, or flame retardants (or their residues). In such cases, the challenge is to take representative samples of such plaques, as tool components can rarely be analyzed directly for obvious reasons. Different techniques can be used, which must be selected empirically depending on the consistency, layer thickness, or composition. "Wipe sampling" has become well-established, in which a sample is taken from the tool surface with a lint-free cloth or cotton swab ("cleanroom swab"). A tried-and-tested procedure is also to absorb deposits with suitable organic solvents, which can then be removed again by evaporation. If this is not possible, the intrinsic spectrum of the absorbing medium can also be taken into account when evaluating the spectrum, for example by subtracting a reference spectrum of a separately measured pure substance from the spectrum obtained. This is possible due to the Beer–Lambert law and is implemented as a function in all common software packages. However, for technical reasons, it is not always guaranteed that a clean subtraction of spectra will be successful, which can make automated evaluation more difficult or even prevent it.

If the sample is of low viscosity, parts of the sample may penetrate the fabric. If the sample is then transferred to the ATR crystal, it is not necessarily guaranteed that the composition of this sample corresponds exactly to that of the coating. However, this circumstance is of secondary importance in most scenarios; often it is only a matter of assigning coatings to expected and therefore known substances or substance mixtures.

In any case, it is advisable to examine more than one sample from different sampling locations, especially as this can provide additional information that is helpful in explaining the origin of the sample composition, if it is found to be location-dependent.

References for Section 4.5

[1] Griffiths, P. R.; de Haseth, J. A.: *Fourier Transform Infrared Spectrometry*, John Wiley & Sons, 2007

[2] Hesse, M.; Meier, H.; Zeeh, B.: *Spektroskopische Methoden in der Organischen Chemie* (7th edn), Thieme, 2005

[3] Frank, H.; Phrood, A.: Elektromagnetisches Spektrum und der Bereich der Infrarotstrahlung, Wikimedia Commons, *https://commons.wikimedia.org/w/index.php?curid=3726606*, 2008

[4] Balster, A.: Schadensanalyse an Kunststoffen [Lecture notes] (1st edn), FH Südwestfalen, 2011

[5] Hediger H.-J.: *Infrarotspektroskopie: Grundlagen, Anwendungen, Interpretationen*, Akademische Verlagsanstalt, 1971

[6] Burns, D. A.; Ciurczak, E. W.: *Handbook of Near-Infrared Analysis* (3rd edn), CRC Press, 2007

[7] Looijmans, S. F. S. P.; Carmeli, E.; Puskar, L.; et al.: Polarization modulated infrared spectroscopy: A pragmatic tool for polymer science and engineering, *Polymer Crystallization*, 2020, 3: e10138

[8] Bienz, S.; Bigler, L.; Fox, T.; Meier, H.: *Spektroskopische Methoden in der Organischen Chemie* (9th edn), Thieme, 2016

[9] Bovey, F. A.; Mirau, P. A.: *NMR of Polymers*, Academic Press, 1996

[10] Holtzclaw Jr, H. F.; Collman, J. P.: Infrared absorption of metal chelate compounds of 1,3-diketones, *Journal of the American Chemical Society*, 1957, 79: 3318–3322

[11] Lawson, K. E.: The infrared absorption spectra of metal acetylacetonates, *Spectrochimica Acta*, 1961, 17: 248–258

[12] Tayyari, S. F.; Milaninejad, F.: Vibrational assignment of acetylacetone, *Spectrochimica Acta Part A*, 2000, 56: 2679–2691

[13] Nicholls, D.: The chemistry of iron, cobalt and nickel, in: *Comprehensive Inorganic Chemistry*, 1973, pp. 1109–1161

4.6 X-Ray Fluorescence Analysis

Dr. Martin Ciaston

4.6.1 Introduction

X-ray fluorescence (XRF) analysis is an established technique for the non-destructive, quantitative and qualitative analysis of solid, liquid, and powder materials. XRF analyses are frequently used in material analysis, quality control, and research, and can be used, for example, to determine the composition of a material, determine its purity, or to characterize coating thicknesses.

XRF is a physical phenomenon that is observed when an X-ray beam hits a material. Some of the higher energy X-ray photons are absorbed by the atoms of the material and then react with the material to produce shorter-wavelength, visible photons. This phenomenon is mainly used in the X-ray analysis of metals and other elements. XRF is one of the foundations of X-ray analysis and is used in many areas of science and technology. Theoretically, it can be applied to elements as light as boron (atomic number 5). However, with the lighter elements (atomic numbers 5–10), the fluorescence is completely absorbed by the environment, so that in practice, measurements can only be carried out for sodium (atomic number 11) and upwards. A distinction can be made between two areas of application for X-ray fluorescence analyses: qualitative and quantitative analyses. Qualitative analyses are based on the determination of the material, while quantitative analyses also take into account the intensity of the X-ray fluorescence and thus allow a statement to be made about the quantity of elements in the sample.

4.6.2 Physical Principles of X-Ray Fluorescence

The following is a brief summary of the basic interactions of X-rays with matter, to provide an understanding of the essential physical properties of an energy-dispersive XRF spectrometer. A more detailed introduction can be found in textbooks such as [1, 2].

XRF analysis refers to the quantitative and qualitative investigation of the characteristic fluorescence radiation of chemical elements, which occurs after a sample has been irradiated by X-rays (generated by X-ray tubes, for example). In the energy range 0.1–100 keV, there are essentially four different interaction mechanisms of X-ray photons with matter: Rayleigh scattering, Compton scattering, the photoelectric effect, and pair formation. Rayleigh scattering describes the elastic scattering of photons by collective excitation of core electrons. The scattering of photons can take place in all spatial directions. Regardless of the spatial direction, however, the photons have the same energy as before scattering. Compton scattering describes the scattering of a single photon with a single electron weakly bound to an atom. Both collision partners are to be regarded as particle-shaped and are subject to the law of conservation of energy and momentum. In contrast to the Compton effect, in the photoelectric effect the full energy of the incident photon is transferred to a shell electron. The photon itself is practically eliminated, as it no longer has any energy: the electron absorbs the energy of the photon as kinetic energy, and it leaves the atomic shell. During pair formation, photons are scattered in the vicinity of atomic nuclei and generate an electron–positron pair. These mechanisms are described by the total atomic interaction cross-section $\mu_{A,}$ which consists of the atomic interaction cross-sections μ_{Ph} for the photoelectric effect, μ_R for Rayleigh scattering, μ_{Pa} for pair formation and μ_C for Compton scattering (Equation 4.18) [3].

$$\mu_A = \mu_{Ph} + \mu_R + \mu_C + \mu_{Pa} \quad \text{with } [\mu_A] = 10^{-24} \cdot \frac{\text{cm}^2}{\text{atom}} \tag{4.18}$$

μ_A = Total atomic interaction cross-section (cm^2/atom)

μ_{Ph} = Photoelectric effect cross-section (cm^2/atom)

μ_R = Rayleigh scattering cross-section (cm^2/atom)

μ_C = Compton scattering cross-section (cm^2/atom)

μ_{Pa} = Pair formation cross-section (cm^2/atom)

The interaction probabilities shown can be used to calculate the attenuation of X-rays in matter. Using the Beer–Lambert law, the attenuation of the intensity I of a collimated X-ray beam passing through a layer of finite thickness x can be calculated, where μ_{lin} is the linear attenuation coefficient and an energy-dependent material constant (Equation 4.19).

$$\frac{dI}{I} = \mu_{lin}\, dx \tag{4.19}$$

I = Intensity (W/cm^2)

μ_{lin} = linear attenuation coefficient (1/cm)

x = Layer thickness (cm)

By integration, the layer thickness x can be accessed using the mass per unit area $F = \rho \cdot x$ (Equation 4.20).

$$I(x) = I_0 \cdot e^{\mu_{lin} \cdot x} = I_0 \cdot e^{-\mu_{mass} \cdot F} \tag{4.20}$$

I_0 = Primary intensity (W/cm^2)

F = Mass per unit area (g/m^2)

ρ = Density (g/cm^3)

μ_{mass} = Absorption coefficient (cm^2/g)

The intensity of the radiation before it enters matter is called the primary intensity I_0, and its intensity after passing through the layer is referred to as $I(x)$, where ρ is the density of the layer. The two absorption coefficients μ_{lin} and μ_{mass} are defined by the Avogadro constant N_A, the density and the molar mass M (see Equation 4.21).

$$\frac{\mu_{lin}}{\rho} = \mu_{mass} = \left(\frac{N_A}{M}\right) \cdot \mu_A \tag{4.21}$$

N_A = Avogadro constant (1/mol)

M = Molar mass (mol)

The mass attenuation coefficients μ_{mass}, μ_{lin}, and μ_A are fundamental parameters that quantitatively describe the interaction of X-rays with matter.

By exposing electrons to radiation with high energy, they can be in the K, L or M shells of an atom into an excited state. If the energy transferred during irradiation is greater than the binding energy of the electron to the nucleus, the electron can be removed from the shell. During this process, an electron gap is created, also known as the initial vacancy. In order to return to the energetically more stable state, an electron from an outer shell fills the vacancy and energy is emitted by X-rays. The energy released corresponds to the energy difference between the energy levels of the different electron shells. This emission of X-ray radiation is referred to as X-ray fluorescence, and leads to the formation of a characteristic line spectrum of discrete X-ray lines. The energy of the emitted X-rays depends on the energy difference between the shell with the initial vacancy and the energy of the electron that fills the shell (Figure 4.59).

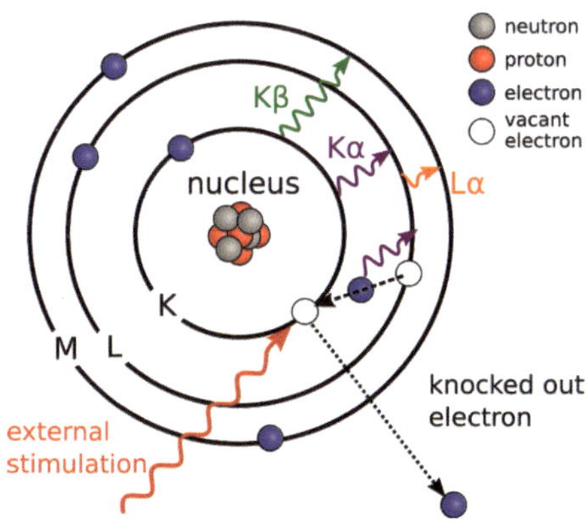

Figure 4.59
Schematic representation of the formation of characteristic X-rays (Image source: *https://de.wikipedia.org/wiki/Röntgenstrahlung*, Wikimedia Commons/Rainer Ziel/Cepheiden, reproduced under the GNU Free Documentation License, Version 1.2, *https://commons.wikimedia.org/wiki/Commons:GNU_Free_Documentation_License,_version_1.2*)

There are different possibilities for electron transitions for each element, which depend on which shell the high-energy electron falls from and in which shell the electron gap is to be filled. These transitions lead to the formation of characteristic X-rays, which are characterized by designations such as K_α, K_β, L_α, etc. The energy of an X-ray fluorescence radiation, i. e. its position in the spectrum, indicates which element it is. The intensity of an X-ray line depends on the quantity of the element in the sample [1–4].

4.6.3 Instrumental Aspects of X-Ray Fluorescence Spectroscopy

In X-ray fluorescence spectroscopy, there are basically two types of devices, whereby a distinction is made between wavelength-dispersive XRF (WDXRF) and energy-dispersive XRF (EDXRF). In both types of device, an X-ray radiation source is used to excite the sample. The difference lies in the way the emitted X-ray spectrum is detected. Typical EDXRF spectrometers, such as those used in laboratories or for quality control in industry, consist of an X-ray tube that generates the primary X-rays, an element to limit the radiation (pinhole or X-ray optics), and a detection system that measures the intensity of the element-specific fluorescent radiation emitted by the sample. This data can be used to draw conclusions about the properties of the sample, such as layer thickness or atomic composition. As a rule, electrons are emitted from an electrically heated cathode. These electrons are accelerated by a high voltage and hit the anode material, which generates primary X-rays and irradiates the sample to be measured. The primary filter is used to optimize the energy distribution of the primary X-rays. Depending on the manufacturer and device, optional safety devices such as a shutter provide additional safety for the user. The shutter is only opened for the duration of the measurement. As soon as the device chamber is opened or there is a fault, the shutter closes. The position to be measured is often illuminated by a light source and the image is projected onto a video camera using a perforated mirror and a lens. The mirror has an opening in the center so that the primary X-ray radiation can pass through. The collimator limits the cross-section of the primary beam in order to create a measuring spot of a defined size (Figure 4.60) [5].

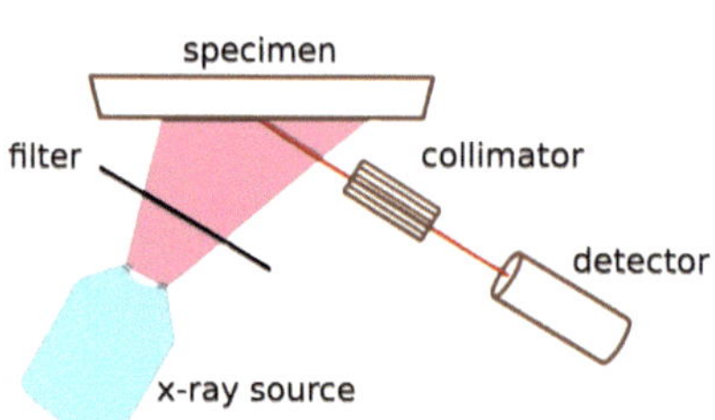

Figure 4.60
Schematic representation of an EDXRF instrument [5]

The energy-dispersive detection of X-rays captures the photons of all elements simultaneously. The entire spectrum is measured at a detector and then processed further. The total radiation of the sample is separated into the radiation of the elements it contains, and information on all elements is obtained in relation to an energy scale.

In the wavelength-dispersive system, the spectrum of the substrate is broken down into individual wavelengths by an analytical crystal in order to obtain the different energies, and fed separately to the detector (Figure 4.61) [5].

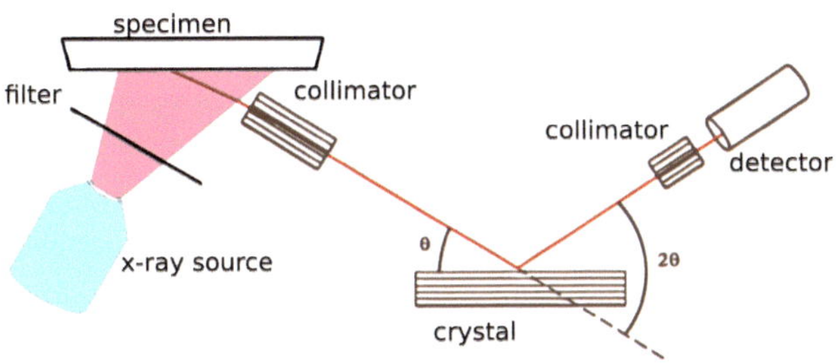

Figure 4.61 Schematic representation of a WDXRF [5]

The individual wavelengths of the elements to be measured are detected separately from each other, which leads to a significant increase in measurement time if several elements are being analyzed.

4.6.4 Applications of X-Ray Fluorescence Spectroscopy in Material Analysis

XRF can be used in many ways to determine the elemental composition of unknown substances. In X-ray fluorescence analysis, the measured intensity of the elements in the sample depends decisively on the surrounding matrix in addition to the concentrations. The matrix effects are based on the absorption, amplification, attenuation, and scattering effects mentioned above. Non-linear effects such as elemental composition, sample thickness, packing density, and particle sizes should not be neglected, as these play a significant role and influence the fluorescence intensity. In general, suitable sample preparation (grinding, melting, digestion, or dilution) of the starting material is necessary to solve these problems. Particles of different sizes have an influence on the fluorescence intensity of an element. A requirement for optimum fluorescence yield is that the chemical composition of the sample is homogeneous and representative. Powdery or particulate samples that show a particle size or grain size distribution can be expected to have an influence on the fluorescence intensity.

Due to its fast and simple application, XRF is used in many areas of industry. In contrast to the use of electron beams as the excitation source in SEM-EDX analyses, XRF uses an X-ray tube. This means that even moist or liquid samples can be analyzed without the need for a vacuum. Another advantage is that the sample does not need to be electrically conductive, which considerably simplifies sample preparation. Insulators can also be analyzed without additional effort. Thanks to its high sensitivity,

XRF is also suitable for screening for RoHS-restricted substances. For example, the solder contacts of components can be checked for freedom from lead and correct coating. Testing plastics for halogen or phosphorus-containing flame retardants is also possible.

4.6.5 Quantitative Aspects of X-Ray Fluorescence Spectroscopy

While qualitative or semi-quantitative determinations may be sufficient in some cases, the actual aim of X-ray fluorescence analysis is often the quantitative determination of major, minor, and trace elements. X-rays can be measured in different ways, e.g. by using excitation lines with different energies or by using filters to absorb the X-rays. Thanks to advances in XRF technology, in particular the use of high-performance silicon drift detectors (SDDs), it is possible, for example, to measure both the nickel and phosphorus content in NiP layers directly, and thus determine the layer thickness and chemical composition simultaneously. In order to achieve the highest precision when measuring nickel–phosphorus alloys (NiP) with XRF, special conditions are required that feature a suitable combination of detector and optimized geometry (minimum distance between sample and detector). This enables the simultaneous determination of several elements in a material and the percentages within a layer, e.g. nickel and phosphorus in electroless nickel layers (Figure 4.62). The different energies of the K_α-lines of the elements are used and the intensities determined [6].

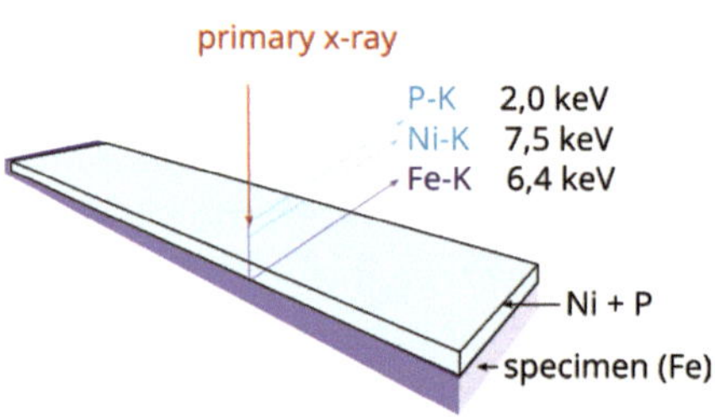

Figure 4.62
Simultaneous determination of nickel and phosphorus in electroless nickel coatings (Image: Based on [10])

In order to achieve high measurement accuracy, calibration standards are used, which contain known concentrations of nickel and phosphorus. The XRF method thus enables fast and precise determination of nickel and phosphorus in electroless nickel layers. By analyzing several elements simultaneously, the efficiency and accuracy of the analysis can be improved. If coating components are are not known, difficulties can arise, for example, due to the absorption of fluorescence radiation from lighter elements. For example, measurements on coatings made of tungsten sulfide at the Kunststoff-Institut Lüdenscheid Table 4.5 were obtained for the percentage of tungsten and sulfur in a sample.

Table 4.5 Percentages of a WS Layer Determined by Means of XRF Measurement

Sample (XRF measurement)	Proportion W (%)	Proportion S (%)
WS0031	65	35

However, further analyses using SEM-EDX have shown that significant amounts of carbon and oxygen are contained in the layers, and therefore that these values were incorrect (cf. Table 4.6). These results clearly show that the lack of measurement of lighter elements and the allocation of residual contents to the measurable elements can lead to incorrect conclusions.

Table 4.6 Percentages of a WS Layer Determined by Means of SEM-EDX Measurement

Sample (SEM-EDX measurement)	Proportion W (%)	Proportion S (%)	Proportion C (%)	Proportion O (%)
WS0031	27	48	18	7

XRF analysis can also be used as a non-destructive and precise method for determining the coating thickness of various materials. Difficulties in the application of XRF often arise due to complex relationships between intensity and concentration. If several elements are present in a sample at the same time, different absorptions of primary and fluorescence radiation as well as secondary excitations can occur, depending on the position in the periodic table. As a result, the relationship between the content of the element to be determined and the intensity of its fluorescence radiation is not always proportional. Furthermore, the measurement accuracy of the coating thickness measurement is primarily dependent on the material to be analyzed and the coating thickness. For example, chromium can still be measured with an accuracy of 10 nm up to a coating thickness of 50 nm, whereas chromium/copper/nickel coatings that are thicker than 40 µm have a high degree of inaccuracy when measuring the copper layer, as the exit depth of the fluorescence to the analyzer is limited. Ultimately, the measurement accuracy depends crucially on the quality of the selected standards. Measurement accuracies must be requested from the device manufacturer, as these are different for each coating system.

If elements cannot be detected by XRF, such as oxygen and carbon, measurements of oxide coatings on substrates such as injection molds are difficult. Using complementary methods on test specimens, which are added to the coating process of the actual substrate, the coating thickness of the coated substrates can be determined non-destructively. One possible method is the crater grinding process, but this of course is not non-destructive (see Section 4.1). The crater grinding method is used to determine the coating thicknesses of the test specimens. If the coating thickness of the specimens is known, XRF can be used to measure the coating thickness non-destructively on the actual substrate. The relationship from Equation 4.20 is used to measure the intensity of the X-ray fluorescence radiation on the substrate, and the density is adjusted in the measurement method until the coating thicknesses of the test specimens and the substrate match. When determining the layer thickness of zirconium oxide layers, for example, it is possible to measure the X-ray fluorescence of zirconium (which has a density of 6.51 g/cm^3) without any problems. However, there is a deviation from the actual coating thickness. By reducing the density to a lower value, which was previously determined by comparing the test specimens with the coating thickness from the calotte grinding process and the intensity signal obtained from the XRF, the coating thickness can be determined using Equation 4.20 and Equation 4.21.

4.6.6 Summary and Outlook

A wide variety of methods are used to analyze materials, both in research and in quality control. The use of scanning electron microscopy (SEM) in conjunction with supplementary methods such as focused ion beam (FIB), energy dispersive X-ray spectroscopy (EDX), and X-ray diffraction (XRD) enables comprehensive investigation of the relevant properties of materials. Using the FIB technique, material can be removed from the sample using a high-energy ion beam, exposing internal structures. These structures can then be analyzed by EDX for elemental composition and XRD for structural analysis. There are also other analysis options such as atomic force microscopy (AFM) and nanoindentation, which can determine material properties such as hardness, brittleness, and modulus of elasticity. In contrast to XRF, however, these methods are sometimes destructive and more time-consuming, and often require sample preparation, meaning that XRF will continue to be used in material analysis, quality control and research in the future.

Micro X-ray fluorescence analysis (μ-XRF) is a method in which an X-ray beam is focused to a few micrometers using X-ray optics in order to achieve a very high resolution (in the micrometer range). An extension of this method, which was developed at TU Berlin in the 2000s, is 3D micro X-ray fluorescence spectroscopy (3D-μ-XRF). This method enables non-destructive, three-dimensional imaging of samples. Both the excitation beam and the detection beam are guided through a polycapillary lens in order to define an examination volume in the micrometer range [7]. The latest results from measurements with XRF show that the practical application of this analysis method still has great potential. For example, Breuckmann et al. recently published a patent for the non-destructive determination of the calorific value of a sample using X-ray spectroscopy [8]. By recording an X-ray spectrum of the sample, they were able to determine the calorific value of the sample on the basis of the recorded X-ray spectrum and the data set. In further investigations, an approach for determining carbon, hydrogen, nitrogen, and oxygen (CHNO) in polymers was researched using wavelength dispersive X-ray fluorescence analysis (WDXRF) in combination with partial least-squares regression (PLS). CHNO quantification was based on the Rayleigh and Compton scattering spectra of 84 polymers using a Rh X-ray tube [9].

The multitude of changes in industrial production will pose a challenge to current process and quality controls in the future. For example, energy-intensive production processes such as the manufacture of steel will have to switch to sustainable energy sources. In the transport sector, in addition to new materials, new processes for possible impurities in the product will also play a role, so that continuous further development of measurement methods using XRF is essential.

References for Section 4.6

[1] Beckhoff, B.; Kanngießer, B.; Langhoff, N.; et al.: *Handbook of Practical X-Ray Fluorescence Analysis*, Springer-Verlag, 2006

[2] Hahn-Weinheimer, P.; Hirner, A.; Weber-Diefenbach, K.: *Grundlagen und Praktische Anwendung der Röntgenfluoreszenzanalyse (RFA)*, Vieweg+Teubner Verlag, 1984

[3] Bremekamp, M.: Energiedispersive Mikro-Röntgenfluoreszenzanalyse technischer Mikrostrukturen [Dissertation], Technical University of Berlin, 2017

[4] Haschke, M.; Flock, J.; Haller, M.: *X-Ray Fluorescence Spectroscopy for Laboratory Applications*, Wiley-VCH, 2021

[5] Schramm, R.: *Röntgenfluoreszenzanalyse in der Praxis*, Fluxana, 2012

[6] Kreiner, M.; Zhu, J.: Der Prozess der stromlosen Vernicklung und seine Bedeutung in der Anwendung, *WOMag*, 2016, issue 07–08

[7] Mantouvalou, I.: Quantitative 3D Micro X-ray fluorescence spectroscopy [Dissertation], Technical University of Berlin, 2009

[8] Breuckmann, M.; Hanning, S.; Kreyenschmidt, M. P.: Verfahren zur zerstörungsfreien Bestimmung des Brennwertes einer Probe mittels Röntgenspektroskopie, Patent DE102022106365A1, 2022

[9] Breuckmann, M.; Wacker, G.; Hanning, S.; et al.: Quantification of C, H, N and O in polymers using WDXRF scattering spectra and PLS regression depending on the spectral resolution, *Journal of Analytical Atomic Spectrometry*, 2022, 37: 861–869

[10] Phosphor-Gehalt in chemisch Nickel direkt messbar (AN002), Helmut Fischer GmbH, Institut für Elektronik und Messtechnik

4.7 Electrochemical Impedance Spectroscopy

Dr. Anatoliy Batmanov

4.7.1 Introduction

Electrochemical impedance spectroscopy (EIS) is a valuable tool for investigating both the transport properties of materials and the electrochemical reactions at interfaces. This measurement method [1, 2] has been used for over 50 years to characterize coatings on metal surfaces. At the end of the 19th century, it was discovered that two electrodes in an electrolyte solution do not exhibit purely ohmic behavior when a sinusoidal alternating voltage is applied, but show clear capacitive and inductive properties [3–5].

EIS has been used to investigate the kinetics of electrochemical processes since the 1960s. As the performance of measuring instruments has improved, the range of applications of EIS has also expanded. This includes investigations of metal dissolution, deposition, and corrosion [6, 7]. Other areas of application are the investigation of barrier/protective layers and the structure of materials used in battery and fuel cells [2, 8, 9].

The cause of aqueous corrosion and high-temperature oxidation of steels are electrochemical processes at the interface [10]. The metal atoms are ionized, which is accompanied by electron transfer. The corrosion resistance of steel in aqueous solution can be assessed by measuring the polarization resistance at the open-circuit potential of the system [11], as shown in Figure 4.63. The surface condition plays a decisive role in the development of the polarization resistance. An oxide and/or hydroxide layer on the surface generally leads to an increase in the polarization resistance, as the oxide and hydroxide layers hinder charge transport [2].

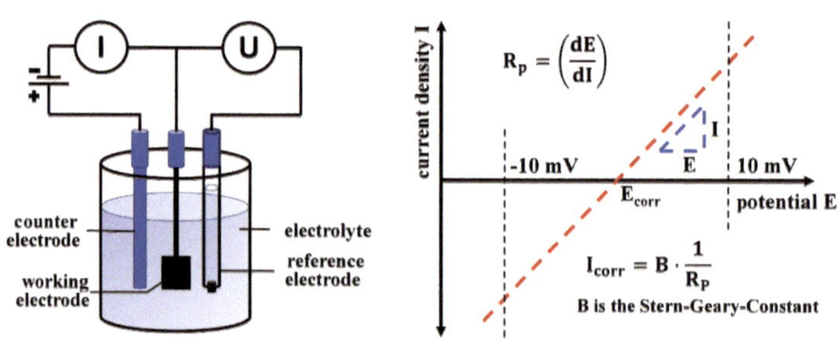

Figure 4.63 Measuring principle of the polarization resistance: a measuring cell (left) and calculation of the polarization resistance (right) [11]

The presence of a protective coating can hinder or even stop the corrosion process of the underlying metal. Therefore, the electrical resistance of the coating serves as an important indicator of the efficiency of the corrosion resistance. In the case of aqueous corrosion and high-temperature oxidation on the surface of metals and alloys, the growth rate of an oxide and/or hydroxide layer depends on the transfer of cations, anions, or their vacancies through the protective coating [10, 12]. If the protective coating is not defect-free and defects such as pores, cavities, channels, or cracks are relatively large, they act as short-circuit paths for the oxidation reaction. The quality of the protective coating therefore has a considerable influence on the electrical resistance of the coating and the corrosion resistance of the metal.

The frequency range of impedance spectroscopy is from 0 Hz (direct current) to 5 MHz, which makes it possible to investigate slower processes such as diffusion or pore filling and fast processes such as chemical reactions [2]. This enables the investigation of high-resistance coating systems and the validation of models for the interpretation of physico-chemical systems. High-resistance coatings are mostly used to protect against corrosion, and serve as a barrier layer against contact of the metal surface with water, air, and other aggressive media. The general requirements for modern corrosion protection coatings include the following points [12, 13]:

- Environmental compatibility
- Low risk to health
- Longer durability
- Lower production costs.

The development and quality testing of such coatings can take several years. In addition to the standard methods for testing the protective properties of anti-corrosion coatings, such as the salt spray test (ISO 9227) [14], the condensation water change climate test (ISO 6270) [15], and the short weathering test (ISO/TR 19402) [16], impedance spectroscopy is very useful for assessing the electrochemical condition of the system and thus shortening the testing time. EIS can be used to examine the ageing process and the occurrence of defects and corrosion.

4.7.2 Basics of EIS

Impedance spectroscopy is a non-destructive method that examines the response of a system to an applied low-amplitude alternating current- or alternating voltage load with low amplitude, as shown in Figure 4.64 [2]. This allows the respective resistance, the impedance of the electrochemical system, and the frequency-dependent time shift (phase shift) between the current and voltage maxima to be determined and evaluated. The correct assessment of the measurement results requires a good understanding of the system under investigation and the possible processes within it, as well as knowledge of the fundamentals of the measurement method.

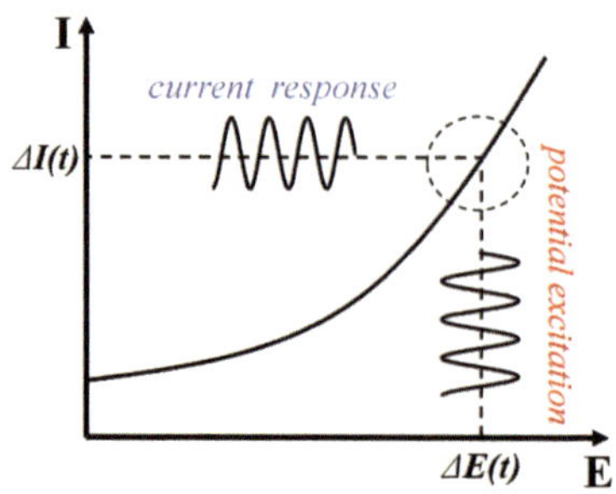

Figure 4.64 Measuring principle of the potentiostatic EIS

The impedance is the complex electrical resistance of a system that is measured when an alternating current is applied to the system. The total resistance of the system is made up of the individual resistances of the various components of the system, including the electrodes and electrolytes [17].

If a system consists only of electrical conductors such as metals or electrolytes, the current–voltage behavior can be calculated using Ohm's law according to Equation 4.22 [2, 17].

$$\text{Direct current (DC)}: R = U/I \quad (4.22)$$

$$\text{Alternating current (AC)}: Z = E/I$$

R = Ohmic resistance (Ω)

U = Electrical voltage (V)

I = Electric current (A)

Z = Impedance (Ω)

E = Potential (V)

Figure 4.65 shows two periodic waves. The blue wave describes the potential signal (excitation) and the red wave the current signal (response). Since one wave causes the other wave, both signals oscillate with the same frequency and intensity. However, there is an important difference between the two signals, namely the constant time shift (phase angle shift φ) at a certain angle, which can vary between 0° and 90°. If the excitation potential $E(t)$ is a sinusoidal wave, the response signal in the form of the current is also measured sinusoidally. The excitation signal as a function of time t can be calculated according to Equation 4.23 [2, 17].

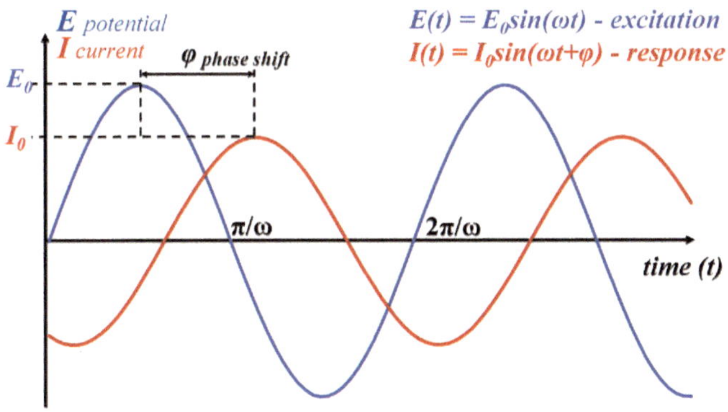

Figure 4.65 Current and voltage as a function of time

$$E\,(t) = E_0 \cdot \sin\,(\omega \cdot t) = E_0 \cdot \mathrm{e}^{j \cdot \omega \cdot t} \tag{4.23}$$

$E(t)$ = Potential at time t (V)

E_0 = Amplitude of the signal (V)

ω = Angular frequency (= 2 π f) (1/s)

j = Imaginary portion ($j^2 = -1$) (—)

The current response can be calculated according to Equation 4.24. The excitation potential and the current response depend on the electrical circuit under consideration.

$$I\,(t) = I_0 \cdot \sin\,(\omega \cdot t + \varphi) = I_0 \cdot \mathrm{e}^{j \cdot (\omega \cdot t + \varphi)} \tag{4.24}$$

I_0 = Amplitude of the current (A)

φ = Phase angle shift (°)

By inserting the excitation signal according to Equation 4.23 and the current response according to Equation 4.24 into Equation 4.18, Ohm's law for the alternating voltage is given by Equation 4.25.

$$Z = \frac{E(t)}{I(t)} = \frac{E_0 \cdot \sin(\omega \cdot t)}{I_0 \cdot \sin(\omega \cdot t + \varphi)} = |Z_0| \cdot \mathrm{e}^{-j \cdot \omega \cdot t} = Z' - j \cdot Z'' \tag{4.25}$$

Z_0 = Absolute value of the impedance (Ω)

Z' = Real part of the impedance (Ω)

Z'' = Imaginary part of the impedance (Ω)

The impedance $|Z_0|$ is specified as an absolute value and defines the ratio between the amplitudes of the voltage and the current. It represents a resistance to the flow of electrons or current in the AC circuit. The phase shift is most pronounced at the frequency at which the ratio of the imaginary part Z'' to the real part Z' of the impedance is greatest [2].

The ohmic resistance is independent of the waveform, so that the phase shift is zero and the peak values of the signal are reached at the same time, as shown in Figure 4.66. This results in the impedance of an ohmic resistor being given by Equation 4.26 [2, 17].

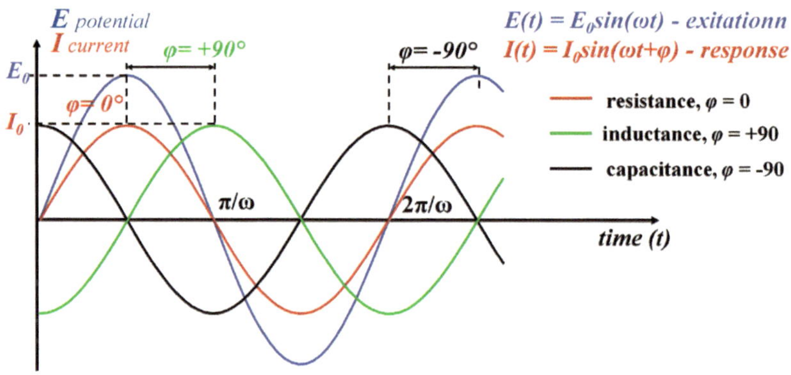

Figure 4.66 Sinusoidal excitation voltage and the current response for a resistor, a capacitor, and an inductor

$$Z_R = \frac{R \cdot I(j \cdot \omega)}{I(j \cdot \omega)} = R = Z' \tag{4.26}$$

Z_R = Impedance of an ohmic resistor (Ω)

In contrast to ohmic resistance, the resistance of reactive elements such as capacitors and inductors is frequency-dependent. In the case of a capacitor, the resistance is extremely high with direct current and very low with alternating current frequencies, meaning that almost no current can be transmitted. In the high-frequency range, the impedance of a capacitor is relatively low, so that the current can flow through the system. However, the peak values of the current and voltage amplitude are offset by –90°, as shown in Figure 4.66. The impedance of a capacitor can be calculated using Equation 4.27 [2, 17].

$$Z_C = \frac{1}{j \cdot \omega \cdot C} = Z_C' \tag{4.27}$$

Z_C = Impedance of a capacitive resistor (Ω)

C = Capacitance of the capacitor (F)

The capacitance can be calculated from the examination of protective coatings according to Equation 4.28:

$$C = \frac{\varepsilon_0 \cdot \varepsilon_r \cdot A}{d} \tag{4.28}$$

ε_0 = Electric field constant (= 8.854×10^{-12}) (F/m)

ε_r = Dielectric constant of the coating material (—)

A = Area of the coating (m^2)

d = Thickness of the coating (m)

The phase shift between current and voltage caused by a coil or inductance can be calculated according to Equation 4.29 [2, 17], and is +90° (see Figure 4.66). The impedance at high frequencies is very high, and at low frequencies and DC voltage it is almost zero.

$$Z_L = j \cdot \omega \cdot L = Z_L' \tag{4.29}$$

Z_L = Impedance of an inductive resistor (Ω)

L = Inductance (H)

When investigating corrosion processes and protective coatings, the inductance generally does not occur, so that the total impedance is defined by Equation 4.30:

$$Z_G = R + \frac{1}{j \cdot \omega \cdot C} \tag{4.30}$$

Z_G = Impedance of a combined resistor (Ω)

In real systems, the corrosion process, the effect of the protective coating, and the electrochemical double layer at the interface between a metal and the electrolyte are usually described by a combination of resistances and capacitances.

The presence of defects in the coating such as pores, cracks, or a rough surface causes a non-ideal capacitive behavior. This can be corrected using the constant-phase element (CPE) [2, 17, 18]. The CPE is a mathematical consideration of a defective capacitor, and also describes the behavior of an electrochemical double layer. The impedance representation of the CPE is given by Equation 4.31 [18].

$$Z_{CPE} = \frac{1}{Y_0 \cdot (j \cdot \omega)^N} \tag{4.31}$$

Z_{CPE} = Impedance of the CPE (Ω)

Y_0 = Admittance of the CPE ($S \cdot s^{-N}$)

N = Value between 0 and 1 (—)

The admittance of the CPE takes into account the conduction paths for water absorption, ion migration, and the penetration of salts into the coatings. The value of N is between 0 and 1. The CPE works as a pure resistor at $N = 0$ and as a pure capacitor at $N = 1$.

Other frequency-dependent processes that can occur during the investigation of metal corrosion and coatings are diffusion processes, which are represented by the Warburg impedance. This is a constant-phase element with a constant phase shift of 45°, which is independent of the frequency, and with a value that is inversely proportional to the square root of the frequency (Equation 4.32) [2, 17, 18].

$$Z_W = \frac{W}{\sqrt{j \cdot \omega}} = \frac{W}{\sqrt{2 \cdot \omega}}(1 - j) \tag{4.32}$$

Z_W = Warburg impedance (Ω)
W = Warburg coefficient ($\Omega \cdot s^{-1/2}$)

Table 4.7 lists examples of the most common elements and their respective equivalents in a system under investigation [2, 17, 18].

Table 4.7 Elements of Impedance Most Commonly Used in the Investigation Of Corrosion Protection Coatings [2, 17, 18]

Element	Impedance	Phase angle (°)	Origin
Ohmic resistance	R	0	Current conductor (resistance of an electrolyte, resistance of the charge transfer)
Capacitor	$\frac{1}{j \cdot \omega \cdot C}$	-90	Electrochemical double layer (adsorption), defect-free coating
Constant-phase element	$\frac{1}{Y_0 \cdot (j \cdot \omega)^N}$	$\frac{(-N) \cdot \pi}{2}$	Non-ideal capacitor (defects in the coating)
Warburg impedance	$\frac{W}{\sqrt{j \cdot \omega}}$	-45	Linear diffusion

4.7.3 Presentation of EIS Measurement Results

In impedance spectroscopy, the impedance spectrum of an entire system under investigation is measured and evaluated in a broad frequency spectrum. Technically, however, current or voltage (depending on the excitation signal), and the phase shift between the excitation and response signal, are measured [18]. The impedance of the entire system is then calculated according to Equation 4.25.

The impedance spectrum can be displayed in different ways. In the Nyquist diagram (Figure 4.67, left), the imaginary part is plotted against the real part of the impedance [2, 18]. For the Bode diagram (Figure 4.67, right), the impedance and the phase shift are plotted as a function of the frequency [2, 18]. The frequency axis and the impedance axis in the Bode diagram are plotted logarithmically, as they vary over a very wide range of values.

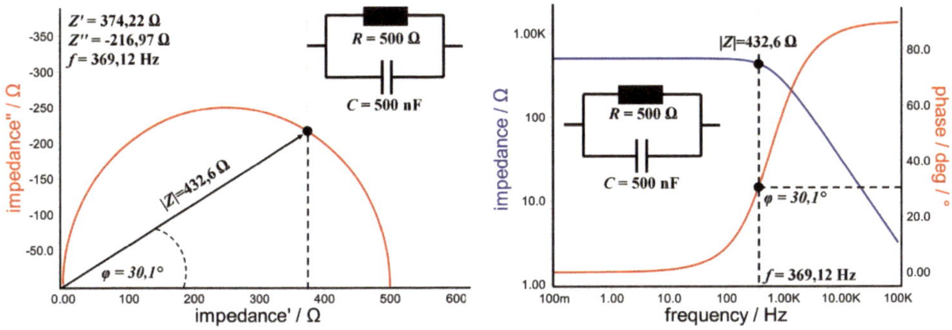

Figure 4.67 Nyquist diagram (left) and Bode diagram (right) of an RC circuit

Both display formats are used to visualize the EIS measurement results. In a Nyquist diagram, small impedances are difficult to recognize in the presence of large impedances. The Nyquist diagram can also be used to resolve individual impedances, but the disadvantage is that the frequency is not explicitly displayed. In contrast, the Bode diagram is frequency-dependent, and small impedances are easily recognizable. This is why Bode diagrams are often used to display and evaluate EIS measurements.

4.7.4 EIS Investigation of Protective Coatings

The sample, e.g. a coated metal plate, is dipped in a liquid electrolyte in order to examine the corrosion resistance of the coating. If there is contact between the electrolyte solution and the substrate material, the metal ions go into solution and free electrons are left behind [2, 19]. As a result, the electrochemical double layer (Helmholtz layer) is formed. In the case of alternating current, they form a capacitor due to polarization processes. This is described by a double layer capacitance C_{dl} and a polarization resistance R_p connected in parallel, as shown in Figure 4.68. The corresponding equivalent circuit diagram is called a "Randles circuit", and describes the corrosion process in such a way that the corrosion resistance can be read directly from the measurement results [2, 8, 19]. The value of the double layer capacitance is normally in the range 10–40 $\mu F/cm^2$. The typical polarization resistance for a bare metal is approx. 5 $k\Omega \cdot cm^2$ [20] and is inversely proportional to the corrosion rate. The double layer capacitance C_{dl} and the polarization resistance R_p must be normalized, as they are dependent on the electrode area. The resistance R_{el} represents the resistance caused by the electrolyte and the leads.

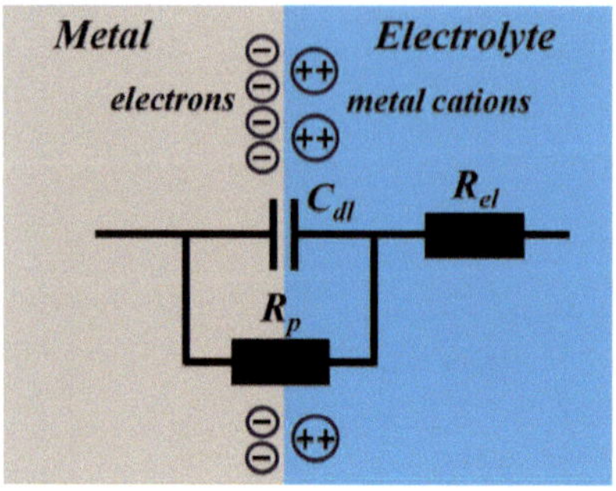

Figure 4.68
Equivalent circuit diagram for a metal–electrolyte (double layer) model (Image based on [19])

The simulated result of the impedance spectroscopy of an electrochemical double-layer system is shown in Figure 4.69. The direct current resistances R_{el} and R_p can be read directly from the diagrams. The capacitance C_{dl} can be calculated using the time constant according to Equation 4.33 [2, 19].

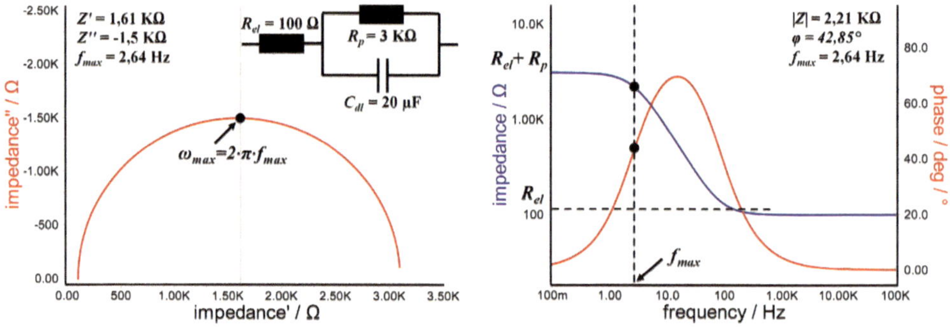

Figure 4.69 Simulated EIS results for an electrochemical double-layer system: Nyquist diagram (left) and Bode diagram (right)

$$\omega_{max} = 2 \cdot \pi \cdot f_{max} = \frac{1}{R_p \cdot C_{dl}} \tag{4.33}$$

F = Frequency (Hz)

R_p = Polarization resistance (Ω)

C_{dl} = Double layer capacitance (F)

A non-conductive coating exhibits a high electrical resistance with direct current, and behaves like a capacitor with the capacitance C_{cpc} with alternating current. In the case of defects (e.g. pores or cracks) in the coating, the resistance of the solution in the pores R_{por} is measured (Figure 4.70).

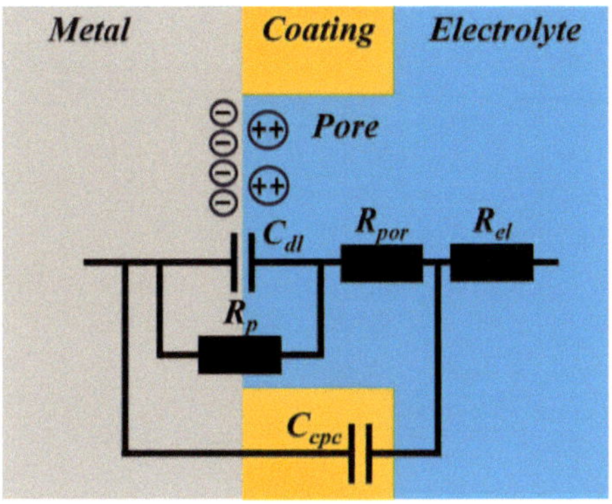

Figure 4.70
Equivalent circuit diagram for a metal–coating–electrolyte model (Image based on [19])

The double layer capacitance C_{dl} is considerably greater than the capacitance of the coating C_{cpc}. It is normally approx. 1–100 nF/cm^2. This means that the double-layer capacitance caused by a small defect in the coating such as a scratch, a crack, or a hole (pinhole) becomes visible in the system response, as shown by the right semicircle in the Nyquist diagram or the left time constant in the Bode plot (Figure 4.71). The left semicircle in the Nyquist diagram and the right time constant in the Bode diagram characterize the coating in terms of thickness and porosity [2, 8, 19]. A coating that adheres strongly to the metal surface does not allow metal–electrolyte contact, so that the double-layer capacitance can sometimes be caused by delamination of the coating [20].

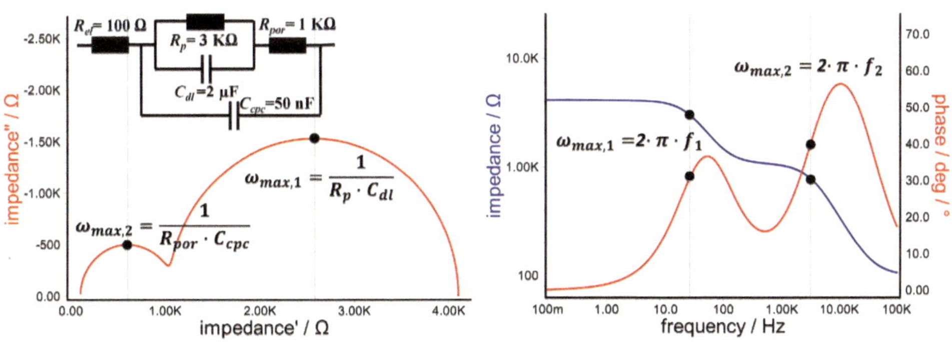

Figure 4.71 Simulated EIS results for a metal coating–electrolyte system with a non-conductive porous layer: Nyquist diagram (left) and Bode diagram (right)

Figure 4.72 shows the EIS measurement results and SEM images of a zirconium oxide layer stabilized with yttrium (YSZ) (left) [21] and an aluminium oxide layer (right) [22], which were deposited at the non-profit KIMW Forschungs-GmbH. The two coatings were applied to metal coins with a diameter of 20 mm made of 1.4301 steel (Cr 18.5%, Ni 8.5%) using a CVD process.

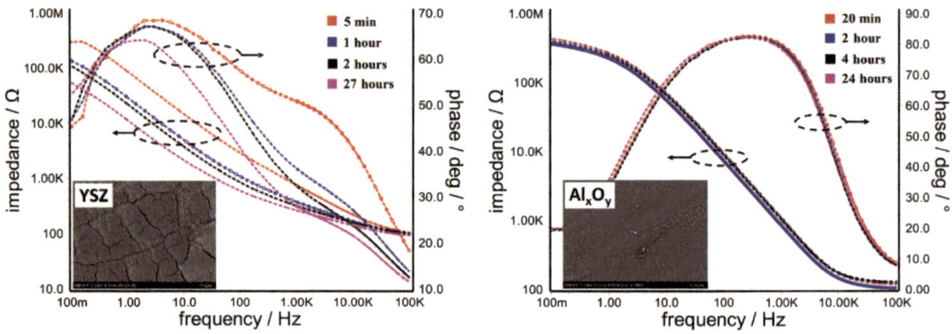

Figure 4.72 EIS measurement results (Bode diagrams) and SEM images of surfaces of the crystalline YSZ (left) and the Al_xO_y coating (right)

Within the first 27 hours, the impedance of the YSZ layer in the low frequency range and the phase angle in the high frequency range decreases. This indicates a flooding of the existing pores and an associated loss of the layer's insulating effect, and thus greater porosity. The Al_xO_y layer shows no change in impedance and phase angle, which indicates low porosity (see Figure 4.72 (right)). Due to the high density of the Al_xO_y coating, this coating can be used for corrosion protection.

An electrical equivalent circuit diagram that can describe the coated 1.4301 steel is shown in Figure 4.73 (left). The resistance of the electrolyte R_{el} is also taken into account here, and is approx. 80–90 Ω at a measuring temperature of 40 °C. The capacitance of the coating is defined by a constant phase element (CPE_{cpc}) and characterizes the dense part of the coating and depends on the permittivity of the coating material, the sample geometry, and the coating thickness. The porous or defective part of the coating can be characterized with an electrical oscillating circuit connected in parallel, consisting of a polarization resistance R_p and a double-layer capacitance C_{dl}, which is connected in series with a pore resistor R_{por}. In order to reduce the effect of the porous or defective part of the layer, the pore resistance must therefore be correspondingly high. The native passivation layer, consisting of iron (Fe), chromium (Cr), and nickel (Ni) oxides [23], is represented by the capacitance CPE_{pl}. The measurement results for both layers after 2 hours in the electrolyte, adjusted (fitted) with the equivalent circuit diagram, are shown in Figure 4.73 (right).

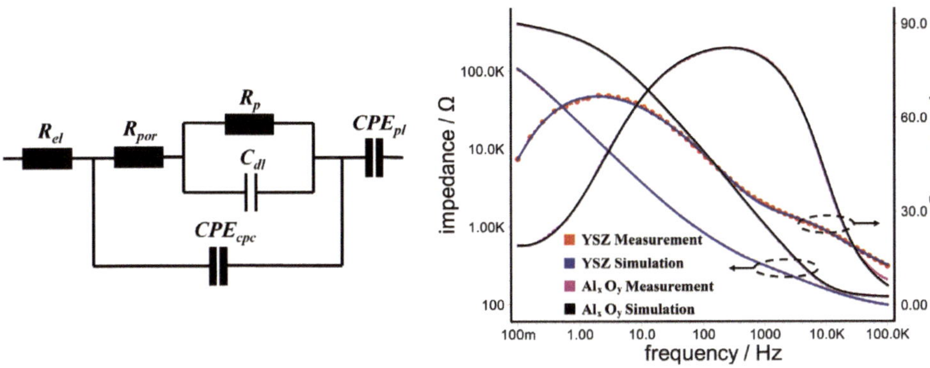

Figure 4.73 Electrical equivalent circuit diagram for the coated 1.4301 steel (left), and EIS measurement results modeled with the equivalent circuit diagram (right)

The values of the elements of the equivalent circuit diagram extracted from the simulation are shown in Table 4.8. The pore resistance R_{por} of the YSZ coating is significantly lower than that of the Al_xO_y coating and amounts to 91.4 Ω, which indicates a very high porosity of the coating structure. The capacitance C_{dl} of the YSZ steel system is three orders of magnitude higher than that of the Al_xO_y steel structure, indicating a large contact area between the electrolyte and the metal. The power N of the CPE_{cpc} of the YSZ layer is significantly lower than the power N of the CPE_{cpc} of the Al_xO_y layer, which indicates a reduced electrically insulating effect of the layer.

Table 4.8 Values of the Elements of the Electrical Equivalent Circuit Diagram for the Coated 1.4301 steel

Element	Unit	YSZ	Al_xO_y
R_{el}	Ω	68.7	70.1
C_{dl}	F	2.49×10^{-6}	12.18×10^{-9}
R_p	kΩ	214	361
CPE_{cpc}	$S \cdot s^N$	8.95×10^{-6}	173×10^{-9}
N_{cpc}	—	0.783	0.99
R_{por}	Ω	91.4	26.2×10^3
CPE_{pl}	$S \cdot s^N$	149×10^{-6}	15.5×10^{-6}
N_{pl}	—	0.4	0.71

4.7.5 The Test Setup for EIS Measurements

The measuring station for electrochemical impedance spectroscopy (EIS) consists of a potentiostat or a galvanostat, a frequency response analyzer (FRA), a measuring cell, and control and evaluation software (Figure 4.74). The EIS is usually recorded in the three-electrode arrangement. In this configuration, the sample to be analyzed is referred to as the working electrode. A counter-electrode is usually made of graphite or platinum. An independent reference electrode is used to accurately measure the applied potential in relation to a stable reference reaction. Commonly used reference electrodes in aqueous electrolytes include the saturated calomel electrode (SCE), the standard hydrogen electrode (SHE), and the AgCl/Ag electrode.

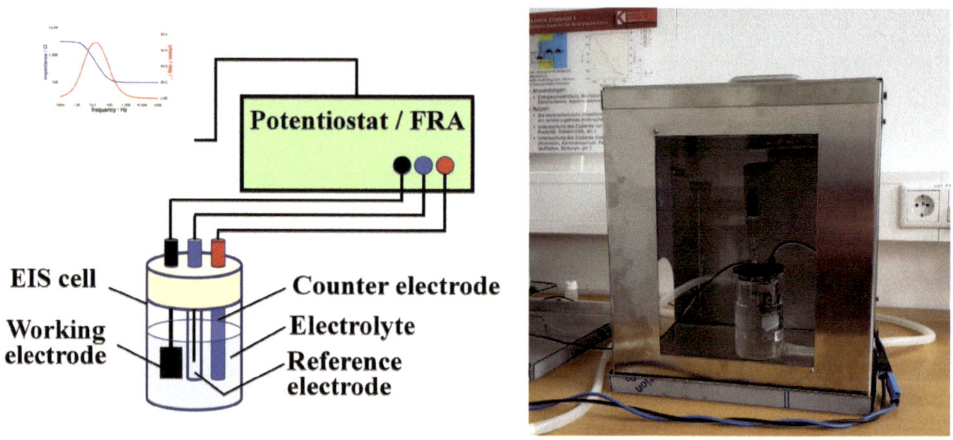

Figure 4.74 EIS measurement setup with three-electrode arrangement (left), and a measuring cell in a Faraday cage (right)

All electrodes must be dipped in the liquid electrolyte to ensure a reliable current flow between all electrodes. The type and concentration of the electrolyte solution have a significant influence on electrochemical experiments, as they usually determine the conductivity in the system. Typical electrolyte solutions consist of the salts Na_2SO_4, Na_2CO_3, NaCl, NH_4Cl, KCl, KNO_3, KNO_2, and $CaCl_2$, with a typical concentration of 10^{-6} mol/L to 5 mol/L [24].

The recording of an impedance spectrum at low frequencies (mHz range) can take several hours, as the measurement time for a measurement point is inversely proportional to the frequency. This must always be taken into account, especially when investigating an unstable system, as the system state can change significantly during the recording of the impedance spectrum [2, 18, 19].

The surface of the working electrode or the coated metal sample should be 3–30 cm^2. The reasons for this are:

- The larger surface area has higher currents, which ensures stable measurement
- A large coating area can be tested in a single measurement, which speeds up the generation of statistics on the quality of the entire coating.

An important note when examining coated metal samples is that the measurements must always be carried out in a Faraday cage. A Faraday cage is an earthed, conductive housing (e.g. a steel box) that completely surrounds the measuring cell and all electrodes, as shown in Figure 4.74 (right). This reduces the current noise recorded by the working electrode and the voltage noise recorded by the reference electrode [18].

4.7.6 Conclusion

Electrochemical impedance spectroscopy is a powerful, non-destructive method to investigate the response of a system to the applied alternating current load. EIS is often used in corrosion technology and science to obtain information about the polarization resistance and time constants of corrosion processes using an equivalent circuit. This requires a deeper understanding of the corrosion and diffusion processes. The EIS measurement technique is relatively simple and easily reproducible. Nevertheless, the measurement conditions (the measurement temperature, the sample geometry, etc.), must be kept strictly the same, as slight deviations lead to a recognizable deviation in the system response.

In order to achieve a fully-fledged characterization of the corrosion protection system, it makes sense to combine the EIS with other examination methods, such as cyclic voltammetry, a polarization method, examination using SEM (scanning electron microscopy) and EDX (energy-dispersive X-ray spectroscopy), and analysis using XRD (X-ray diffraction) or LICT (local ion concentration technique) [10, 11, 24, 25].

References for Section 4.7

[1] Ferse, A.; Paul, M.: Capacitance measurements of the system metal-polymer film-electrolyte solution, *Journal of Physical Chemistry (Leipzig)*, 1973, 252: 198–208

[2] Barsoukov, E.; Macdonald, J.; R.: *Impedance Spectroscopy: Theory, Experiment, and Applications*, Wiley, 2005

[3] Kohlrausch, F.: Über die elektromotorische Kraft sehr dünner Gasschichten auf Metallplatten, *Annalen der Physik*, 1873, 148: 143–154

[4] Wien, M.: Über die Polarisation bei Wechselstrom, *Annalen der Physik*, 1896, 58: 37

[5] Warburg, E.: Über das Verhalten sogenannter unpolarisierbarer Elektroden gegen Wechselstrom, *Annalen der Physik*, 1899, 67: 493–499

[6] Gabrielli, C.: Identification of electrochemical processes by frequency response analysis (Technical Report Number 004/83), Solartron, 1980

[7] McAdams, E.; T.; Jossinet, J.: Tissue impedance: A historical overview, *Physiological Measurement*, 1995, 16: A1–A13

[8] Sauer, D. U.: Impedanzspektroskopie – Eine Methode, viele Anwendungen – Definitionen, physikalische Bedeutungen, Darstellungen, *1st Symposium on Impedance Spectroscopy*, 2006, 1/3: 7–11

[9] Radtke, M.; Hess, C.: Easy-made setup for high-temperature (up to 1100 °C) electrochemical impedance spectroscopy, *Journal of Materials Engineering and Performance*, 2022, 31: 6980–6987

[10] Fontana, M. G.: *Corrosion Engineering* (3rd edn), McGraw-Hill, 1987

[11] Schmitt, G.: Sophisticated electrochemical methods for MIC investigation and monitoring, *Materials and Corrosion*, 1997, 48: 586–601

[12] Jones, D. A.: *Principles and Prevention of Corrosion* (2nd edn), Prentice Hall Publishing, 1996

[13] Montemor, M. F.: Functional and smart coatings for corrosion protection: A review of recent advances, *Surface and Coatings Technology*, 2014, 258: 17–37

[14] DIN EN ISO 9227:2017: Corrosion tests in artificial atmospheres – Salt spray tests, Beuth-Verlag, 2017

[15] DIN EN ISO 6270-2:2017: Paints and varnishes – Determination of resistance to humidity – Part 2: Condensation (in-cabinet exposure with heated water reservoir), Beuth-Verlag, 2017

[16] DIN CEN ISO/TR 19402-02:2022: Paints and varnishes – Adhesion of coatings, Beuth-Verlag, 2018

[17] Ende, D.; Mangold, K.-M.: Impedanzspektroskopie, *Chemie in Unserer Zeit*, 1993, 27: 134–140

[18] Zahner analysis [Manual], Zahner-Elektrik GmbH, *https://doc.zahner.de/manuals/zahner_analysis.pdf*, accessed on 20 May 2023

[19] Knoblauch, C.: Impedanzspektrokopie – Ein Überblick von der Theorie bis zur Anwendung, *WOMag*, 2015, issue 09

[20] Loveday, D.; Peterson, P.; Rodgers, B.: Evaluation of organic coatings with electrochemical impedance spectroscopy: Part 2: Application of EIS to coatings, *JCT CoatingsTech*, 2004, 6: 88–93

[21] Batmanov, A.; Frettloeh, V.; Schlutter, R.; et al.: Zirconium dioxide as a thermo-insulating coating for molds in plastic injection molding, *Proceedings of ANTEC 2022*, 2022

[22] Ciaston, M.: Neue MOCVD-Schichtsysteme zur elektrischen Isolierung von Werkzeugoberflächen, 5th web-treff EFDS, 2021

[23] Lazar, A. M.; Yespica, W. P.; Marcelin, S.; et al.: Corrosion protection of 304L stainless steel by chemical vapor deposited alumina coatings, *Corrosion Science*, 2014, 81: 125–131

[24] Ochs, H.: Physikalisch-chemische Untersuchungen von Grenzschichten auf metallischem Untergrund [Dissertation], Institute for Physical Chemistry, University of Stuttgart, 2003

[25] Elektrochemische Untersuchungsmethoden für Beschichtungen, Fraunhofer Institute for Manufacturing Engineering and Automation, *https://www.ipa.fraunhofer.de/en/competences/coating--systems--and--painting--technology.html*, accessed on 20 May 2023

4.8 Nanoindentation

Dr. Ruben Schlutter

4.8.1 Introduction

The hardness of a material can be defined as its resistance to permanent penetration by another harder material. Due to this comparatively vague definition, various test methods have been developed in which the hardness is measured with a defined indenter [1].

The most important hardness test methods are the Rockwell, Vickers, and Brinell methods. A defined test force is applied perpendicular to the surface to be tested. The hardness values are then calculated after the test force has been removed by evaluating the projected surface (Brinell and Vickers) or measuring the penetration depth (Rockwell). The elastic deformation below the indenter cannot therefore be taken into account [2]. Table 4.9 summarizes the test specifications for the most important classical hardness test methods.

Table 4.9 Summary of the Most Important Hardness Test Methods [3–5]

	Brinell	Vickers	Rockwell
Standard	DIN EN ISO 6506	DIN EN ISO 6507	DIN EN ISO 6508
Test specimen	Polished hard metal ball	Equilateral diamond pyramid with an opening angle of 136°	Diamond cone with an opening angle of 120° and a radius of curvature at the tip of 0.2 mm, or a hard metal ball
Evaluation form	Measuring of the indentation calotte diameter	Calculation of the surface of the indented pyramid based on the two pyramid diameters	Final penetration depth under the effect of a pre-load after removal of the test load
Limitations	Thickness of the specimen must be at least eight times the indentation depth Minimum thickness of the specimen of 0.12 mm	No elastic recovery of the substrate Minimum thickness of the specimen or layer >1.5 times the diagonal length of the indentation Test load to be selected depending on the expected hardness value	Minimum thickness of specimen or layer to be tested >0.1 mm (HRC scale) Thickness must be at least 10 times the penetration depth High demands on the clamping of the specimen

Table 4.9 Summary of the Most Important Hardness Test Methods [3–5] *(continued)*

	Brinell	Vickers	Rockwell
	During the measurement, the specimen must not visibly deform	Polishing necessary for microhardness measurements	No elastic deformation of the specimen during measurement
Permissible penetration depths	Diameter of the indentation 24–60% of the ball diameter	Indentation diagonals 0.02–1.40 mm Indentation depths 4–280 µm	Not defined

The problem with the classic methods of hardness testing is the fact that it is not possible to analyze thin coatings, since the penetration depths are so great that the coating is either penetrated or the substrate is affected by the measurement. Both invalidate the results of the hardness test. In addition, surfaces are measured optically in Brinell and Vickers hardness testing. If the penetration depth is reduced in order to measure only the hardness of the coating, the impressions made by the test specimens may be beyond the resolution limit of the optical measuring device. In addition, higher demands must be placed on the tips of the test specimens in order to obtain valid test results. The detection of elastic processes, such as springback, plastic phase transitions, or crack formation, can also not be detected with the classic methods of hardness testing. It must therefore be possible to achieve penetration depths of approx. 100 nm when testing thin layers [6].

Therefore, hardness testing using nanoindentation has been developed. Nanoindentation can be used to determine local mechanical properties, such as hardness and elasticity [7]. The measurement of the penetration depth and the test force are carried out throughout the entire duration of the test, so that the plastic and elastic deformation can be detected. By evaluating the complete test cycle, hardness values can also be determined that are comparable to the hardness values of the classical test methods. In addition, various other properties of the material, such as its indentation modulus and elastic-plastic hardness, can be obtained without having to measure the indentation optically [2].

The elastic and plastic properties of a coating are crucial for its effect and functionality. Many coatings are specially developed to minimize wear and therefore have high hardness, with hardness often being used as a quality characteristic. The mechanical properties must also be known if further calculations are to be carried out taking into account an applied coating.

While the determination of the hardness or mechanical properties of solid materials is comparatively simple, in the case of coatings these values are often influenced by the properties of the substrate, the thickness of the coating, and the applied force [8].

4.8.2 Experimental Setup for Measurement using a Nanoindenter

The setup and performance of nanoindentation tests are standardized in ISO 14577-1. A wide variety of tests can be carried out, from simple loading and unloading to complex experiments such as the investigation of strain rate dependence or creep properties [2].

Figure 4.75 shows various basic load types. The combination of different loading cycles is also possible. Segments with a loading rate can be used to analyze the hardness and elasticity of the sample. By varying the loading rate and the unloading rate, the strain rate-dependent deformation of the specimen can be investigated. In addition, creep properties can be obtained by the holding segments. Several loading and unloading cycles can be used to draw conclusions about the fatigue of the specimen. By varying the load and the indenter, over 60 different characteristic values can be obtained [9].

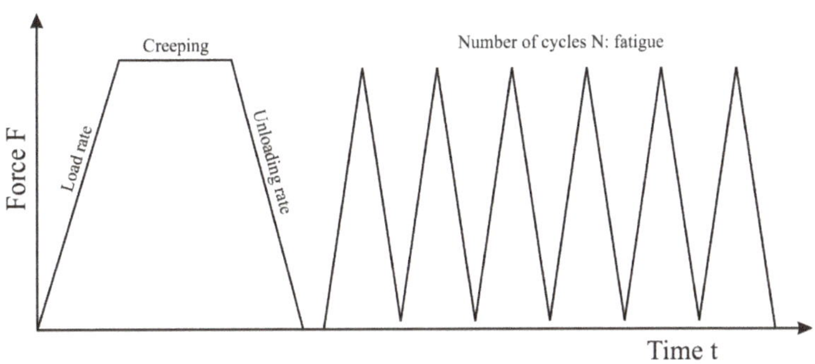

Figure 4.75 Schematic representation of some possible load types (Image based on [7])

The indenter used is a standardized test specimen that is harder than the surface to be tested. Depending on the type of test, different indenters can be used. The most important ones, listed in ISO 14577-1 [2], are:

- Diamond indenter in the form of a pyramid with a square base, similar to the specimen used in Vickers hardness testing
- Diamond pyramid with triangular base (modified Berkovich pyramid)
- Carbide ball (determination of the elastic behavior of materials)
- Diamond cone with a spherical tip.

The investigation of various material properties is possible by the appropriate choice of test specimens. The three-sided *Berkovich pyramid* is most frequently used for nanoindentation. It is similar to the Vickers pyramid, but does not have a "roof edge" at the tip. This eliminates tip defects at small penetration depths, maintaining the self-similarity of the pyramid. If higher strains are to be generated in the sample, the *cube-corner pyramid* is used. This has the geometry of the tip of a cube and therefore a much smaller opening angle than the Berkovich pyramid, allowing the testing of individual small phase components in the sample. The *hard metal sphere* is another important test specimen geometry. It is not self-similar, which means that it causes a continuous transition from purely elastic to fully plastic deformation in the specimen material, allowing determination of the yield stress. *Conical indenters* are mainly used to analyze the effects of crystal symmetry, but they are complicated to manufacture [7].

Figure 4.76 is a schematic showing indentation during the test. Either the test force *F* or the corresponding indentation depth *h* can be used as the control variable. Both variables are recorded time-dependently during the entire test cycle so that the resulting data set can be evaluated. Figure 4.77 shows an SEM image of a indentation made using a nanoindenter, to determine the hardness of a coating.

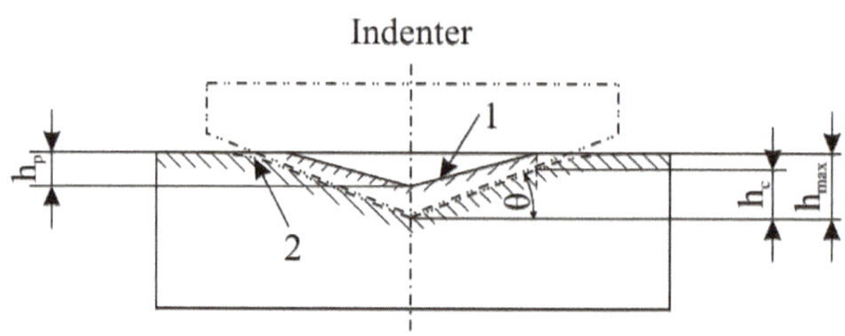

1) Surface of the remaining plastic indentation in a specimen with a "perfectly plastic behavior".
2) Surface area of the indentation at maximum indentation depth and test load

θ largest angle between the specimen surface and the indenter

Figure 4.76 Schematic representation of an indentation test (Image based on [3])

In order to guarantee the reproducibility of the individual measurements, the zero point for the force-penetration depth measurement must be determined individually for each test. During the test, the force and the penetration depth of the test specimen are recorded continuously. The corresponding diagram is shown in Figure 4.78.

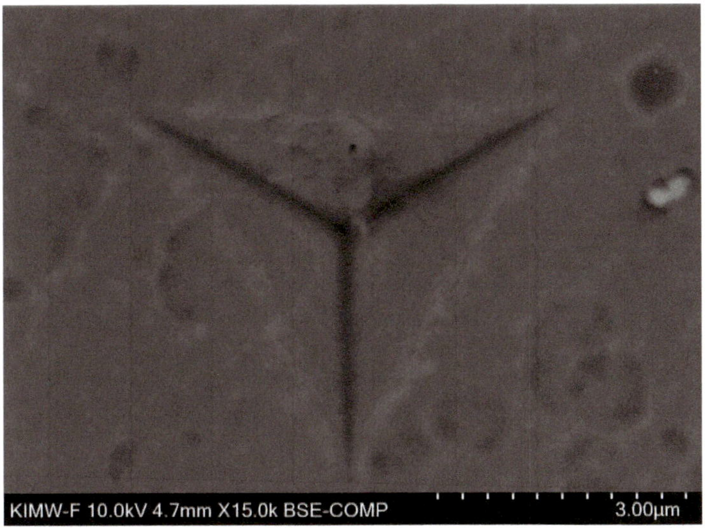

Figure 4.77 SEM image of an indentation made using a modified Berkovich pyramid nanoindenter (Image source: Non-profit KIMW Forschungs-GmbH)

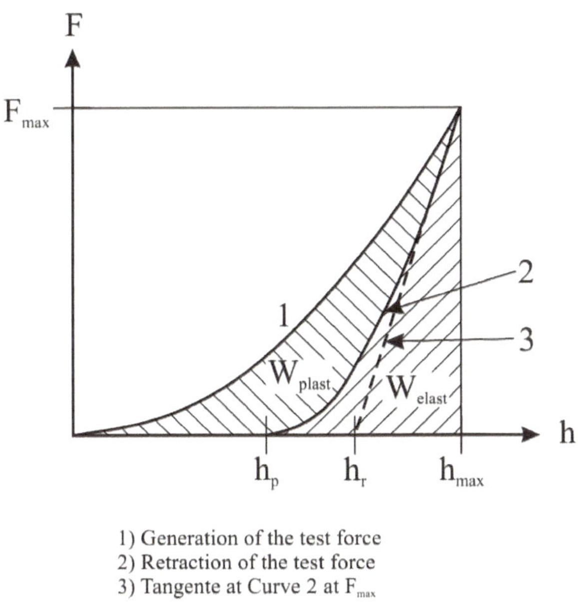

Figure 4.78 Schematic representation of the test procedure (Image based on [3])

Depending on the amount of force to be applied and the penetration depth, the three ranges in Table 4.10 are defined [2].

Table 4.10 Test Ranges for the Instrumented Indentation Test According to ISO 14577-1 [3]

Test range	Test force (kN)	Penetration depth (μm)
Macro-scale	2–30	—
Micro-scale	<2	>2
Nanoscale	—	<2

Especially at the nanoscale, the mechanical deformation of the specimen and the surface to be tested is strongly dependent on the real geometry of the tip of the indenter. The calculated material parameters are therefore significantly influenced by the indenter used. This influence means that it is necessary to carefully calibrate both the testing machine and the indenter geometry to obtain acceptable comparability between different testing machines [2].

At smaller penetration depths, microphysical effects can be observed. These include the indentation size effect and so-called "pop-in" behavior. Both are due to the micromechanical behavior of crystalline materials. In practice, a correlation between the local mechanical properties, the chemical composition, and the microstructure of crystalline layers is possible. It is also possible to analyze the mechanical properties and chemical composition of thin films and microparticles [7].

In addition, the surface roughness has greater influence at small penetration depths. In principle, the interaction cross-section in Equation 4.18 should be adhered to in order to limit the influence of surface roughness. The uncertainty of the penetration depth is then less than 5% [2].

$$h \geq 20 \cdot R_\mathrm{a} \tag{4.34}$$

h = Penetration depth (μm)

R_a = Arithmetic center roughness value (μm)

If the requirement from Equation 4.18 cannot be met, the number of tests should be increased. The surface roughness should then be determined and the surface examined microscopically [3].

If several tests are carried out on one surface, the surface to be tested must not be influenced by interfaces or indentations from previous tests. Therefore, the minimum distance between two indentations must be five times the largest indentation diameter. If cracks occur on an indentation, these are added to the indentation diameter in order to determine the distance to the next measuring point [2].

4.8.3 Common Test Methods

4.8.3.1 Determination of the Indentation Hardness

Indentation hardness is defined as the resistance of a surface to damage or permanent deformation. The first approximation of the indentation hardness is calculated according to Equation 4.35 [2].

$$H_{IT,0} = \frac{F_{max}}{A_p\,(h_c)} \tag{4.35}$$

H_{IT} = Penetration hardness (MPa)

F_{max} = Maximum effective test force (N)

$A_p(h_c)$ = Projected contact area between the indenter and the sample (mm^2)

The contact depth h_c describes the depth of the indentation under an applied test force (see Figure 4.76), and is calculated from the force recovery curve (see Figure 4.78). Accordingly, it depends on the maximum penetration depth and the tangent depth, and is calculated according to Equation 4.36. There are various standardized approaches for determining the tangent depth based on straight-line approximations or the approximation of power curves. Which method is used in each case depends on the type of surface tested and the quality of the data. A straight-line approximation is often used for highly plastic materials with low elastic recovery. If the load-deflection curve is not linear or the results are time-dependent, approximations of power curves are used [2].

$$h_c = h_{max} - \varepsilon\,(m) \cdot (h_{max} - h_r) \tag{4.36}$$

h_c = Contact depth of the indenter with the sample at F_{max} (µm)

h_{max} = Maximum penetration depth at F_{max} (µm)

ε = Correction value ($0.6 < \varepsilon < 0.8$) (—)

h_r = Intersection of the tangent to the force reduction curve at F_{max} with the path axis (see Figure 4.78) (µm)

The maximum indentation depth is the indentation depth at maximum test force, and is determined at the beginning of the force reduction. Therefore, all additional displacements are included in the maximum indentation depth, which occurred, for example, due to creep during the holding periods.

The correction factor ε depends on the geometry of the indenter and the extent of plastic deformation. When using a linear fit, the correction factor can be estimated according to Table 4.11. If a power approach is used, the correction factor can be calculated [2].

Table 4.11 Correction Factor ε for Different Indenter Geometries When Using a Linear Fit According to ISO 14577-1 [2]

Indenter geometry	Correction factor
Flat stamp	1
Conical	$\frac{2(\pi-2)}{\pi} = 0.73$
Rotationally symmetrical paraboloid (including spherical)	¾
Berkovich, Vickers	¾

Depending on the type of indenter, the indentation hardness when using a Vickers pyramid according to Equation 4.37 or when using a Berkovich indenter according to Equation 4.38 can be calculated. From an indentation depth of 6 μm, the projected area is given by the theoretical shape of the indenter.

$$H_{IT,0} = \frac{F_{max}}{A_p\,(h_c)} = 4 \cdot h_c^2 \cdot (\tan\alpha)^2 \tag{4.37}$$

H_{IT} = Indentation hardness (MPa)

$A_p(h_c) = 24.5 \cdot h^2$ (mm^2)

α = Half the opening angle of the Vickers pyramid (68) (°)

$$H_{IT,0} = \frac{F_{max}}{A_p\,(h_c)} = 3 \cdot \sqrt{3} \cdot h_c^2 \cdot (\tan\alpha)^2 \tag{4.38}$$

H_{IT} = Indentation hardness (MPa)

α = Opening angle of the Berkovich pyramid (68°) (°)

Depending on the shape of the Berkovich indenter, the projected surface area and the opening angle can be determined from Table 4.12. The modified form is used more frequently [2].

Table 4.12 Standard Values for the Projected Contact Area and the Opening Angle of the Berkovich Indenter ISO 14577-1 [2]

	Projected contact area between the indenter and the sample, $A_p(h_c)$ (mm^2)	Opening angle of the Berkovich indenter, α (°)
Original Berkovich indenter	$23.97 \cdot h^2$	65.03
Modified Berkovich indenter	$24.49 \cdot h^2$	65.27

Since all indenters do not have a perfect tip, but are rounded at the tip, the area function of the indenter cannot be assumed to have the perfect theoretical shape at small contact depths. A contact indentation depth <6 µm serves as a limiting value. As a rule, the rounding at the tip is also not uniform, so that the exact surface function must be determined according to ISO 14577-2 [2, 10].

Finally, the first approximation of the indentation hardness must be corrected with respect to the radial displacement of the surface. The correction is iterative. The final indentation hardness contains a value n for the number of iterations performed ($H_{IT,n}$). For metallic materials, the radial correction is usually small (<0.5%). For highly elastic materials, it can assume values up to 5% [2].

4.8.3.2 Determination of the Indentation Modulus

The contact depth and the projected indentation area are also used to determine the indentation modulus. The first approximation of the reduced indentation modulus is calculated according to Equation 4.39. This is then also corrected by the radial displacement [2].

$$E_{r,0} = \frac{\sqrt{\pi}}{2 \cdot C_S \cdot \sqrt{A_p\,(h_c)}} \tag{4.39}$$

$E_{r,0}$ = Reduced modulus for penetrating contact (MPa)

C_S = Contact compliance, i. e. dh/dF of the force reduction curve (mm/mN)

$A_p(h_c)$ = Projected contact area (mm^2)

The projected contact area depends on the geometry of the indenter used:

- Vickers and modified Berkovich indenter: $\sqrt{A_p} = 4.950 \cdot h_c$
- Berkovich penetrator: $\sqrt{A_p} = 4.895 \cdot h_c$

From the reduced indentation modulus, the biaxial strain modulus according to Equation 4.40 can be calculated. This is then used to determine the indentation modulus according to Equation 4.41. To do this, the Poisson ratios of the indenter and the sample must be known. These can either be estimated or must be known.

$$E^* = \frac{1}{\frac{1}{E_{r,n}} - \frac{1-\nu_i^2}{E_i}} = \frac{E_{IT}}{1-\nu_s^2} \tag{4.40}$$

E^* = Biaxial strain module (MPa)

$E_{r,n}$ = Reduced modulus at indentation contact after correction of the radial displacement (MPa)

E_i = E-modulus of the indenter (1140 GPa for diamond) (MPa)

ν_i = Poisson ratio of the indenter (0.07 for diamond) (—)

E_{IT} = Penetration modulus (MPa)

ν_s = Poisson ratio of the sample (—)

The indentation modulus Equation 4.41 is comparable to the modulus of elasticity of the sample material, so long as no bulging or sinking occurs during the measurement, when significant differences can occur [2].

$$E_{IT} = \frac{1 - \nu_s^2}{\frac{1}{E_{r,n}} - \frac{1-\nu_i^2}{E_i}} \qquad (4.41)$$

4.8.3.3 Determination of the Penetration Creep

Indentation creep is the change in indentation depth over time at a constant test load. The creep process is shown in Figure 4.79.

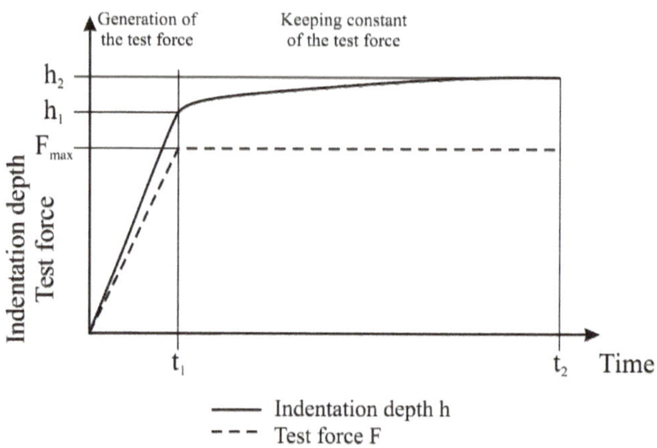

Figure 4.79 Schematic representation of penetration creep (Image based on [3])

The penetration creep is calculated according to Equation 4.42. It can be influenced by thermal drift [2].

$$C_{IT} = \frac{h_2 - h_1}{h_1} \cdot 100 \qquad (4.42)$$

C_{IT} = Penetration creep (%)

h_2 = Penetration depth at time t_1 of reaching the test force, which is kept constant (mm)

h_1 = Penetration depth at time t_2 while holding the test force constant (mm)

4.8.3.4 Determination of the Penetration Relaxation

If the test is performed under control of the penetration depth, the indentation relaxation can be determined. To do this, the indentation depth is kept constant over time and the change in test force over time is measured. The relaxation process is shown in Figure 4.80 [2].

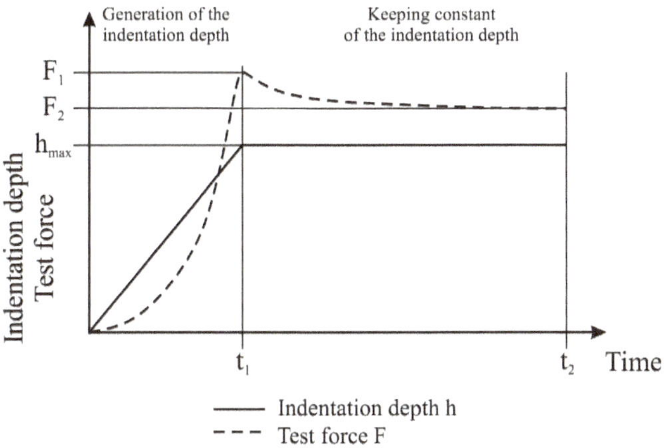

Figure 4.80 Schematic representation of penetration relaxation (Image based on [3])

The penetration relaxation can then be calculated according to Equation 4.43.

$$R_{IT} = \frac{F_1 - F_2}{F_1} \cdot 100 \tag{4.43}$$

R_{IT} = Penetration creep (µm)

F_1 = Force on reaching the penetration depth, which was kept constant (N)

F_2 = Force after the time during which the penetration depth was kept constant (N)

4.8.3.5 Determination of the Plastic and Elastic Part of the Indentation Work

Mechanical work is performed during the application of force and the penetration process. This is partly converted into plastic deformation work and partly stored in elastic work which is released again during re-deformation after the test force is removed. The proportion of elastic and plastic work is shown in Figure 4.78 as an area in the force–displacement diagram [2].

The total indentation energy can be determined by integrating the curve during the test load application. It is the sum of elastic and plastic indentation work (Equation 4.44).

$$W_{total} = W_{elast} + W_{plast} \tag{4.44}$$

W_{total} = Total mechanical work during penetration (N·m)

W_{elast} = Elastic recovery work during penetration (N·m)

W_{plast} = Plastic part of the penetration work (N·m)

The elastic indentation work can be determined directly from the diagram by integrating the curve of the test force reduction, and expressing it as a proportion of the indentation work according to Equation 4.45.

$$\eta_{IT} = \frac{W_{elast}}{W_{total}} \cdot 100 \quad (4.45)$$

η_{IT} = Elastic part of the penetration work (%)

The plastic part of the indentation work can then be calculated according to Equation 4.46 [2].

$$\frac{W_{plast}}{W_{total}} = 100\% - \eta_{IT} \quad (4.46)$$

4.8.4 Test Methods for Coatings

Usually, coatings are not thick enough for the substrate to remain unaffected during the test. If the coating remains unaffected during the test, it can be considered a solid material. As a limit, it is assumed that the penetration depth does not exceed one tenth of the coating thickness [11].

In addition, only individual coatings can be examined. In the case of multi-layer coatings or gradient coatings, a cross-sectional examination is recommended, whereby the individual coatings must be thick enough for the indentation process to be limited to a single coating. The examination of metallic and non-metallic coatings is possible. The coatings or individual coatings must be homogeneous along the indentation depth, since the properties of the coating are assumed to be homogeneously distributed [8].

During the test, plastic deformation must occur within the coating without plastic deformation occurring in the substrate. According to Figure 4.81, the maximum of the main shear stress is then in the coating. Ideally, the yield point of the substrate is not reached. The tip of the indenter must not be too sharp, in order to prevent the coating from breaking.

The plastic deformation of the substrate can be problematic, especially with very hard coatings on a soft substrate. In this case, the best compromise must be found between the level of the test force and the penetration depth. The test force must then be high enough to prevent the substrate from flowing. The penetration depth should be small enough so that the substrate is not affected by the penetration, and the influence on the measurement is minimized.

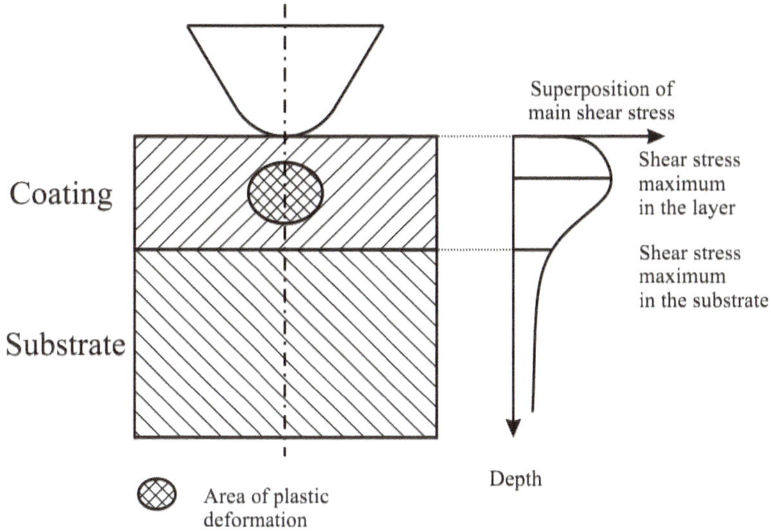

Figure 4.81 Main shear stress as a function of the depth under the indenter (Image based on [10])

Bulging or sinking during the penetration test cannot be taken into account in the evaluation. If bulging or sinking occurs or is suspected, the indentation must be examined microscopically to confirm the situation. In the case of bulging, the surface effects will be underestimated, whereas in the case of sinking, the surface effects will be overestimated [8].

The indenter used is a pyramid or a cone whose tip rounding is small enough to achieve plastic deformation within the coating. Berkovich pyramids or Vickers pyramids are often used, and spherical indenters or cube corners are also possible. If plastic properties of the coating are to be determined, pointed indenters are used. The thinner the coating, the larger the tip of the indenter should be. If only the elastic properties of the coating are to be investigated, elastic indenters are sufficient, since the problems described with regard to bulging or fracture of the coating can be avoided here. In this case, an indenter with a large radius or a ball should be selected. These allow elastic indentations over a wide range of forces.

Preparation of the sample should be kept to a minimum. The surface roughness R_a should be less than 5% of the maximum penetration depth, as otherwise, the variation of the results increases too much. Measurements must be made away from surface disturbances such as lumps or scratches. If possible, polishing should be avoided, especially with metallic surfaces, as this can lead to a hardening of the surface. With ceramic surfaces, the hardening is less significant. It may be necessary to determine the coating thickness again after polishing, since polishing reduces the coating thickness, so that the influences of the substrate can be increased. Cleaning of the surface is usually not necessary. If contamination is present on the surface, it is removed with an oil-free gas stream, sublimed CO_2, or a solvent that is inert to the coating. The use of ultrasonication should be avoided [8].

The indentation tests are evaluated on the basis of the measured force and displacement data, whereby the displacement data must be corrected for thermal drift (C_S) and the compliance of the testing machine frame (C_F). During the analysis, a distinction must be made between brittle and ductile layers [10].

It is not possible to determine the coating properties directly, as the substrate always deforms at least elastically. All measured and evaluated characteristic values must therefore be regarded as composite values. In parallel, it is advisable to determine the characteristic values of the substrate in order to be able to carry out a suitable evaluation.

The coating properties can only be estimated by extrapolation. For this purpose, the characteristic values are plotted as a function of a dimensionless ratio. This is calculated either from the quotient of the contact radius a (Equation 4.47) or the penetration depth h_c and the coating thickness t_c. It is only necessary to measure the coating thickness if different coating thicknesses are to be compared with each other, or if the coating thickness varies considerably. As the indenters have different geometries, the contact radius a is approximated by the radius of a circle that has the same area as the projected contact surface, according to Equation 4.47 [8].

$$a = \sqrt{\frac{A_p}{\pi}} \tag{4.47}$$

a = Contact radius (mm)

A_p = Projected contact surface with the indenter (mm^2)

4.8.4.1 Penetration Modulus of the Coating

If the response of the material composite to the penetration is unknown, preliminary tests must be carried out with different, widely separated test forces in order to be able to estimate the ideal test force. The aim of the actual tests is to produce indentations over a wide range of penetration depths without cracks or fractures occurring in the coating.

When softer coatings are analyzed on a harder substrate, a sufficient number of tests must be carried out in the range $a/t_c < 1.5$. The indentation modulus of the coating is then determined by linear extrapolation of the measurement points over a/t_c towards zero (Figure 4.82, left).

If harder coatings are examined on a soft substrate, the test range is at $a/t_c < 2$. Here too, the indentation modulus of the coating is determined by linear extrapolation of the indentation modulus over a/t_c towards zero (Figure 4.82, right).

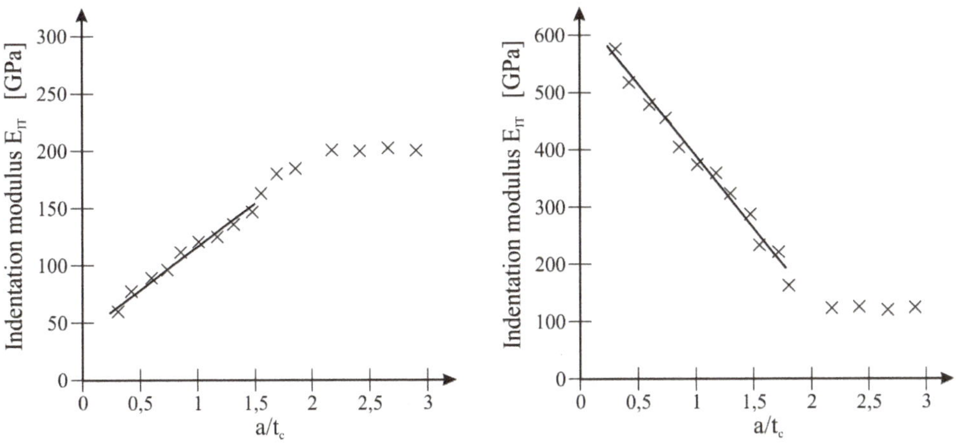

Figure 4.82 Indentation modulus over the standardized contact radius:
Left: Ductile coating on harder substrate; Right: Hard coating on softer substrate
(Image based on [8])

The linear extrapolation is only an approximation, since the occurring non-linear effects cannot be captured in a model. An application over the mentioned ranges is therefore not possible.

In principle, at least 15 individual measurements must be carried out, whereby at least five different values of a/t_c should be examined. To reduce the uncertainties in linear extrapolation, it is recommended that 50 individual measurements with 10 different values of a/t_c are carried out. The tendency is to analyze the impressions at different values of a/t_c rather than repeating fewer impressions with the same values of a/t_c [8].

4.8.4.2 Penetration Hardness of the Coating

The normalized parameter a/t_c can also be used to evaluate the results of the indentation hardness. However, the determination of the hardness requires indenters with a self-similar geometry, i.e. with a defined tip, so that the dimensionless parameter h_c/t_c, i.e. the ratio of the contact depth to the coating thickness, can also be used. The tip radius should be as small as possible to ensure plastic deformation in the layer.

If the indentation hardness of soft, ductile layers on a harder substrate is to be investigated, the indentation hardness is determined in the range $0 < h_c/t_c < 1$. However, the requirement is always that the substrate only deforms elastically.

When evaluating the indentation hardness, a constant minimum value can result, as shown in Figure 4.83 (left). This plateau describes the indentation hardness of the coating. It is also possible that no constant minimum value can be determined for the indentation depth. In this case, the indentation hardness of the coating is extrapolated linearly towards $h_c/t_c = 0$, as shown in Figure 4.83 (right). If possible, the determination of the indentation hardness should be verified on several layers of different thicknesses. To identify the plateau, it is recommended to carry out at least 50 measurements with 10 values of h_c/t_c [8].

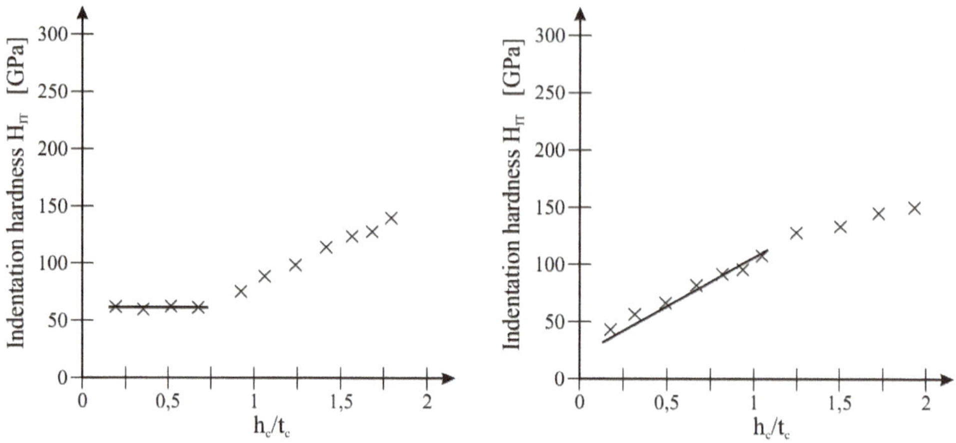

Figure 4.83 Indentation hardness above the normalized contact depth of a ductile layer on a harder substrate: Left: Constant plateau at minimum indentation hardness; Right: Without constant plateau (Image based on [8])

When testing hard coatings on a soft substrate, only sharp indenters with a small tip may be used to determine the indentation hardness in order to force plastic deformation of the coating without plastic deformation of the substrate. If the behavior of the coating–substrate bond is unknown, a preliminary test with a spherical indenter is useful to determine whether the coating first or the substrate first will be plastically deformed. If the substrate is plastically deformed first here, it is not possible to determine the hardness of the coating.

In parallel, the hardness of the substrate should also be determined. The hardness of the substrate can also be used to detect exfoliation or fracture of the coating, since the hardness values of the coating approach the substrate value for small values of h_c/t_c.

The parameters of the indentation test, consisting of the indentation force or depth and the geometry of the indenter, must be chosen so that the maximum of the indentation hardness can be detected. Usually, the maximum of the indentation hardness is in the range of $0 < h_c/t_c < 0.5$.

During evaluation, a constant highest possible value may result. This is then the indentation hardness of the coating (Figure 4.84, left). However, the maximum can also consist of only a single measured value (Figure 4.84, right). In this case, the indentation hardness must be verified on further test specimens with greater coating thicknesses.

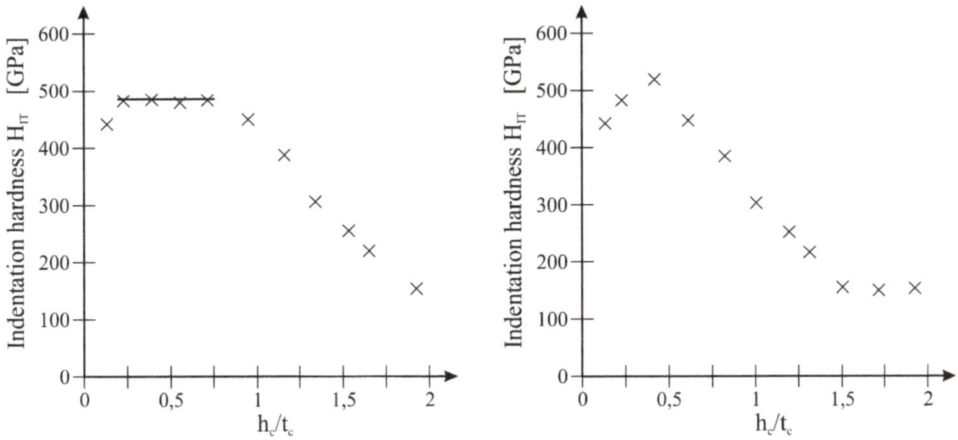

Figure 4.84 Indentation hardness above the standardized contact depth of a hard, brittle coating on a softer substrate: Left: Constant plateau at maximum indentation hardness; Right: Single maximum value of indentation hardness (Image based on [3])

In general, at least 10 individual measurements should be carried out to identify the plateau or the maximum individual value. Preference should be given to increasing the values of h_c/t_c rather than repeating individual measurements with the same values of h_c/t_c.

References for Section 4.8

[1] Domke, W.: *Werkstoffkunde und Werkstoffprüfung* (10th edn), Girardet, 2001

[2] ISO 14577-1:2015: Metallic materials – Instrumented indentation test for hardness and materials parameters – Part 1: Test method, 2015

[3] ISO 65061:2014: Metallic materials – Brinell hardness test – Part 1: Test method, 2015

[4] ISO 65071:2018: Metallic materials – Vickers hardness test – Part 1: Test method, 2018

[5] ISO 65081:2016: Metallic materials – Rockwell hardness test – Part 1: Test method, 2016

[6] Kurze Einführung: Nanoindentation, *https://qd-europe.com/de/de/produkt/kurze-einfuehrung-nano indentation/*, accessed on 01 November 2022

[7] Durst, K.; Göken, M.: Nanoindentierung – Eine Sonde für die lokalen mechanischen Eigenschaften, *Metallography Conference Special Volume*, 2004, 36 : 319–328

[8] ISO 14577-4:2017: Metallic materials – Instrumented indentation test for hardness and materials parameters – Part 4: Test method for metallic and non-metallic coatings, 2017

[9] ZwickRoell nanoindentation, *https://www.zwickroell.com/de/branchen/hochschulen/nanoindentation/ #c1180*, accessed on 01 November 2022

[10] ISO 14577-2:2015: Metallic materials – Instrumented indentation test for hardness and materials parameters – Part 2: Verification and calibration of testing machines, 2015

[11] ISO 4516:2002: Metallic and other inorganic coatings – Vickers and Knoop microhardness tests, 2002

4.9 Determination of the Thermal Diffusivity of Coatings

Patrick Engemann

4.9.1 Influence of the Mold Wall Temperature on the Injection Molding Process

The temperatures present during injection molding are important from both a qualitative and economic point of view. The quality of the molded parts can be decisively influenced by the temperature of the injection molding process. For example, the use of metals with high thermal conductivity, such as copper, can influence the injection molding process in favor of shorter cycle times or to counteract warpage of the molded part, as the heat can be dissipated more quickly (Figure 4.85) [1, 2].

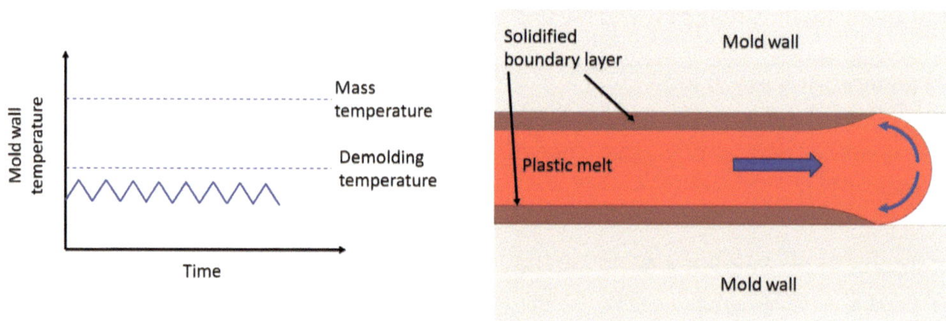

Figure 4.85 Formation of the surface layer during isothermal tempering (Image based on [3])

In contrast, thermally insulating materials can be used to conceal defects on the surface of plastic molded parts or to achieve longer flow paths of the plastic melt [4, 5]. Within the limits set by the plastic material, such effects can also be achieved by raising the inlet temperature of the tempering medium at the injection mold, although this always leads to an undesirable increase in the cycle time of the injection molding process. The heat carried along in the plastic can no longer be dissipated to the same extent at a higher cavity temperature, which means that the necessary demolding temperature is reached later. For this reason, the market offers various options for introducing the necessary heat into the mold just before the plastic is actually injected. As only the surface of the molded part cavity is heated, the cycle time remains largely unaffected. These technologies include infrared heating systems, which use radiation to heat the cavity in the injection molding tool when it is open, or variothermal temperature control systems, which can switch between temperature control media with different flow temperatures. Alternatively, heating elements are also inserted into the relevant areas of the injection mold under the cavity [5]. The consequence of the methods listed is that the heat transfer of the plastic melt to the injection mold at the time of contact can be reduced. As a result, the formation of a solidified surface layer between the mold wall and the molded part core is delayed, as shown in Figure 4.86. As a result, the molding of the mold cavity is improved. The temperature to be exceeded is usually the glass transition temperature of the polymer to be processed, above which the plastic is viscous [6].

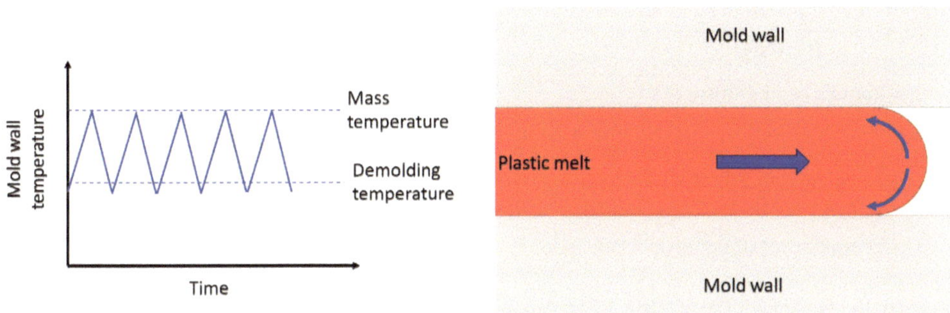

Figure 4.86 Formation of a surface layer with variothermal process control (Image based on [3])

Due to their thermally insulating effect and their time-limited influence, ceramic coatings on the surface of the cavity are also suitable for this purpose, in order to temporarily delay the release of heat from the plastic melt without having to resort to technical equipment.

4.9.2 Contact Temperature

The thermal effect of a ceramic coating on the injection molding process can be described by the contact temperature. The contact temperature is the temperature that is reached when two bodies come into contact until the temperature is equalized [7]. In the case of the injection molding process, the contacting partners are the plastic melt and the mold surface.

The contact temperature (T_C) is derived from the heat penetration coefficients (b) according to Equation 4.48 and the temperature (T_1, T_2) of the materials in contact. If the heat penetration coefficient is not known, it can be calculated using the thermal conductivity (λ), the density (ρ) and the specific heat capacity c [8].

$$b = \sqrt{\rho \cdot \lambda \cdot c} \tag{4.48}$$

b = Heat penetration coefficient ($W \cdot s^{1/2}/(m^2 \cdot K)$)

λ = Density (kg/m^3)

ρ = Thermal conductivity ($W/(m \cdot K)$)

c = Specific heat capacity ($J/(kg \cdot K)$)

If the heat penetration coefficients of the materials are known, a statement about the contact temperature can be made using Equation 4.49.

$$T_C = \frac{b_1 \cdot T_1 + b_2 \cdot T_2}{b_1 + b_2} \tag{4.49}$$

T_C = Contact temperature (K)

b_1 = Heat penetration coefficient of material 1 ($W \cdot s^{1/2}/(m^2 \cdot K)$)

T_1 = Temperature of material 1 (K)

b_2 = Heat penetration coefficient of material 2 ($W \cdot s^{1/2}/(m^2 \cdot K)$)

T_2 = Temperature of material 2 (K)

It can be inferred from this that the material on the surface is important for the period before the temperature equalization occurs. As the thermal characteristics of thin layers may differ from those of solid bodies, they must be determined by measurement if they are not available [9–11].

4.9.3 Time-Domain Thermoreflectance (TDTR)

Time-domain thermoreflectance (TDTR) is a purely optical method that can be used to determine the thermal conductivity of thin layers on substrates. In order to determine the thermal conductivity, a defined laser beam is directed at the sample. To carry out the measurement, this is split into a pump beam and a sample beam (see Figure 4.87), with the sample beam hitting the sample with a time delay. The pump beam that arrives first is used to heat the sample. As the reflectivity is dependent on the temperature of the sample, the sample beam is directed onto the sample in such a way that it can be reflected by the sample and falls onto a measuring system. By evaluating the beam, a statement can be made about the temperature of the sample [12]. Metallic surfaces in particular show a strong correlation between temperature and the reflectivity of the sample surface [13, 14]. To enable measurement over a longer period of time, a continuous-wave (CW) laser can be used as the sample beam, which is continuously reflected by the surface. This modified form of TDTR measurement is referred to as transient thermoreflectivity (TTR) [15].

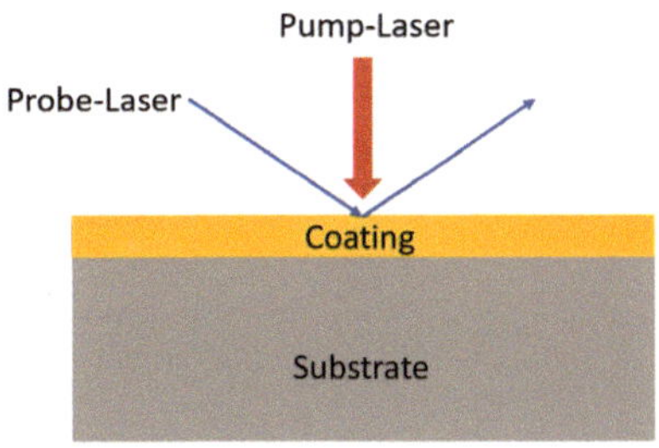

Figure 4.87
TDTR measurement setup (Image based on [16])

4.9.4 3-Omega

The 3-omega measurement for determining the thermal conductivity of materials was developed by Cahill [17]. The basis of the measurement is the dependence of the electrical resistance on the temperature of an electrical conductor. However, the method developed for solid materials is now also used to determine the thermal conductivity properties of thin coatings. In order to determine the thermal conductivity of the coating, the thermal conductivity of the substrate material must be determined by means of an additional measurement. Fourier's law, which describes the heat diffusion in a solid body, can be used to draw conclusions about the thermal properties of the coating, as it acts as an additional thermal insulator. A thin metallic conductor track is applied to the surface of the material to be measured, and serves as both a measuring device and heating element (Figure 4.88) [18, 19].

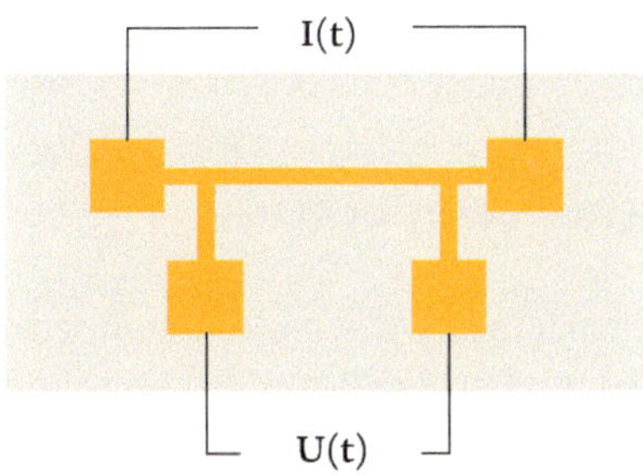

Figure 4.88
3-Omega measurement (Image based on [20])

To measure the thermal conductivity, an alternating current $I(t)$ with the frequency ω flows through the metallic conductor (Equation 4.50). As a result of the electrical resistance of the conductor, a temperature $T(t)$ is periodically impressed on the sample.

$$I(t) = I_0 \cdot \cos(\omega \cdot t) \tag{4.50}$$

I = Alternating current (A)

I_0 = Maximum of the sinusoidal curve (A)

ω = Frequency (1/s)

t = Time (s)

The temperature of the sample and the electrical conductor changes with a frequency of $2 \cdot \omega$. The amplitude of the temperature change around the zero position $T_0(t)$ depends on the thermal properties of the sample and the conductor, as does the shift φ of the temperature amplitude to the amplitude of the introduced current (Equation 4.51)

$$T(t) = T_0(t) + \Delta T \cdot \cos(2 \cdot \omega \cdot t + \varphi) \quad (4.51)$$

T = Temperature (K)

T_0 = Zero position of the temperature (K)

φ = Phase shift (—)

The power introduced for the measurement via the heat-sealing band can be calculated using the current introduced and the electrical resistance of the band. The calculation is based on Equation 4.52.

$$P(t) = I_0^2 \cdot R \cdot \cos^2(\omega \cdot t) \quad (4.52)$$

P = Contributed power (W)

R = Resistance of the sample (Ω)

As the electrical resistance is dependent on the temperature of the electrical conductor, it follows the change in temperature in an oscillating manner. If the voltage is measured, it can be seen that the frequency of the current and the change in resistance over time result in a superimposed frequency in the voltage. This has a frequency of $3 \cdot \omega$, which gives the method its name. This superimposed oscillation can be detected via a lock-in amplifier [20].

4.9.5 Test Setup for Measuring the Contact Temperature

Previous technologies have taken the approach of calculating the contact temperature via heat conduction. In order to compare the thermal properties of coatings and solids, a test setup can be used in which a constantly heated temperature sensor is pressed onto the surface of the sample to be measured. If the temperature sensor comes into contact with the colder sample surface, the temperature at the sensor drops. Depending on the heat penetration coefficient, the measured temperature drops to varying degrees. Materials with a lower heat penetration coefficient lead to a smaller drop in the measured temperature curve of the sensor.

Figure 4.89 shows a setup described in this way, in which the sensor is fixed to a copper rod, which can be heated to the defined temperature. In the graph shown, the temperature on the rod and on sensor 1 ($T_{Sensor1}$) is 80 °C at the start of the measurement (t = 0). Sensor 2 ($T_{Sensor2}$), which is located on the back of the coated substrate, records the sample temperature, which corresponds to the ambient temperature at t = 0. If sensor 1 comes into contact with the sample, a change in temperature ($T_{Sensor1}$ to $T_{Sensor2}$) can be measured over time ($t > 0$), as the thermal energy from the copper rod is transferred to the sample [21].

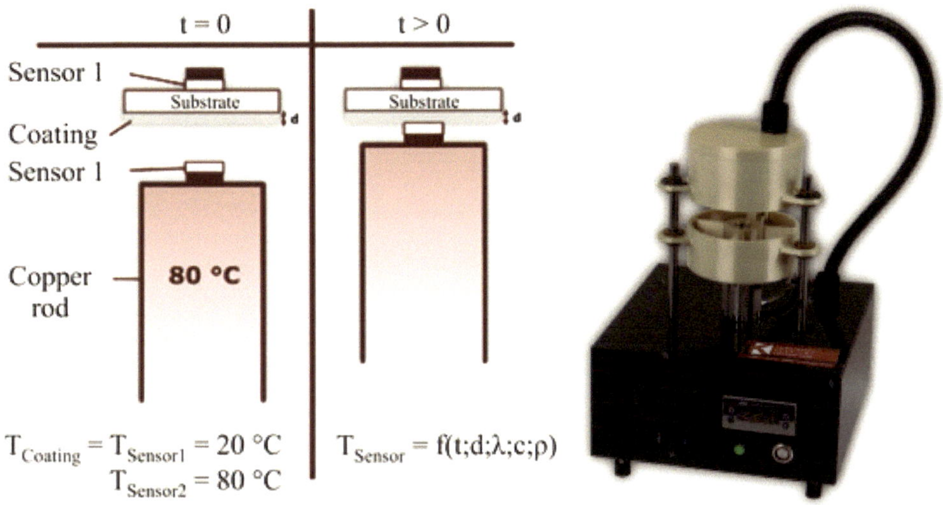

Figure 4.89 Left: Structure of the device for measuring the thermal diffusivity (Image based on [21]). Right: Measuring device for determining the thermal diffusivity at non-profit KIMW Forschungs-GmbH

Sensor 1 is brought into contact with the sample or with the coating on the sample, the measurement is started, and the time-dependent temperature change on the copper rod is detected. The slower the temperature drop at the sensor 1, the higher the thermal insulation effect of the coating. The low thermal conductivity of the applied thermal insulation coating reduces the heat flow from the copper block into the metal substrate of the sample. On the back of the sample, another thermocouple is in contact with the sample (sensor 2), which records the temperature increase caused by the temperature conduction through the sample.

4.9.6 Test Procedure for Measuring the Contact Temperature

Figure 4.90 shows an example of the temperature–time curves of undoped and lanthanum-doped thin coatings of zirconium oxide on 1.4301 steel, and an untreated sample of steel, obtained using this method. The initial temperature at the sensor at the time of measurement is 80 °C, that of the sample 20 °C [4]. After the measured temperature at the sensor drops to the contact temperature, the measured temperature increases again. This is due to the temperature equalization which occurs.

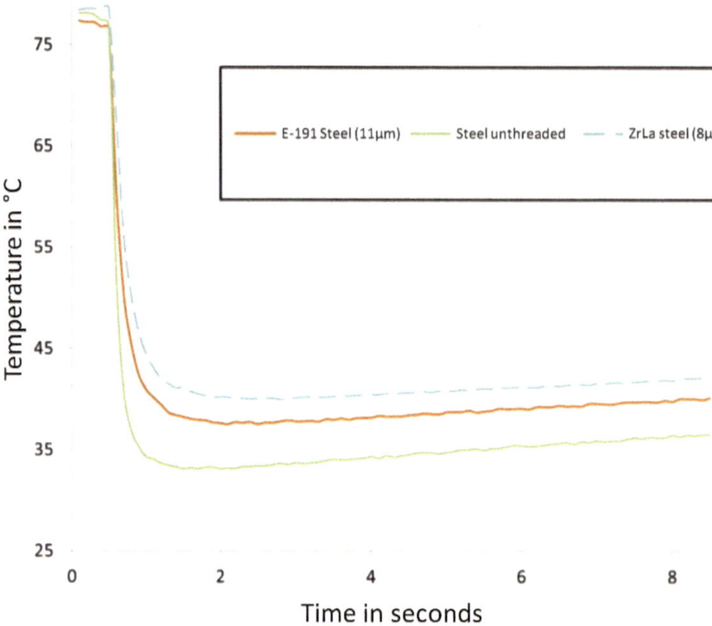

Figure 4.90 Representation of three temperature curves for a zirconium oxide coating on 1.4301 steel. Blue: 8 µm thick coating doped with lanthanum; Orange: 11 µm thick coating without doping; Green: Untreated 1.4301 steel [4]

As the temperature that occurs on contact is dependent on both the thermal properties of the sample and those of the temperature sensor, these must be known in order to be able to calculate the heat penetration coefficient of the sample from the measurement. For this purpose, the values given in Section 4.9.2 can be used for this purpose. The heat penetration coefficient of the sensor must be obtained using literature values corresponding to the materials processed in the sensor.

References for Section 4.9

[1] Menning, G.: *Werkzeuge für die Kunststoffverarbeitung – Bauarten, Herstellung, Betrieb*, Hanser, 1995.

[2] Zöllner, O.: Optimized mold temperature control, Bayer AG, 1999

[3] Tewald, A.: Entwicklung und Untersuchung eines schnellen Verfahrens zur variothermen Werkzeugtemperierung mittels induktiver Erwärmung [Dissertation] (Institute Report no. 13), University of Stuttgart IKFF, 1997

[4] Tilsner, C.; Mumme, F.; Süß, P.: Dünnwandspritzgießen von PET-Vorformlingen zur Gewichtsreduktion und Energieeinsparung zur Herstellung von PET-Getränkeflaschen (Final report on a cooperative research project), Deutsche Bundesstiftung Umwelt, 2020

[5] Giessauf, J.; Pillwein, G.; Steinbichler, G.: Mold temperature control: Variotherm temperature control is fit for production, *Kunststoffe International*, 2008, 08: 87–92

[6] Thienel, P.: Verfahren und vorrichtung zur beseitigung von fließ- bzw. bindenahteinkerbungen in spritzgießteilen aus thermoplast, German Patent 3621 379 A1, 1988

[7] Böckh, P.; Wetzel, T.: *Wärmeübertragung*, Springer Berlin Heidelberg, 2009

[8] Hopmann, C.; Menges, G.; Michaeli, W.: *Spritzgießwerkzeuge*, Hanser, 2018

[9] Linseis V.; Völklein, F.; Reith, H.; et al.: Thickness and temperature dependent thermoelectric properties of Bi87Sb13 nanofilms measured with a novel measurement platform, *Semiconductor Science and Technology*, 2018, 33: 085014

[10] Linseis, V.; Völklein, F.; Reith, H.; et al.: Thermoelectric properties of Au and Ti nanofilms, characterized with a novel measurement platform, *Materials Today: Proceedings*, 2019, 8: 517–522

[11] Li, D.; Wu, Y.; Kim, P.; et al.: Thermal conductivity of individual silicon nanowires, *Applied Physics Letters*, 2003, 83: 2934–2936

[12] Koh, Y. K.; et al.: Comparison of the 3ω method and time-domain thermoreflectance for measurements of the cross-plane thermal conductivity of epitaxial semiconductors, *Journal of Applied Physics*, 2009, 105: 054303

[13] Guidotti, D.; Wilman, J. G.: Novel and nonintrusive optical thermometer, *Applied Physics Letters*, 1992, 60: 524–526

[14] Döring, F.: Minimizing thermal conductivity in laser deposited multilayers [Doctoral thesis], Georg-August-Universität Göttingen, 2017

[15] Meyer, D.: Spektrale Kontrolle von elastischer Dynamik und Wärmetransport in periodischen Nanostrukturen [Doctoral thesis], Georg-August-Universität Göttingen, 2021

[16] Mitterhuber, L.; Kraker, E.; Defregger, S.: Structure function analysis of temperature-dependent thermal properties of nm-thin Nb_2O_5, *Energies*, 2019, 12: 610

[17] Cahill, D. G.; Pohl, R. O.: Thermal conductivity of amorphous solids above the plateau, *Physical Review B*, 1987, 35: 4067–4073

[18] Cahill, D. G.; Katiyar, M.; Abelson, J. R.: Thermal conductivity of a-Si:H thin films, *Physical Review B*, 1994, 50: 6077–6081

[19] Borca-Tasciuc, T.; Kumar, A. R.; Chen, G.: Data reduction in 3ω method for thin-film thermal conductivity determination, *Review of Scientific Instruments*, 2001, 72: 2139–2147

[20] Jaber, W.; Chapuis, P.-O.: Non-idealities in the 3ω method for thermal characterization in the low- and high-frequency regimes, *AIP Advances*, 2018, 8: 045111

[21] Atakan, B.; Khlopyanova, V.; Mausberg, S.; Mumme, F.; Kandzia, A.; Pflitsch, C.: Chemical vapor deposition and analysis of thermally insulating ZrO_2 layers on injection molds, *Physica Status Solidi C*, 2015, 12: 878–885

4.10 Determining the Demolding Force during Injection Molding

Dr. Ruben Schlutter

4.10.1 Introduction

Economical injection molding production in combination with high quality requirements absolutely requires a smooth and trouble-free production process. For this purpose, all functional complexes of an injection mold, such as the temperature control and gating system, the mechanical stiffness and strength, and necessary mold movements, must be adapted to the respective requirements [1].

In the area of using large-area molds in conjunction with different plastics, demolding problems are increasingly occurring, which hinder smooth large-scale production. Highly polished surfaces additionally increase static friction and thus also the force required to demold the components. Ejector marks, deformation of the component, or demolding grooves on the component are the result [2].

4.10.2 State of the Art

After the cooling time has expired, the molded part has solidified and must be demolded from the injection mold. In an ideal world, the molded part would be released from the cavity or mold core solely by the influence of gravity. Due to internal stresses, adhesive forces and undercuts on the molded part, however, this type of demolding is not possible, so the molded part must be demolded by an ejector system [3, 4].

In addition to the mold material used, the design of the molded part, and the set injection molding process, the surface structure of the of the molded part cavity has a decisive influence on the necessary demolding force. The tribological properties that act during demolding result from the interaction of these influences. The demolding itself is usually a parallel relative movement. Figure 4.91 summarizes the influences on the demolding force [1].

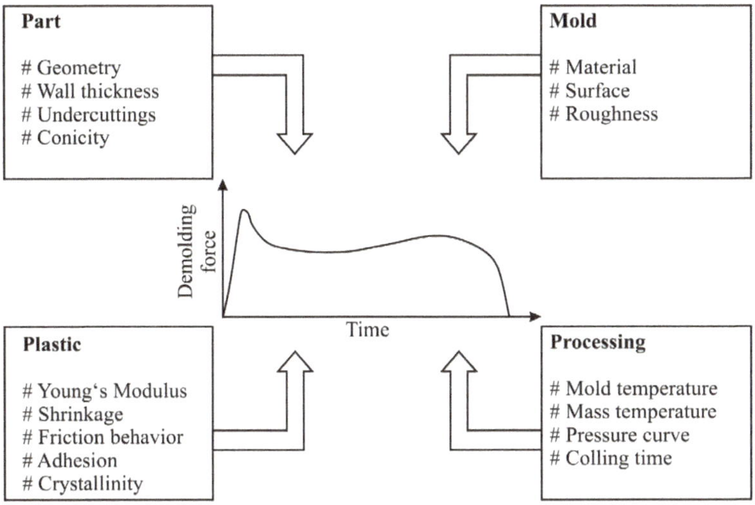

Figure 4.91 Influences on the demolding force (Image based on [1, 5]

The demolding forces can be divided into two parts. The first is the breakaway force. During filling, holding pressure and cooling, the molded part shrinks onto the mold core. Shrinkage is also possible with thin, slim rips with a low conicity. After tearing loose, shear forces act. These occur when the mold core has insufficient conicity, so that there is frictional contact between the mold core and the molded part during demolding [3].

A large number of test methods have already been defined in standards in order to define the tribological conditions between two friction partners (e.g. [6–9]). However, the standardized tests cannot represent the states during demolding in injection molding processes, so that the measured characteristic values cannot be used [10, 11].

Due to this reason, various molded part geometries were developed in order to quantitatively assess the demolding behavior. The first principles for estimating and calculating the demolding force on cylindrical parts are presented in [10]. In [5] and [12] a cylindrical part is also described. In [13] a cup-shaped part is used for the investigations. In [14], a box-shaped molded part is developed. In all the methods described, the demolding force required to demold the part from a core is determined. The problem is that a change or adaptation of a conicity is only possible to a limited extent (cf. Design rule 7: Provide sufficient conicities [3, 15]). In general, a conicity of 1.5–3° for amorphous thermoplastics and 0.5–3° for semi-crystalline thermoplastics should be used as a reference value. In addition, the structuring of the surface of the molding cavity must be taken into account. Here, the recommendation is that the angle of inclination is 1° per 25 µm structure depth [3]. Due to the comparatively complex internal geometries of the described three-dimensional moldings for determining the demolding forces, finishing or uniform coating of the cores is also more difficult to achieve than with a flat test geometry.

In [11], the determination of the coefficient of static friction directly in the injection mold has been described using a vertically molded plate-shaped component. This was pressed hydraulically against a test surface during demolding so that the friction could be measured translationally. The influence of various process parameters was investigated, and the results were very promising. However, the adaptation of the test surface to required machining, coatings, or structures would be comparatively complex because it is located on a "sliding jaw".

Another promising approach is described in [16]. Here, a disk-shaped rotationally symmetrical component is manufactured, which is fixed in its position by teeth on the outer geometry. The ejector side is equipped with a torque transducer and a standard force transducer and can be rotated relative to the gate side. Through the rotation, both the coefficient of static friction and the coefficient of dynamic (sliding) friction of the plastic–metal pairing can be determined. The adhesive bond between the test specimen and the mold block is broken when the maximum torque is applied, and thus corresponds to the coefficient of static friction.

4.10.3 Experimental Setup for Analyzing Static and Dynamic Friction

Previous tribological laboratory tests are based on the determination of the coefficient of static and dynamic friction by friction of a defined test surface on a flat surface [6]. The coefficients of friction of the tested friction partners are then measured. This procedure is primarily designed for the tribological conditions of the friction pair during service life, and provides corresponding friction coefficients. These tribological tests cannot adequately represent the condition in an injection mold, so that the friction coefficients cannot be used to estimate the demolding of a part from an injection mold.

Experiments carried out so far to map the demolding of parts from an injection mold provide various influences on the demolding process [5, 17]. However, the results obtained cannot be transferred directly to any shaped plastic parts, since, for example, the different shrinkage of the plastics of the plastics is not taken into account. The contact of the plastic part with the mold surface cannot be determined precisely either, due to the manufacturing tolerances, so that the actual friction situation remains unclear [5, 17]. It is also possible to determine the demolding force using load cells on the ejectors. In this case, it is particularly important to ensure that the friction conditions of the ejectors in the ejector bore do not change due to the formation of plaque or abrasion, as this would invalidate the test results. It is also difficult to transfer the results obtained to other injection molds and other parts, since comparative measurements are mainly carried out with different surface modifications of the mold surface [12].

4.10.3.1 Experimental Setup

In order to be able to model and quantitatively measure the friction conditions directly in the injection mold, a test mold was manufactured with a rotationally symmetrical part geometry (transparent part in Figure 4.92). The ribbed structure on the upper side serves to fix the test specimen on the nozzle side of the injection mold in order to prevent the specimen from tearing loose. This ensures that the coefficient of static and dynamic friction between the plastic used and the mold surface is actually determined.

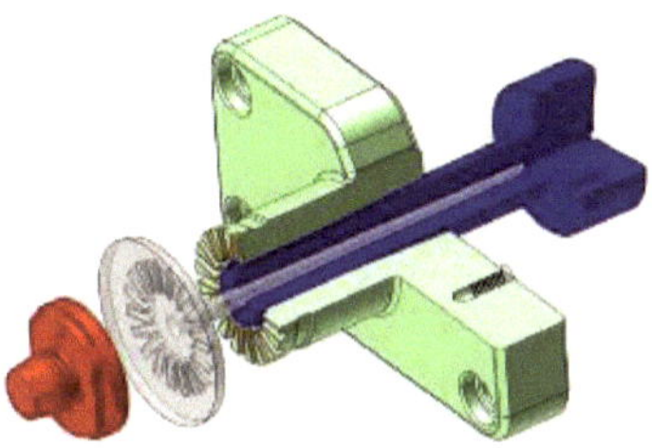

Figure 4.92
Geometry of the test specimen and general test setup (Image source: Non-profit KIMW Forschungs-GmbH)

The red component is a test blank made of 1.2343 steel, the surface of which can be modified according to the requirements of the demolding process. In its basic state, the test surface of the blank is hardened to 50–54 HRC and has a mirror finish. Depending on the problem, the test surface can be modified and functionalized using various manufacturing processes in order to be able to reproduce the conditions in the series production process. Modification of the test surface includes, among other things:

- High-gloss polishing/diamond polishing
- Sandblasting
- Chemical etching
- Die sinking
- Coating by chemical vapor deposition (CVD)
- Coating by physical vapor deposition (PVD).

Figure 4.93 shows a selection of different test blanks that have undergone different modifications of the test surface.

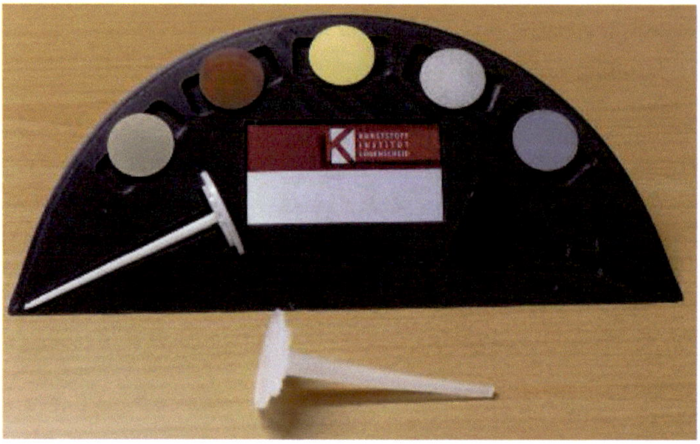

Figure 4.93 Selection of test blanks with different surface treatments (Image source: Non-profit KIMW Forschungs-GmbH)

The test blanks can be exchanged during the tests so that different surface modifications can be tested with the same plastic without any time delay.

The ejector side of the injection mold is equipped with a device for measuring the torque in order to determine the adhesive torque and the dynamic frictional torque between the test specimen and the test surface of the test blank. The design of the ejector side of the injection mold is shown in Figure 4.94.

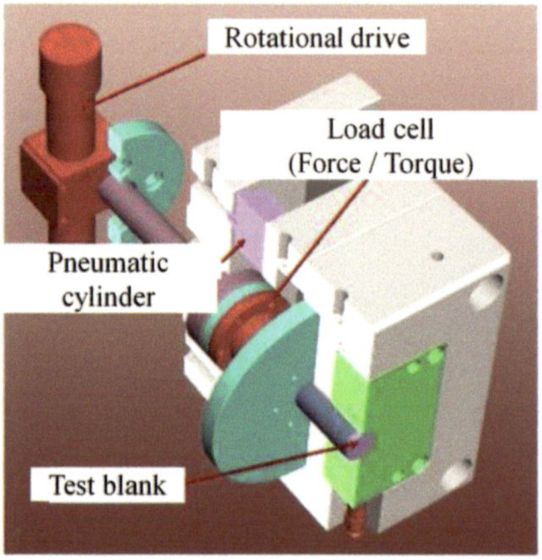

Figure 4.94 Ejector side of the injection molding tool (Image source: Non-profit KIMW Forschungs-GmbH)

4.10.3.2 Carrying out the Experiment

The aim of the test is to map the tribological conditions during the injection molding process. Accordingly, a standard injection molding process with injection phase, holding pressure phase, and cooling phase is defined to produce the test specimen. During the holding pressure phase, the part is pressed against the test surface to prevent shrinkage. At the end of the cooling phase, the injection mold is opened a few millimeters. At the same time, the test specimen and the test round are held together pneumatically with a defined force to prevent the friction partners from loosening and to ensure comparability of the measurements, regardless of the plastic used and the injection molding process set up. The actual measurement of the adhesion torque and the sliding (dynamic) friction torque follows. For this purpose, the test blank is rotated by means of a hydraulic rotary drive, while the test specimen remains fixed by the rib structure in the nozzle side of the injection mold.

During the test, the curve in Figure 4.95 is recorded by the torque load cell. The first maximum of the torque describes the adhesion torque. Up to this point, the test specimen adheres to the test blank. The subsequent plateau of the torque curve then describes the sliding torque, which is formed by the constant friction between the specimen and the test blank. It is evaluated as an integral over time, with the evaluation time depending on the plastic.

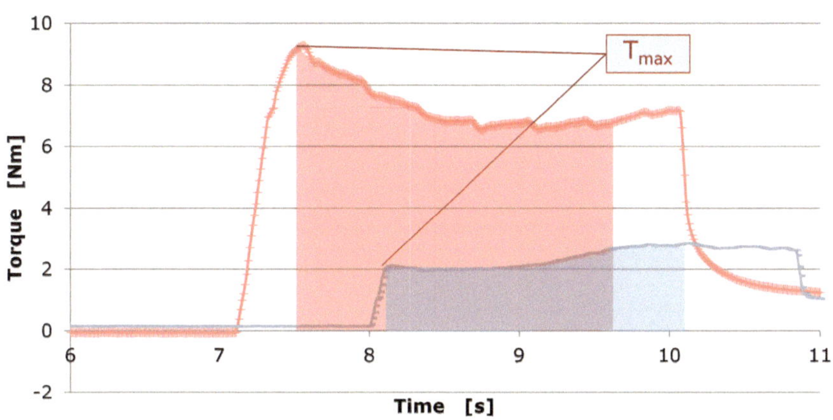

Figure 4.95 Evaluation of the measured torque curves of different materials (red and blue lines) for one demolding (Image source: Non-profit KIMW Forschungs-GmbH)

After finishing the test, the injection mold is opened and the test specimen hanging in the nozzle side is removed. During a test, the first 30 injection molding cycles are usually considered as start-up scrap. During these injection molding cycles, the torque is not evaluated. If the test specimens are manufactured reproducibly with a stable injection molding process, the actual measurement follows. As a rule, this comprises 15 specimens in order to ensure statistical reliability of the static and dynamic frictional torques determined and to exclude outliers.

4.10.3.3 Qualification of the Injection Mold in the Long-Term Test

In order to exclude long-term effects that could corrupt or negatively influence the measurement results, the test mold was subjected to several long-term runs. In each case, over 300 individual measurements were carried out to characterize initial run-in effects and long-term stability. The cleaning of the blanks also has a significant influence on the measured values.

Figure 4.96 shows the final long-term tests to qualify the measurement method. The long-term test was carried out twice: once with cleaning of the test blank after each cycle, and once without cleaning of the test blank. Acetone was used as the cleaning agent. Since part removal and cleaning take place in stages, there is a greater variation in the cycle time and in the quality of the cleaning. Both have an influence on the maximum torque, so that the values in this test series fluctuate more strongly (blue curve in Figure 4.96). In addition, a significantly lower maximum torque can be measured compared to the long-term test without cleaning (blue curve in Figure 4.96). In this test series, only the part was removed manually, and the cavity was not cleaned. It can be assumed that a plaque from the abrasion of the previous measurements forms on the test blank. This covers the surface roughness of the test blank, so that it is no longer steel that rubs against plastic, but plastic that rubs against plastic. The result is an increase in the specific adhesive bonds and thus an increase in the maximum torque. The cleaning agent acetone has a further influence, as it cools the surface of the cavity while vaporizing, so that the plastic on the blank can freeze more quickly if cleaning is carried out between cycles. As a result, the blank can no longer be molded accurately, which also leads to a smaller maximum torque [1].

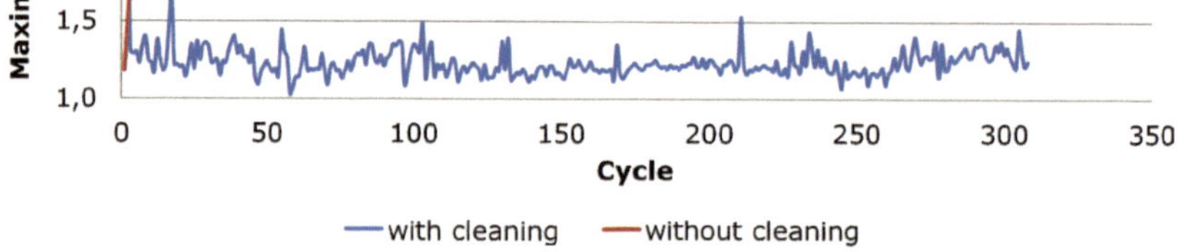

Figure 4.96 Evaluation of the measured torque curves (Image source: Non-profit KIMW Forschungs-GmbH)

Figure 4.97 shows the statistical evaluation of the final series of long-term tests. To detect outliers, the measured values were balanced with a moving average and checked for outliers. A deviation of the moving average of >0.01 N·m was defined as an outlier criterion. From the 14th cycle onwards, no outlier can be measured. The first 15 measured values are therefore considered as start-up scrap and are not included further in the evaluation.

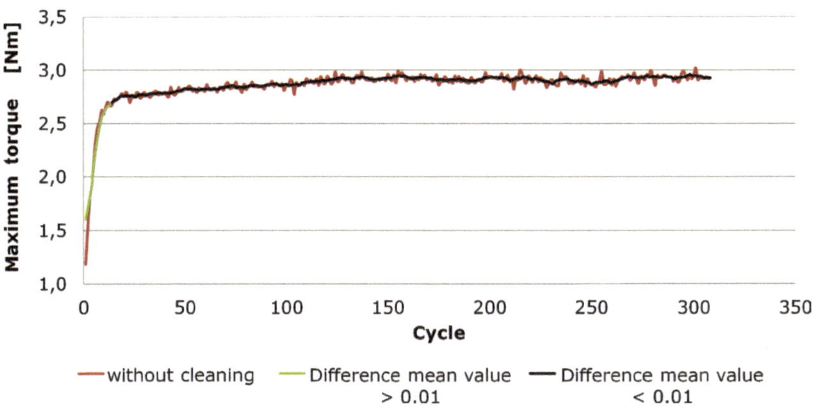

Figure 4.97 Evaluation of the obtained torque curves (Image source: Non-profit KIMW Forschungs-GmbH)

4.10.4 Summary

As a result of this study, it was possible to achieve accurate and reproducible modeling of the friction conditions in the injection mold during demolding under near-series conditions. Thus, the previous deficiencies that occur in the modeling of the friction conditions during demolding could be overcome. As a result, a simple and robust test was established to map the demolding process. There are no limits to the modification of the mold surface in terms of material selection, the introduction of structures, or the application of coatings through the use of standardized demolding probes. The plastic material can also be chosen as freely as possible to carry out the tests. The only restrictions are the maximum mold wall temperature, which is 140 °C, and the minimum hardness of the plastic, which must be a minimum of 30–40 Shore A.

References for Section 4.10

[1] Korres, M.: Optimierung eines Messwerkzeuges für das Entformungsverhalten im Spritzgießprozess [Bachelor thesis], South Westphalia University of Applied Sciences, 2016

[2] KIMW: Guide to defects in moulded thermoplastic parts (13th edn), Kunststoff-Institut für die mittelständische Wirtschaft N. R. W. GmbH, 2017

[3] Hopmann, C.; Menges, G.; Michaeli, W.; et al.: *Spritzgießwerkzeuge* (7th edn), Hanser, 2018

[4] Baur, E.; Brinkmann, S.; Osswald, T.; et al.: *Saechtling Kunststoff Taschenbuch* (30th edn), Hanser, 2007

[5] Burkard, E.; T.; Schinköthe, W.: Einfluß von Werkzeugbeschichtungen auf das Entformungsverhalten beim Spritzgießen, *17th Stuttgart Plastics Colloquium*, 2001

[6] ISO 8295-10: Plastics – Film and sheeting – Determination of the coefficients of friction, Beuth-Verlag, 1995

[7] ISO 7148-2:2012: Plain bearings – Testing of the tribological behaviour of bearing materials – Part 2: Testing of polymer-based plain bearing materials, Beuth-Verlag, 2012

[8] DIN 50324-07:1992: Tribology; testing of friction and wear model test for sliding friction of solids (ball-on-disc system), Beuth-Verlag, 2017

[9] ISO 16047-01: Fasteners – Torque/clamp force testing, Beuth-Verlag, 2013

[10] Kaminski, A.: Messen und Berechnen von Entformungskräften an geometrisch einfachen Bauteilen, *Kunststoffe*, 1976, 66: 208–214

[11] Menges, G.; Bangert, H.: Messung von Haftreibungskoeffizienten zur Ermittlung von Öffnungs- und Entformungskräften bei Spritzgießwerkzeugen, *Kunststoffe*, 1981, 71: 552–557

[12] Schattka, G.; Burkard, E.; T.; Schinköthe, W.: Entformungskraftuntersuchung beim Spritzgießen Neuer messtechnischer Ansatz zur Ermittlung der Adhäsions- und Gleitreibungskräfte bei der Entformung, *23rd Stuttgart Plastics Colloquium*, 2013

[13] Pönisch, N.; Hering, W.: Anforderungen an Werkzeugbeschichtungen zur Verbesserung des Entformungsverhaltens im Spritzguss, *TECHNOMER'26*, 2019

[14] Michaeli, W.; Hopmann, C.; Erler, I.: Mold design: Virtual demolding forces, *Kunststoffe International*, 2012, 102: issue 06

[15] Brinkmann, T.: *Handbuch Produktentwicklung mit Kunststoffen* (1st edn), Hanser, 2011

[16] Kaminski, A.; Salewski, K.: Device and method for determining a friction coefficient, Patent specification WO/2002/082058, 2002

[17] Molekulare Beschichtungen von Formen und Werkzeugen für die Kunststoff-Verarbeitung, *https://gvt.org/Forschung/IGF_Forschungsprojekte/Abgeschlossene+GVT_Vorhaben/17993+N-p-954.html*, accessed on 04 September 2019

4.11 Determination of Emissions in Plastics Processing

Dr. Andreas Balster and Matthias Korres

4.11.1 Introduction

The qualitative and quantitative determination of emissions is standard in the optimization and assessment of technical processes in industry. For process optimization, this information is of decisive importance for the effectiveness of the design of systems, tools, and peripherals, but also for the protection of system elements against increased wear and tear; it thus serves to reduce maintenance costs.

Due to the variety of polymers, additives, fillers, reinforcing materials, process parameters, and batch fluctuations, precise analysis of emitted substances is a difficult task in itself. The fact that the material emissions during injection molding consist of very complex gas mixtures and that the gaseous products can themselves be reactive at the prevailing temperatures and pressures, and form secondary products with other components, further complicates the investigations.

A no less complex method, gas chromatography/mass spectrometry (GC/MS), is used to analyze the gas mixtures. This is a combination of analytical methods that is used to characterize mixtures of organic chemical compounds. As the name suggests, the principle of GC/MS consists of two main parts: gas chromatography is used to separate the mixture, while the mass spectrometer identifies the components by determining their mass and charge. Due to the enormous range of variations available, its high sensitivity and today's robust and simple handling, GC/MS is one of the most important tools for clarifying the composition of scientifically or technologically relevant samples, e.g. in the food, environmental, forensic, pharmaceutical, and petrochemical industries.

4.11.2 Gas Chromatography/Mass Spectrometry (GC/MS)

In order to make qualitative or quantitative statements about the individual components of an unknown mixture of substances, the sample to be analyzed must be separated into its components as clearly as possible. After this step, initial statements about the complexity of the mixture and first indications of the chemical similarity of the sample components can be made.

What all chromatographic methods have in common is that mixtures of substances are exposed to the interaction of a stationary phase and a mobile phase. The samples can be solid, liquid, or gaseous in their initial state. In the case of gas chromatography – hence the name – the separation takes place under conditions in which the analytes are gaseous and can be transported stably in a carrier gas stream for the duration of the analysis.

Sample allocation

The analysis procedure can be divided into the sample preparation, the actual separation, and the subsequent detection or identification of the components. The first step in substance separation is sample introduction, which is often carried out by injecting a few microliters of a solution of the mixture into an organic solvent such as acetone, methyl ethyl ketone, 2-propanol or others into the column. This tiny amount is split once again; only a fraction of the injection volume enters the column, which would otherwise be "overloaded", severely disrupting the separation process. Common split ratios are 1:60 or 1:70. The sample quantities analyzed after this dilution cascade are in the nanogram range, which illustrates the enormous detection capacity of the method. The exact sample quantities and concentrations can vary depending on the area of application. It is therefore important to determine the optimum sample quantity and concentration for a particular analysis in order to ensure reliable and accurate measurement.

There are numerous variants of sample feeding which are used for a wide variety of analytical questions. In the field of emission analysis, the samples, which are gaseous under standard conditions, are first captured on a carrier material. This is an adsorbent with a high specific surface area made of special adsorber resins such as poly(2,6-diphenyl-*p*-phenylene oxide) (trade name Tenax®) [1]. The compounds on the adsorbent, which are usually of a semi-volatile nature (SVOCs, semi-volatile organic compounds), must first be separated from the carrier by thermal desorption (TD) (also known as thermodesorption), collected again (cryofocusing), and then analyzed as simultaneously as possible. As this form of sample introduction is of great interest for the analysis of pollutants in the air, there are pre-configured sample introduction systems that can be integrated into the analysis system as a functioning unit. This form of analysis is referred to as TD–GC/MS [1, 2].

Substance separation

The stationary phase of a GC system is called the chromatographic column. An analytical GC column is usually a polyimide-coated glass capillary with a length of typically 30 m and an inner diameter of 250 µm. It is wound into a roll with a diameter of approx. 20 cm and is located in a temperature-controlled oven whose temperature behavior can be controlled very precisely. The stationary phase is located on the inner wall of the column or capillary in the form of a coating with a layer thickness of 0.1–5 µm. The mixture of substances to be separated is transported through the chromatographic column in a laminar flow at flow rates of a few milliliters per minute with the aid of the carrier gas, for example helium. This results in permanent interactions (adsorption and desorption) between the introduced components and the stationary phase. Differences in the size, composition, structure, and polarity of the molecules lead to differences in the residence time of the substances. The greater affinity a substance has for the stationary phase, the more time it needs to pass through the column. The least interacting components require only a little longer than the unloaded carrier gas, which has passed through the column after 1–2 min at the dimensions and flow rates mentioned. The time after which a substance leaves the column is called the retention time (t_R), and is substance-specific for the system under consideration and the selected test parameters. Common retention times range between 3 min and 30 min (Figure 4.98).

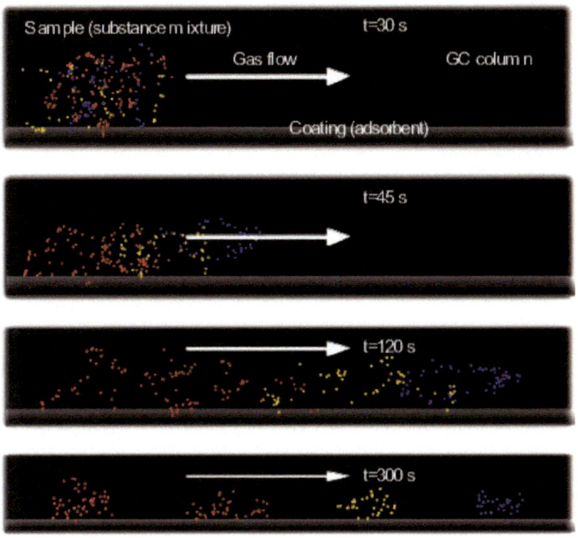

Figure 4.98
Principle of substance separation due to adsorptive interactions in the GC. The times given are examples

In this context, it is important to mention that the substances in the process interact with the stationary phase exclusively due to physical effects and not through chemical processes. A chemical reaction would cause undesirable side-effects that would prevent the planned separation process, and it would make the next step, namely the identification of the components using MS, impossible. The basic separation behavior can be influenced by the type of loading, primarily by determining the polarity of the adsorbent [3]. Typical column loadings are e.g. squalane, OV-101 (non-polar) or Carbowax (polar, PEG-based) [4]. The exact nature of the loading can be adapted for special separation problems.

The analytical question posed must inevitably deal with the problem of selectivity. In addition to the choice of column configuration, the following parameters are essentially available for optimizing the separation of substances:

- Feed temperature and split ratio (sample quantity)
- Type of carrier gas and its flow rate
- Column or oven temperature, including the temperature program and its gradients.

At the end of the column, the individual components are are discharged in the gas flow according to their characteristic retention times, i. e. separated by time. For the species of interest, an overlap of the time intervals must be avoided in order to achieve a high degree of selectivity, which is of decisive importance for the subsequent step, identification.

Detection and identification

If the components of a sample are already known in advance, they can often be identified based on their retention times, which can be determined by injecting the expected pure substances under the same conditions. The retention time is determined by registering the exit of the respective pure substance at the end of the GC column using a suitable detector. Non-substance-specific detectors are, for example, thermal conductivity detectors, which register the presence of a substance through differences in the thermal conductivity of the pure and the temporarily substance-loaded carrier gas, or a flame ionization detector (FID), which generates ions such as CHO^+ or H_3O^+ from the exiting sample packets through combustion, which in turn can be measured in an electric field. The prerequisite here is the use of a combustible carrier gas. Hydrogen is used for this purpose, which increases the requirements for laboratory safety. Another method, which is hardly used on a broad scale, is the so-called electron capture detector (ECD), with a detection sensitivity that is orders of magnitude higher [5]. What all the systems mentioned have in common is that they provide a quantity-dependent signal, which is the basic requirement for quantifying the substances.

The most common combination for routine analyses today consists of a mass spectrometer (MS) downstream of the GC (Figure 4.99) [6]. In the mass spectrometer, the individual components are ionized and then separated in an electric field based on their mass-to-charge ratio.

The different ions are then registered on a detector and converted into a mass spectrum, which represents a characteristic pattern for the compounds present in the sample. As these processes take place in a high vacuum, the system must in turn significantly reduce the gas flows from the GC column. Modern mass spectrometers are capable of generating signals with just a few thousand to millions of ions.[5)] Re-sampling is therefore rarely a problem.

[5] For comparison: 1 ng (nanogram, 10^{-9} g) of the solvent acetone has about 10 trillion (10^{13}) particles.

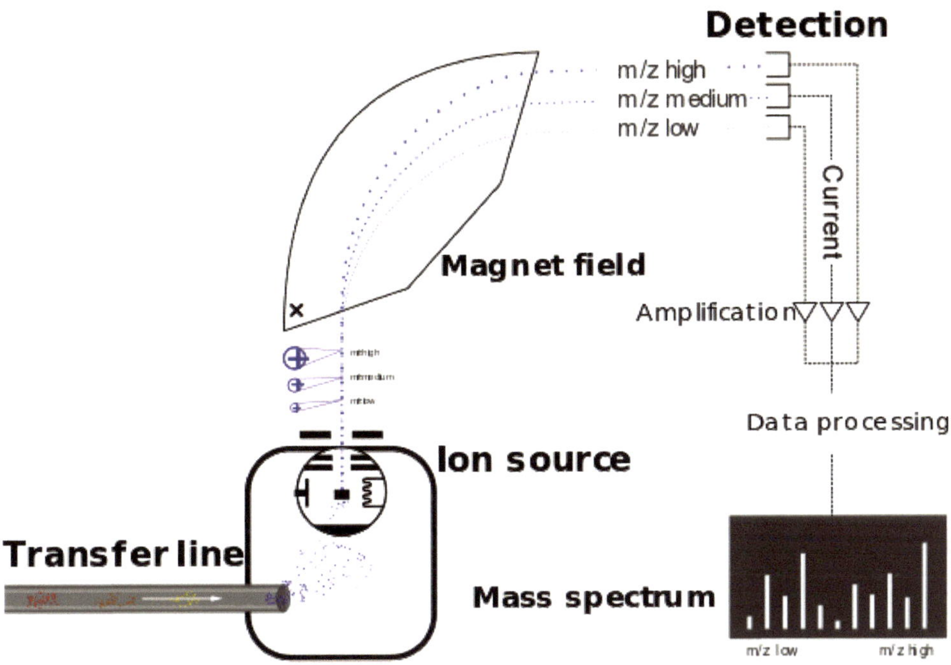

Figure 4.99 Ionization of the separated fraction and mass-selective splitting of the ions in a magnetic field. Different techniques exist for splitting; the sector field method is shown here as an example

In the first step, the introduced molecules are ionized. This can be done in various ways, such as electron impact ionization (EI), chemical ionization (CI), photo-ionization (PI), field ionization (FI), electrospray ionization (ESI), fast atom bombardment (FAB), inductively coupled plasma (ICP) ionization, and other methods.

It should be noted that each of these methods has its advantages and disadvantages for different substance classes and areas of application. The EI method is often used for routine analysis. The aim of ionization is to obtain charged debris (fragments) from the substances, whereby the ratio of the fragment mass to its charge m/z represents the subsequent measured variable in the characterization.

Organic compounds are, simply explained, composed according to a modular principle, whereby certain "building blocks" (sequences, functional groups, or residues) repeatedly occur in different molecules and can be regarded as a unit that determines the chemical character of a substance and also makes it assignable to certain substance classes. During the fragmentation of organic substances, more or less stable fragments can be formed which often correlate or correspond to known functional groups. As a result, when the molecules are fragmented in doses, certain groups often remain intact and can be detected as a unit or "debris" with a precisely defined and known mass. So what emerges from ionization is a list of fragments with specific mass-to-charge ratios of which the largest fragment is called the molecular ion (M^+) – no larger fragments are possible (Figure 4.100). The stability of a fragment also determines its formation frequency. If the frequency of the fragments is plotted as a function of *m*/*z*, this histogram represents a so-called mass spectrum, which is characteristic of the pure substance under the measurement conditions. It can therefore be used to identify the substance.

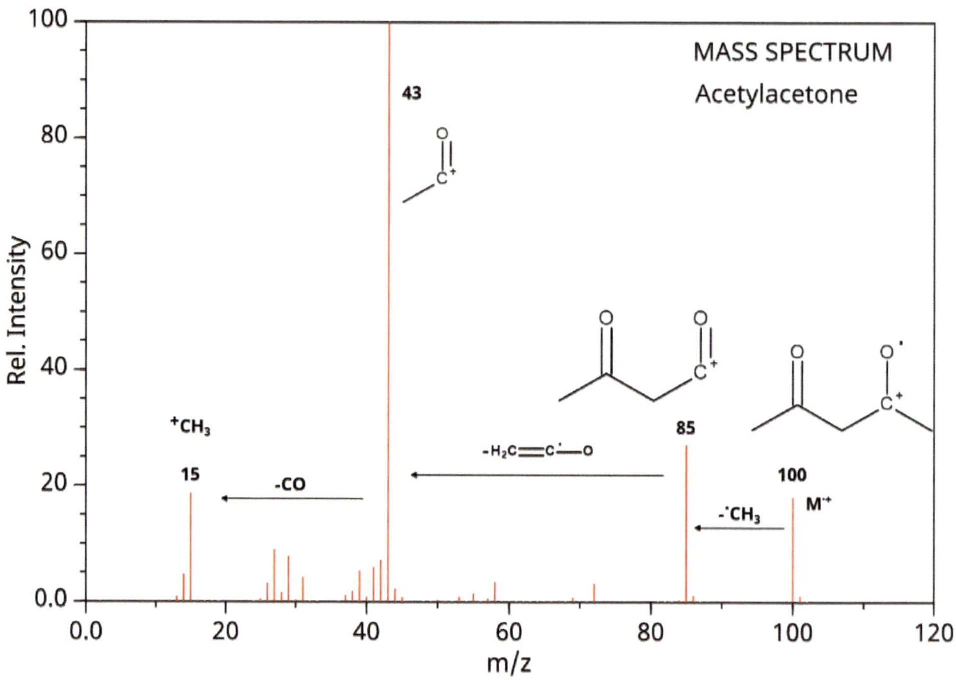

NIST Chemistry WebBook (https://webbook.nist.gov/chemistry)

Figure 4.100 Mass spectrum of 2,5-pentanedione (acetylacetone), an important complexing agent. The primary fragmentation patterns are shown [7]

In order to generate such a fragment pattern, it is necessary to separate the resulting charged fragments from each other after the ionization process in order to be able to count them. For this purpose, the generated beam of charged particles is directed into an electric field, where the different mass inertias of the fragments allows selection according to the mass-to-charge ratio. As the reaction to the electric field is also proportional to the charge of the particle due to Coulomb's law, a fragment ion with a given mass and a single positive charge delivers the same signal as one that is twice as heavy but twice as positively charged.

Common techniques that allow separation by *m*/*z* are quadrupole (QMS) or time-of-flight (ToF) mass spectrometers. A QMS works by selectively passing a single fragment with a defined m/z ratio at a given time, as only this has a stable trajectory between the four quadrupole rods. By scanning a high-frequency alternating field, different m/z ratios can pass through the quadrupole one after the other, and the total ion count (TIC) is used as a measured variable for quantification. The ToF technique, which is also widely used, accelerates the ions in a static field, with ions of different masses arriving at the detector at different times due to their inertia, resulting in a time-dependent m/z distribution. Figure 4.99 shows another variant, a sector field mass spectrometer.

Although it is possible to evaluate the mass spectrum "manually", i.e. on the basis of plausible assumptions about the possible fragmentation pathways, and make an identification in this way, instead it is usually compared with databases of known mass spectra. Such databases are partly publicly available [7] and partly commercially available. Since each GC/MS system has its own individual catalogs of expected compounds in combination with the laboratory-specific questions, self-created databases are often used in practice. While prepolymer and granulate samples rarely contain unknown substances in terms of their composition and the components they contain (additives, excipients, or oligomers, which can also be obtained and measured individually), the situation changes when it comes to the determination of products resulting from thermo-oxidative degradation, pyrolysis, or contamination. In such cases, the reaction products can usually be chemically derived and understood, but they are not necessarily available as references.

The intensities of the signals in the mass spectrum can also be used to determine the relative concentrations of the different compounds in the sample. In this case, the method is used to quantitatively determine the species of interest. Calibration is always required to obtain meaningful results. For this purpose, samples of different, known concentrations are added to the GC/MS system and the signal strengths obtained are plotted against the concentration. If the target compounds are not available, at least semi-quantification can be achieved by including an internal standard in the analyses, the identity and concentration of which can be easily controlled [8].

In the case of emission analysis, it cannot be granted that the substances emitted are stable, easily accessible compounds. The special conditions in the gas space of a just-filled cavity during injection molding generate a gas mixture with a wide variety of reaction products whose occurrence and distribution depends on the prevailing pressure and temperature conditions, the starting materials, the process parameters, and even the part and mold design. For this reason, so-called sum parameters are sometimes used to quickly classify a product mixture by summarizing the detected signals and expressing their quantity in equivalents of an accessible substance. For example, in VDA 277 [9], a widely used automotive standard, the signal peaks detected in the course of a so-called headspace GC analysis are integrated and totalled. This sum is expressed in acetone equivalents, i.e. a calibration is performed with increasing amounts of the readily available solvent acetone. The result of the analysis is expressed as if the total of all emitted substances consisted of acetone. This allows at least a comparison of different samples with regard to the quantity of emitted compounds, but not with regard to the composition of the gas mixture. However, such a sum parameter is not always sufficient when evaluating emission samples from the point of view of the medical assessment of plastic formulations or in the course of process optimization.

4.11.3 Emission Formation in Plastics Processing

In the industrial processing of plastics, especially thermoplastic molding compounds, the properties of plastics are changed by various process steps such as material conditioning, plasticizing, processing, and downstream machining due to the many influencing variables. These include thermal fluctuations, shear effects, pressure changes and contact with external media, which cause direct or indirect material degradation. Under the effective temperature and pressure conditions of the process, substances can escape from the plastic, some of which are unstable or reactive and in turn react with other species. The resulting products (Figure 4.101) are as varied as the materials and processes used, both in terms of their quantity and their chemical structures [10].

Figure 4.101 Plaque formation on a parting line due to the processing of PBT [11]

The resulting precipitate is deposited in the molded part cavity, but also in the venting areas and the parting line of the mold. This can lead to the formation of plaque in the cavity and on the molded parts, as well as to the formation of nasty smells and corresponding stress for employees. The reduction of such behavior often conflicts with the parameters used in the process to ensure part quality. Particularly in the case of thin-walled parts or long flow paths, strict requirements are placed on the temperature and flow behavior of the plastic, which, however, also support the formation of emissions from the respective material. Accordingly, this behavior must be replicated as accurately as possible when analyzing the emissions. Emissions from plastic parts are measured at various stages of production. Some of these are supported by standards that are widely used in testing technology. The investigation of process-related material emissions has so far been largely excluded from this, and plays more of a role in directly ensuring production stability. In the following, the measurement in the injection molding process will be highlighted in particular.

4.11.4 Process-Dependent Emission Formation

Despite its many variants and various additional peripherals, the injection molding process is characterized by a clear process structure. This can be divided into three process areas, in the following order:

- Material preparation and processing
- Processing, plasticizing and/or molding
- Other process steps, if applicable.

Downstream process steps can include, for example, tempering or joining with other materials by painting, printing, or the assembly of model groups by gluing, welding, and other joining processes. Many of these steps can produce emissions that can affect product quality. For the injection molding process, material drying, compounding and/or plasticizing, and the injection phase are interesting for an investigation with regard to possible emission formation.

4.11.4.1 Material Drying

The majority of thermoplastics require pre-drying for optimum processing due to their hydrophilic properties. Even at these low drying temperatures (compared to the processing temperatures), volatile organic components can be released. Dust pollution caused by very small particles and the air circulation used in dry-air dryers can make measurement difficult. For a successful measurement, a vacuum applied via a Venturi nozzle should be used to ensure a constant flow rate and thus optimum absorption of the emission (Figure 4.102). Since the generally low emission formation requires a correspondingly long measurement duration, an additional measurement strategy for the flow into a sample holder is helpful to ensure the quality of the results. Using a dry-air dryer as an example, this setup can be placed at the head end of the dryer with a hose connection [11].

Figure 4.102
Recording of process emissions during the drying of various thermoplastics [11]

The small variance in drying temperatures and drying times produces a correspondingly small variance in the emission patterns. Particularly when comparing different critical materials or material batches, the tests should be carried out and recorded precisely. Multiple exposures using thermal desorption tubes can also help to exclude the influence of residues and foreign material.

Such measurements are useful for recording values close to industrial series production in the case of defects, such as traces of deposits in the dryer housing, or thermally critical materials. This procedure can also reveal additional findings when using biodegradable materials, which are specifically conditioned for processing in the case of cellulose-based polymers, for example. An alternative to this measurement setup for testing plastics is the use of a headspace GC analysis with the material samples in a conditioned state, whereby the circulating air in the dryers is not taken into account. A material change observed in a sample in a dry-air dryer should serve as an example here.

The chromatogram in Figure 4.103 shows at higher retention times (from approximately 24 minutes) an overlap in the collected material mixture for less volatile compounds, which remained in the dryer after cleaning with compressed air by the responsible employee and were introduced into the subsequent process with the new material in production. Here they caused streaks on the surface of the molded part. Any deposits on the wall in the dryer were not visually recognizable and could not be detected with a headspace spectrogram.

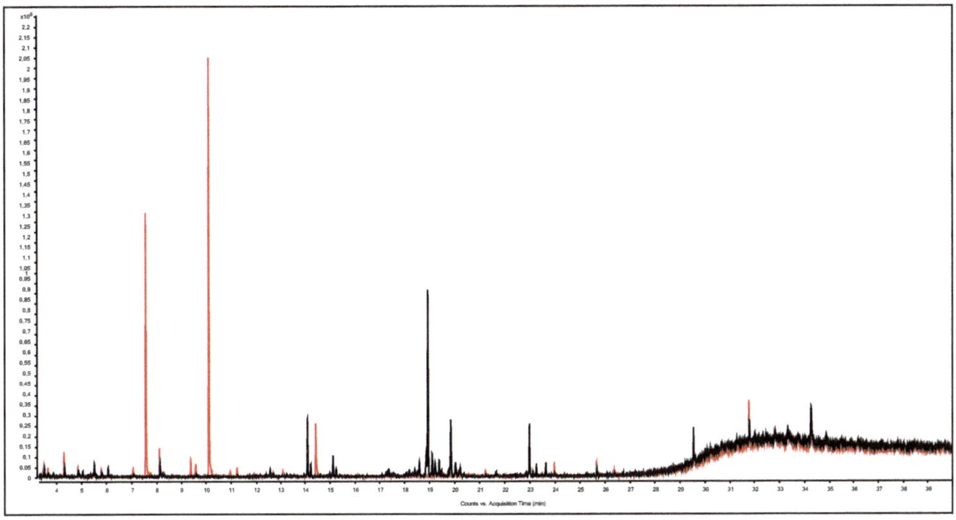

Figure 4.103 GC analysis of emissions from a dry-air dryer with polystyrene (red) and an image after cleaning and filling with polypropylene (black) [11]

As an action to minimize such error patterns, in addition to the material-specific separation of dryer units, the cleaning of smaller dryer units (e.g.: up to 40 mL) with microfiber cloths or with the addition of solvents is also suitable in addition to blowing-out with compressed air.

4.11.4.2 Material Processing

Emission measurement in the manufacturing process shows clear differences to measurement on the granulate or on the finished part. It also depends on which interface is measured in the current process step. Emissions can be in degassing zones of the extruder (Figure 4.104), in the plasticizing cylinder, or in the venting area of an injection molding tool. Appropriate test tools are available for this purpose, which are used to specifically record emissions. A recording can also be made directly on the tool in the series process, which enables a very precise localization of error sources, as there is no change in the process environment and all real influencing factors of the process under consideration can be included.

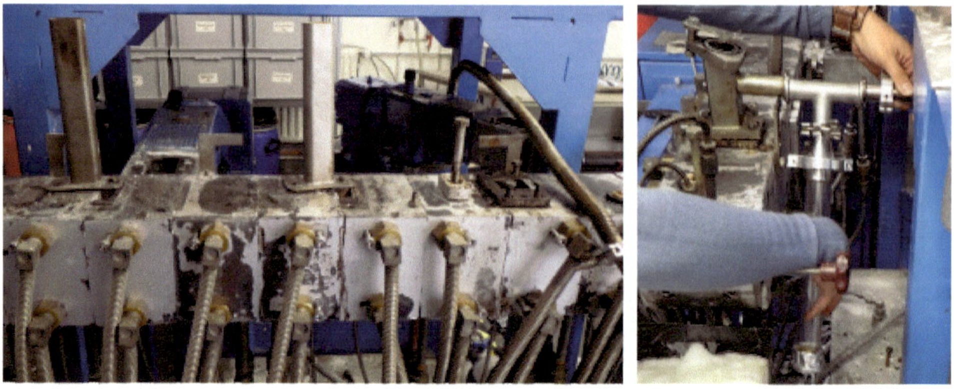

Figure 4.104 Extraction areas for emissions at a compounding plant at the level of the mixing: pressure zones (left) or degassing zones (right)

Regardless of the process step under consideration, a system-specific option is required to achieve reproducible recording of the emissions. The aim is to achieve a constant, defined, and reproducible flow rate through a sample holder that is characteristic of the respective process step. For this purpose, thermal desorption tubes from gas chromatography, possibly with the use of additional filter elements, can be used for this purpose (see Section 4.11.2).

When compounding or extruding plastics, it is possible to place the sampling within a degassing zone (Figure 4.104, right). The absorption of emissions and sample accumulation take place parallel to the continuous process. In contrast, sampling in the injection molding process must be carried out during a sufficient number of cycles due to the discontinuous process.

In the field of extrusion, the amount of escaping emissions depends not only on the material, but also on the position of the sampling device. If possible, this should be located downstream of mixing, shearing, or degassing zones, in order to capture the influence of the additives introduced or to utilize the gas flow that is generated anyway. As the sample intake is both volume- and time-dependent, appropriate control and regulation fittings and monitoring of the volume flow must be provided in addition to the intake through filter systems or thermal desorption tubes, for example. In addition to the qualitative evaluation of the emission, this procedure can also produce a quantitatively comparable result. For example, the thermal desorption tubes cannot be subjected to high pressures, as the resulting measuring section has a significantly smaller diameter compared to the compounding cylinder, and the control fittings and thermal desorption tubes also represent corresponding bottlenecks. It cannot be excluded that the measuring section could be blocked, for example by dust particles from free-flowing additives. Filter units can therefore be installed upstream to protect the measuring section in the event of high expected emission formation or particle formation, such as with flame retardants or similar [11].

Recording emissions in an injection molding tool makes it possible to investigate the dependency of a large number of process parameters, but also requires a precise definition of the interfaces for measurement recording. For example, a high degree of comparability with other injection molding processes can be achieved by translating dosing speeds and other injection molding parameters. This approach means that it is not absolutely necessary to set up a corresponding sample interface in an industrially used injection mold. Targeted venting in a sample mold designed for this purpose can lead to reliable recording of process emissions. The disadvantage of this approach is that the emissions produced during the plasticizing process are superimposed by the substances formed due to shear, pressure, and temperature changes during the injection phase. The test mold of the Kunststoff-Institut Lüdenscheid (KIMW) in Figure 4.105 serves as an example.

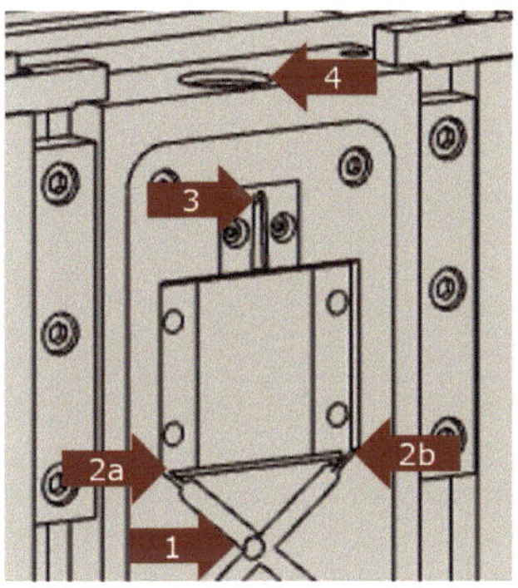

Figure 4.105
Schematic of the tool for recording emissions at the Kunststoff-Institut Lüdenscheid [11]

The injection mold has two cavities. The cold runner manifold (1) connects the respective cavity via two gating points (2a, 2b) on the corners of the molded part. The melt forms a weld line which flows together at the venting module. This directs the air via a hole (3) to the thermal desorption tube which can be attached to the bore hole (4). The sprue and the geometry of the molded part are designed in such a way that the material experiences minimal shear in order to measure the effects of plasticizing without requiring a more complex structure on the plasticizing cylinder. Depending on the viscosity properties of the thermoplastic, the transition from the cavity to the venting channel can be changed. The emissions are recorded with the aid of thermal desorption tubes, which are preceded by a so-called "victim tube" with a coarse filter in order to minimize any soot and condensate contamination of the finer filter areas downstream. This design makes it possible to change the sample holder with little effort and only short process interruptions. A double measurement, achieved by the simultaneous recording of emissions in two thermal desorption tubes at both cavities, reduces the susceptibility to sources of error or can reveal asymmetrical behavior of the cavities.

The measurements shown in Figure 4.106 were taken on a problematic PA12 batch that formed heavy plaques. This defect pattern could not be detected on earlier batches. As no major differences had previously been detected on the pellets and the molded part, emission measurements were carried out on the injection mold of the KIMW. In addition to a large number of highly volatile emissions, which evaporated through the venting gaps in the mold and during the subsequent storage of the parts, systematic differences between the problematic batch and a problem-free batch were recorded. These are released, among other things, by the special process conditions of the injection molding process, which cannot be reproduced in the laboratory. In addition, there are degradation reactions due to different process parameters in different productions, and thus potentially further variations or quantity distributions of the emitted compounds.

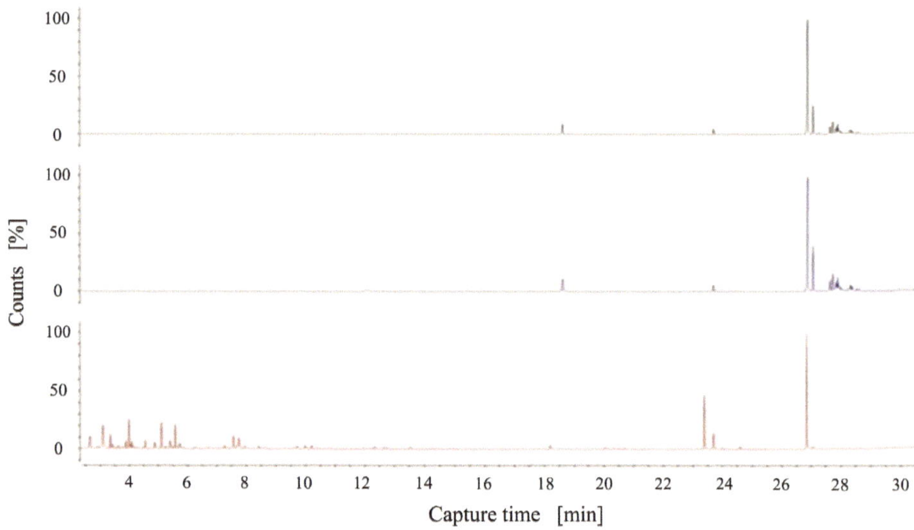

Figure 4.106 Comparison of the measurement results for pellets (top), molded part (middle) and emission tool (bottom) using the example of PA12 [12]

In order to be able to transfer the results to the conditions in the series process, it is important to copy the parameters of the original process as accurately as possible. Table 4.13 shows an excerpt from the semi-quantitative MS evaluation of the emissions of two defective batches (1 and 2) and an older, defect-free batch (3), each recorded with a double determination in the emission tool [12].

Table 4.13 Relative MS Signal Intensities in Scan Mode (33–650 *m*/*z*)

t_R (min)	Substance	Batch 1-1	Batch 1-2	Charge 2-1	Charge 2-2	Charge 3-1	Charge 3-2
20.1	Cyclododecene	++	+++	++++	++++	–	–
22.0	1,4,8-Dodecatriene	–	–	++++	++++	–	–
23.4	Di-*tert*-butylphenol	–	–	++	++	+++	++++
23.7	Cyclododecanone	–	++	++	++	+++	++++
26.8	Laurolactam	++	+++	–	–	++++	++++
27.1	Long-chain nitrile	++	+++	++++	++++	–	–
44.7	Irgafos 168	–	–	+++	+++	–	–

Legend:
- Not detectable
+ Low intensity
++Medium intensity
+++ High intensity
++++ Very high intensity

The substances, e.g. cyclododecene, are synthetic precursors and indicate inadequate material quality. In the case of problematic batches, this causes plaque problems in the cavity. These components were not found in sufficient form on the granulate samples or the molded part to make a reliable statement in this regard. The special circumstances of the process environment offer clear advantages here. In addition, the differences between the batches show the sometimes considerable fluctuations within the series of measurements. Such differences must be taken into account in the process if the injection molding process itself is not stable enough and additional molding defects occur. In this case, the fluctuations only had an influence on the intensity of the plaque formation [12].

Another example (Figure 4.107) shows the influence of the injection molding parameters on the emission chromatogram. By optimizing the process parameters, it is possible to optimize the injection molding process in terms of emissions, which can also lead to a reduction in corrosion and plaque formation.

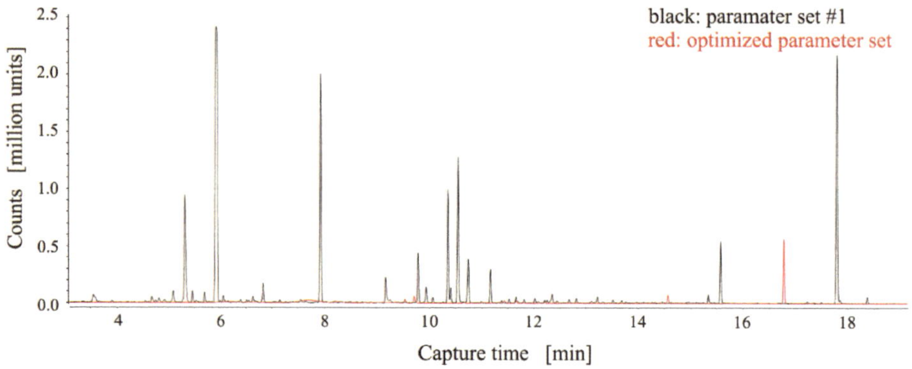

Figure 4.107 Influence of the shaping process parameters on the emission chromatogram [11]

Optimization can be carried out in different ways and is discussed in detail in Section 5.4. In the present case, PA6 was treated purely by process engineering. The parameters were adjusted with 30 run-in cycles and then 20 measurement cycles, which were measured with the emission molding tool at KIMW (Figure 4.105) with attached thermal desorption tubes. First, the processing temperatures (melt temperature and mold wall temperature) were determined and then several series of tests were carried out within this adjustment, each with different dosing parameters and decompression settings. The quantitative comparison of the substances present in the chromatograms then provides information on the reduction in emissions as an indicator of the thermal and rheological material load [11].

Taking into account the deviating geometric conditions of the series tool, the optimization can be carried out in the direction of minimized emissions as an indicator of the thermal and rheological material load, which leads, among other things, to a reduction in deposits on the tool surfaces.

4.11.5 Summary

Emission measurement represents a logical extension of previous laboratory analysis. This means that problems that were solved using conventional tests on samples, e.g. on the granulate, the molded part, or traces of coating from the mold, can now be expanded to include a sampling method that is close to the process.

This approach offers the advantage of simple and effective sample acquisition with minimal material requirements during production. On the other hand, there is an increased need for evaluation, which requires in-depth knowledge of polymer chemistry and a meticulously defined experiment design. When using the system, the question of proportionality arises, i.e. whether and to what extent the effort to record the emissions is necessary. For example, the information collected can be used to directly optimize the process parameters or materials, or to determine indirect solutions such as coatings and steel types to protect the tools from corrosion problems.

References for Section 4.11

[1] Sakodynskii, K.; Panina, L.; Klinskaya, N.: A study of some properties of Tenax, a porous polymer sorbent, *Chromatographia*, 1974, 7: 339–344

[2] de Jong, A. M.; Niemantsverdriet, J. W.: Thermal desorption analysis: Comparative test of ten commonly applied procedures, *Surface Science*, 1990, 23: 355–365

[3] Schmid, T.: Lectures in Analytical Chemistry, ETH Zurich, 2010

[4] Recommended applications for stationary phases, *https://www.chem.agilent.com/cag/CABU/Recappss.htm*, accessed on 24 May 2023

[5] Krejci, M.; Dressler M.: Selective detectors in gas chromatography, *Chromatographic Reviews*, 1970, 13: 1–59

[6] Gross, J. H.: *Massenspektrometrie: Ein Lehrbuch*, Springer, 2013

[7] NIST Standard Reference Database, DOI: 10.18434/T4H594, accessed on 10 June 2023

[8] McNaught, A. D.; Wilkinson, A.: *IUPAC Compendium of Chemical Terminology* (2nd edn), Blackwell Scientific Publications, 1997 (online version, S. J. Chalk)

[9] VDA 277 (version 01): Nichtmetallische Werkstoffe der Kfz-Innenausstattung – Bestimmung der Emission organischer Verbindungen, 1995

[10] Korres, M.: Belag und Entformung analysieren, *Kunststoffe*, 2020, 110: issue 09

[11] Schwenke, M.: Final report ZIM project "eMission", 2022

[12] Korres, M.: Emissionsmessung am KIMW [Presentation], Speedkongress, 2020

4.12 Wear Tests in Plastics Processing

Marko Gehlen

4.12.1 Introduction

Injection molding is an important process in plastics processing. It offers a cost-effective way of producing high-quality plastic parts. A particular advantage over other manufacturing processes is the high dimensional accuracy of the molded parts as well as the haptically and visually high-quality surface appearance. The molding tool steels used are exposed to various stresses, which have a significant influence on the lifetime of machines and maintenance intervals. The stresses on molding tools primarily include material-damaging processes such as corrosion and wear. There is a higher risk of the latter in particular with the use of plastics that exhibit abrasive properties due to their fillers (e.g. glass fibers and beads, and carbon fibers) in conjunction with the conditions prevailing during injection molding processes. Increased wear is to be expected, particularly in the area where the melt enters the cavity and due to frictional forces during demolding.

As a result, the demands placed on the molding tool steels used are becoming ever higher, but the possibilities of combining different properties in one and the same material are limited. Wear on the steel or the surface structure is therefore a major problem for toolmaking. This is also related, for example, to the increasingly tight tolerances in toolmaking. If specified structures are damaged by wear-induced abrasion, they can only be restored by producing a new mold insert, due to the tight dimensional tolerances. In addition to the resulting costs, there is also the corresponding loss of production, which is always accompanied by high monetary losses and a reduction in productivity for a plastics processing company.

4.12.2 Definition of Wear

In plastics processing, there are continuous (e.g. in extrusion) and cyclical (e.g. injection molding) processes. What they have in common is that the molten plastic is in contact with the tool surface or moves relative to it. The constant stress on the surface leads to a gradual removal on the side of the mold, which can be described as wear. On injection molds, this abrasion manifests itself, for example, in an increase in the gloss level of the surface over time. The material removal corresponds to a loss of mass or a loss of volume, which is often specified in surface technology in an abrasion comparison value. In addition to purely mechanical wear, corrosion can also wear or wear down the surface of an injection mold, for example. In many cases, both mechanisms overlap.

The process of hydroabrasion is an interaction between the properties of the abrasive substances in the mixture, the material properties of the material surfaces, the corrosive environmental conditions, and the various wear mechanisms. Depending on the material, mechanical damage can occur through micro-chipping, micro-plowing and micro-fatigue in addition to corrosive processes [1].

4.12.3 The Importance of Wear for the Industry

Molding tool surfaces and machine parts are subject to particular stresses as soon as they interact with other materials. The ever-increasing productivity of industry in particular has led to a significant increase in processing speeds, for example. Projected onto plastics processing, the aim is to reduce cycle times in the respective processes in order to increase the output quantity per unit of time. Furthermore, the demands on materials are increasing, requiring a balancing act between the desired functionalization of the components and their mechanical resilience on the one hand and a perfect external appearance on the other. In this case, it is the steel tool surfaces that are stressed by the polymer materials and, depending on many factors, wear out or are damaged by abrasion over time. According to [2], the economic damage caused by friction and wear is estimated at around 2.7% of the annual gross domestic product.

To ensure that a tool can be used reliably for a long time, it must be suitable for the production conditions. This also includes its resistance to wear and corrosion.

4.12.4 State of the Art and Measurement Methods

The materials can be divided into two groups depending on their microstructure and alloying elements, with regard to wear behavior: brittle and ductile materials. While ductile materials are mainly damaged by micro-chipping and micro-plowing, brittle materials are primarily damaged by micro-fracturing. Cyclic loads, e.g. with impacting particle loading, lead to fatigue processes on the surface (surface disintegration). Typical characteristics are chipping, cracking, and cratering. There are various ways of estimating the resistance of surfaces or surface coating systems to wear.

Estimating the removal rate using mechanical material properties is an attractive alternative to time-consuming laboratory tests, as the material properties are available for many materials, and this is a cost-effective and simple alternative. In [3], the material properties and the properties of the abrasive are combined with the mechanical wear rate. In general, the material hardness is regarded as the characteristic value whose increase is accompanied by a reduction in wear [4]. This applies in particular to abrasive wear.

As with all tribological tests, wear processes in hydroabrasive applications can also be divided into two groups, “laboratory tests” and “tests on real processes”.

Laboratory tests offer the opportunity to obtain quick results under controlled conditions by defining and quickly changing the load, speeds, and other parameters. Two differently designed laboratory test stands have proven themselves to date for investigating mechanical stresses: the wear pot test rig, and the free-jet test rig.

In the wear pot test stands, the liquid–solid mixture is homogenized in a cylindrical pot with a stirrer. At the same time, the material samples are rotated through the bath on holders. It is important that good mixing is ensured even at low motor speeds [5]. Methods for simulating high speeds and large abrasive particle sizes are also possible [6]. The main advantages of the wear pot test stand are its simple design and ease of manufacture and operation. The wear behavior of different materials can be evaluated easily and cost-effectively in a short time, with the results providing comparatively realistic values for different applications [7]. However, the uneven distribution of the solids in the liquid–solid mixture and turbulence in the mixture are disadvantageous [8]. This type of test stand is particularly suitable for analytical modeling. Both the average incidence angles of the particles and their average velocity can be easily controlled [7].

The free-jet test stand is another widely used test device for hydroabrasive wear tests. This design makes it possible to precisely control test parameters such as the impact angle of the abrasive particles and their impact velocity [9, 10]. As with other models, hydroabrasive wear is regarded as a superposition of micro-chipping and micro-plowing. The good reproducibility in particular is a major advantage of this type of test stand [11]. Further advantages are possible high particle velocities. The controllability of the impact velocity and the angle of incidence are also advantageous. A disadvantage of the free-jet test stand is the fact that the impact angle and the impact velocity do not remain constant during the test procedure [12]. The impact velocity must be measured and adjusted regularly or continuously, as the nozzle itself is also affected by abrasion. In addition, the abrasive particles usually cause very localized stress on the surface of the test specimen [13]. The test duration for free-jet test stands is shorter than for wear pot test stands, and is in the range of minutes to hours. Such test stands are often used to generate and prove CFD simulations and semi-empirical modeling.

Other wear test stands are known as the Miller test [14], Coriolis test stand [15], closed-loop test stands [16], or variants of the free-jet test stand such as the submerged impinging jet test stand [17].

To determine the abrasion resistance near-edge layers, a wear apparatus was developed by the company BAQ [18] (Figure 4.108). Here, a steel ball rubs against a material sample under standardized conditions. The penetration depth of the ball into the surface layer of the substrate is determined as a function of time. Based on the diameters determined (see red line in Figure 4.108), the time-dependent wear volume can be determined using mathematical formulae. This parameter indicates a comparable abrasion resistance of different thin-film systems and materials.

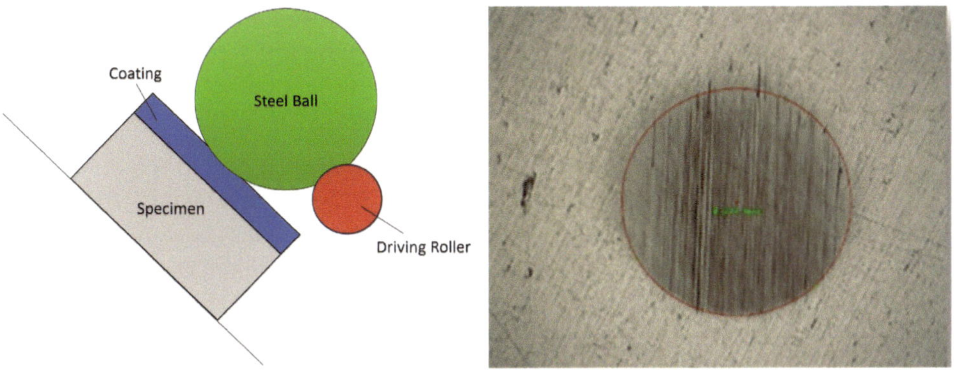

Figure 4.108 Schematic representation of the abrasion measurement (left) and the wear track to be measured (right) [18]

4.12.5 Wear during Injection Molding and in the Injection Molding Tool

There is a clear separation of functions in injection molding production. While the injection molding machine is primarily responsible for melting the material, dosing it, and building up pressure so that the plastic melt can be injected into the injection molding tool, the molding tool finally determines the geometry of the part and defines its surface quality. However, the machine must also ensure that the molding tool remains closed under high pressure during injection molding so that no burr is created, particularly in the separation plane. The clamping force, which must be adjusted to the molding tool, is crucial for this. In other words, the injection molding machine and the injection molding mold must fit together.

As the melt flows into the cavity under high pressure and at high speed, the surface is stressed with each injection molding cycle. Particularly in the case of unprotected mold surfaces, this can result in an increase in the gloss of the part or molding tool after just a few tens of thousands of shots. The reason for this increase in gloss is that the tips of the often eroded structures are rounded, which is reflected in a reduction in roughness.

4.12.6 Investigation of the Wear Behavior in Injection Molding

The injection molding process is therefore suitable as a practical test method for investigating the surface and coating properties in a production-related environment.

The molding tool is designed for the comparative investigation of the wear behavior of different coatings and must fulfill several requirements. Firstly, the maximum shear load should be applied to the area under investigation. Secondly, the inserts should be easy to coat, interchangeable, and finally designed for optimum analysis.

The result is a tool insert (Figure 4.109, right), which can gather two wear inserts in each cavity so that four inserts can always be examined simultaneously. The wear inserts are filled from the central injection point. The flow channels within the cavity are as narrow as possible in the measuring areas (on the inclined flanks). The computed simulation in Figure 4.110 confirms that the highest shear rates (up to 50,000/s) are achieved in the area of these narrow points. The individual wear inserts are first polished to a mirror finish in the area that is subjected to stress in the injection molding process, and then blasted evenly so that the surface of the inserts has a uniform roughness. Finally the inserts receive their coating.

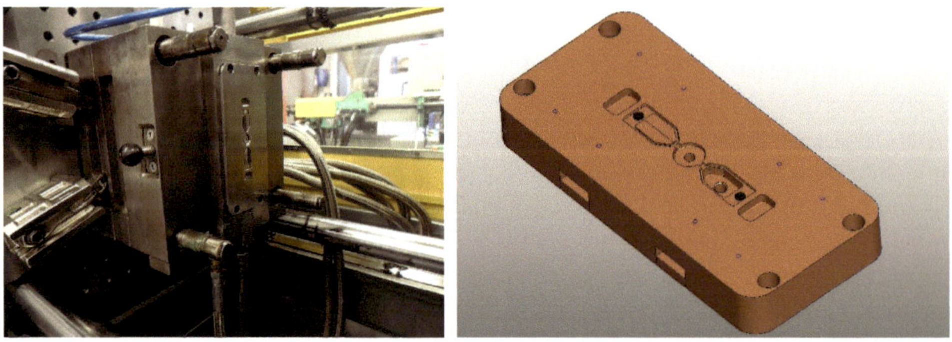

Figure 4.109 Left: Injection molding tool for holding four wear inserts (two per cavity). Right: Tool insert showing one cavity with two wear inserts at the top, and one cavity without wear inserts at the bottom

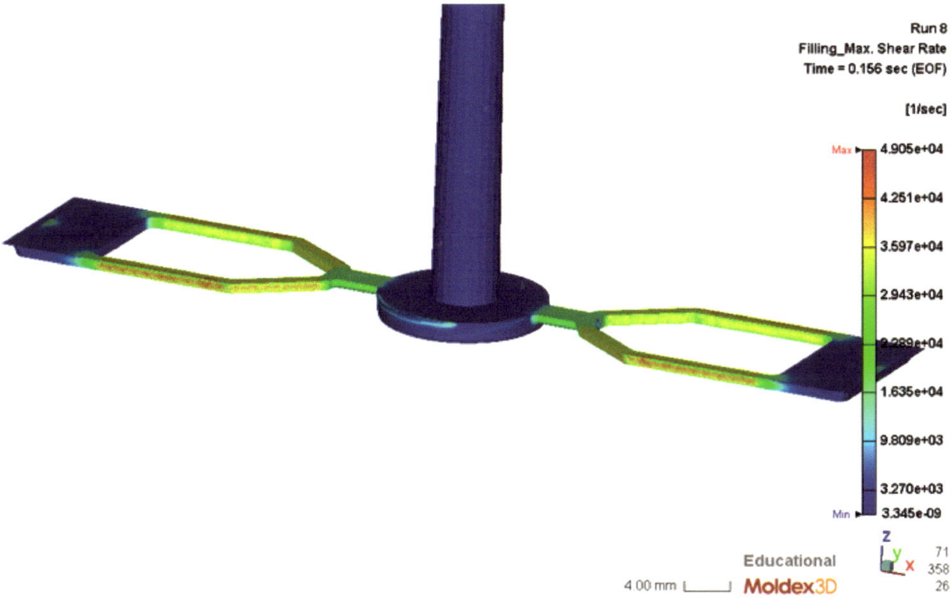

Figure 4.110 Tool for determining wear: Calculation of the maximum shear rates

Before the injection molding tests begin, the prepared wear inserts are measured for surface roughness in the measuring area using a white-light interferometer. Measurements are taken on the lateral inclined flank of the wear inserts, where the melt flows along the mold wall and rubs. The wall thickness of the test specimens is 1 mm. In order to evaluate the roughness with the optical measuring method over the widest possible range, images are taken at 5× magnification with the white-light interferometer. This results in a measurement window of approx. 1.2 mm × 0.9 mm, within which the evaluation range must lie. It is important with this procedure that the same areas are measured again to determine any change in roughness after injection molding. Characteristic points on the surface also help with fine alignment. Laser marks or body edges, for example, can be used for a previous rough alignment.

MountainsMap® software is used to evaluate the images from the white-light interferometer. The surface roughness S_a is a meaningful value for assessing the surface quality.

The result of four differently coated wear inserts is shown in Figure 4.111. This shows the change in roughness ΔS_a that occurs during the same 25,000 injection molding cycles.

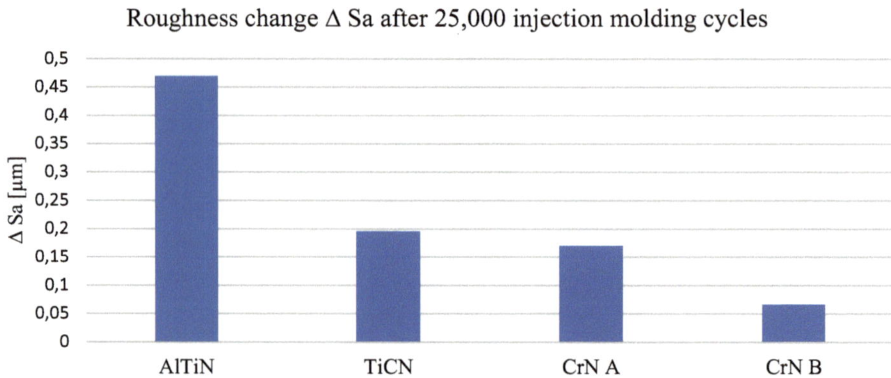

Figure 4.111 Roughness change of the surface of the coated mold inserts after 25,000 injection molding cycles

In this comparison, the CrN B coating shows the best behavior in terms of roughness change. The difference between CrN A and CrN B shows that the process in which the coatings are applied also plays a role, and not just the metals themselves. The AlTiN coating shows the highest roughness level overall and also the greatest change.

Overall, the injection molding process is a very complex method for measuring the influence of the plastic melt on the tool surface. Many thousands of injection molding cycles have to be done, especially with coated tools or tool inserts, in order to be able to determine the behavior of the coatings. However, the process described is unique in that the inserts are kept simple and can be procured cheaply. The molded parts can also be processed in a fully automated process so that the machine can run unattended. A disadvantage of this investigation is the consumption of resources such as material, energy, and time.

4.12.7 Outlook

There are many models and methods for assessing wear on molding tool surfaces, but either mechanical wear (in particular abrasion) or pure corrosion is the focus. In particular, there is a great lack of knowledge about damage processes with low stress intensities that occur in the area close to the surface (where no material parameters are available), despite their great importance (functionality, appearance, and haptics) [19]. It is therefore desirable to use a new test method to determine characteristic values for the combined stress profile, and to derive manageable wear models from this.

With the help of such wear models, tools, machines, and systems can be designed more reliably for a wide range of industries (e.g. plastics, textile, or paper processing), which leads to longer service lives and thus to lower production downtimes and higher effectiveness and productivity.

The models described and the associated test methods could become very important in the future for service providers participating in the processes, such as contract coaters, toolmakers, steel manufacturers, and suppliers.

With a test method for predicting the wear behavior of tools or machines, many cross-industry developments could be made more application-specific. Extending tool life and the associated reduction in maintenance repairs and production downtime has a direct impact on productivity, regardless of the industry under consideration.

If such a test method can be developed, this would have further advantages in terms of sustainability, as materials, time, and energy could be saved.

4.12.8 Summary

Wear is a major problem in many areas of industry. In particular, downtimes of systems and machines lead to very high costs for companies, in addition to the actual repair of tools or assemblies.

Today, there are a large number of methods and models for predicting material damage and material failure. The systematic investigation in a practical test such as the injection molding test described here requires many resources, but provides the opportunity to compare different layers and surfaces under real conditions.

Furthermore, as described in the outlook above, it would be welcome if in future there were a method that could objectively evaluate material combinations and surfaces, taking into account all phenomena involved in wear, be they mechanical and/or corrosive in nature.

References for Section 4.12

[1] Clark, H. M.; Wong, K. K.: Impact angle, particle energy and mass loss in erosion by dilute slurries, *Wear*, 1995, 186–187: 454–464

[2] Reibung und Verschleiß durch mikrostrukturierte Oberflächen steuern [Annual Report], Fraunhofer Institute for Material and Beam Technology IWS, 2016

[3] Oka, Y. I.; Okamura, K.; Yoshida, T.: Practical estimation of erosion damage caused by solid particle impact: Part 1: Effects of impact parameters on a predictive equation, *Wear*, 2005, 259: 1–6

[4] Oka, Y. I.; Yoshida, T.: Practical estimation of erosion damage caused by solid particle impact: Part 2: Mechanical properties of materials directly associated with erosion damage, *Wear*, 2005, 259: 1–6

[5] Gadhikar, A. A.; Sharma, A.; Goel, D. B.: Fabrication and testing of slurry pot erosion tester, *Transactions of the Indian Institute of Metals*, 2011, 64: 4–5

[6] Ojala, N.; et al.: High speed slurry-pot erosion wear testing with large abrasive particles, *Tribologia – Finnish Journal of Tribology*, 2015, 1: 36–44

[7] Franek, F.; Badisch, E.; Kirchgaßner, M.: Advanced methods for characterization of abrasion/erosion resistance of wear protection materials, *FME Transactions*, 2009, 37: 61–70

[8] Madsen, B. W.: Measurement of erosion-corrosion synergism with a slurry wear test apparatus, *Wear*, 1988, 123: 127–142

[9] Rodríguez, E.; Flores, M.; Pérez, A.: Erosive wear by silica sand on AISI H13 and 4140 steels, *Wear*, 2009, 267: 2109–2115

[10] Lindsley, B. A.; Marder, A. R.: The effect of velocity on the solid particle erosion rate of alloys, *Wear*, 1999, 225–229: 510–516

[11] Frosell, T.; Fripp, M.; Gutmark, E.: Investigation of slurry concentration effects on solid particle erosion rate for an impinging jet, *Wear*, 2015, 342–343: 33–43

[12] Abd-Elrhman, Y. M.; Abouel-Kasem, A.; Emara, K. M.: Effect of impact angle on slurry erosion behaviour and mechanisms of carburized AISI 5117 steel, *Journal of Tribology*, 2014, 136: 011106

[13] Ukpai, J. I.; Barker, R.; Hu, X.: Exploring the erosive wear of X65 carbon steel by acoustic emission method, *Wear*, 2013, 301: 370–382

[14] Neville, A.; Wang, C.: Erosion-corrosion of engineering steels – Can it be managed by use of chemicals?, *Wear*, 2009, 267: 2018–2026

[15] Hawthorne, H. M.; Xie, Y.; Yick, S. K.: A new Coriolis slurry erosion tester design for improved slurry dynamics, *Wear*, 2003, 255: 170–180

[16] Wong, C. Y.; Solnordal, C.; Graham, L.: Slurry erosion of surface imperfections in pipeline systems, *Wear*, 2015, 336–337: 72–85

[17] Tuzson, J. J.: Laboratory slurry erosion tests and pump wear rate calculations, *Journal of Fluids Engineering*, 1984, 106: 135–140

[18] Crater grinder kaloMax NT II [Operating manual], BAQ, 2013

[19] Stoyanov, P.; Chromik, R. R.: Scaling effects on materials tribology: From macro to micro scale, *Materials (MDPI)*, 2017, 10: 550

4.13 Adhesion Assessment of Coatings

Dr. Orlaw Massler

A wide range of methods are used to evaluate the adhesion of electroplated, chemical, and PVD (physical vapor deposition) coatings. What all of these have in common is that a load is applied to the coated surface, which then leads to delamination effects with limited adhesion leads. One thing that all *well-established* adhesion assessment methods have in common is their destructive character on the component. This can be avoided by carrying out the test on a moving test piece, or at a point on the test piece that has no function in the application. In this way, the influence of the production and pre-treatment of the tool or component, which is often decisive for the adhesion, is included, which would not be the case with a running test specimen.

In addition to the physical bond between the coating and the base material, the mechanism of adhesion of coatings to a substrate also includes the mechanical bonding of the coating to the topography of the coated surface.

The selection of the appropriate method for evaluating adhesion depends on various factors such as the type of coating, the substrate material, and the requirements of the application. Table 4.14 shows an overview of common methods and their areas of application.

Table 4.14 Overview of Various Liability Assessment Methods

Method	Thin-film PVD	Layers deposited from the bath <700 HV	Coatings deposited from the bath >700 HV	Remark
Rockwell test	Yes	No	Possible	Brittle layers
Cross-cut	No	Yes	Yes	—
Thermal shock	No	Possible	Yes	—
Scratch test	Yes	Possible	Possible	Qualitative
Pull-off shear strength test	No	Possible	Possible	Adhesive often too weak
File test	No	Possible	Yes	—
Scratch test	Yes	No	No	Results difficult to transfer
Cross-section method	Yes	Yes	Yes	High expenditure

Adhesion tests

Adhesion tests are performed to evaluate the bond strength between the coating and the substrate. These tests usually involve applying a load to the coating to see how well it adheres to the substrate. Examples of adhesion tests are the cross-cut test, the scratch test, and the pull-off test.

Cross-cut test

The cross-cut test is a test method for evaluating the adhesion of coatings to a substrate. The test consists of applying a grid pattern to the coated surface with a cutting tool by scratching a series of horizontal and vertical lines on the surface. The grid pattern is typically spaced approximately 1 mm to 5 mm apart. Once the grid pattern has been applied, a defined adhesive tape is placed over the grid pattern and pressed down firmly. The adhesive tape is then removed quickly and firmly.

If the grid pattern remains intact and no parts of the coating are peeled off, this is an indication that the coating adheres well and has sufficient strength. However, if parts of the coating are peeled off, this is an indication of insufficient adhesion and strength of the coating. The resulting grid image is compared with a reference sample and categorized. The test is used to check whether the coating adheres well and whether it has sufficient strength to withstand the loads.

The cross-cut test is a simple and quick method to check the adhesion of coatings. However, it should be noted that this test is not always conclusive and further tests should be carried out to fully assess the quality of the coating.

Scratch test

The scratch test is another test for evaluating the adhesion of coatings. Similar to the cross-cut test, a hard scratching stick or a hard tip is used to scratch the coating on the substrate and assess how strong the adhesion is. The evaluation is purely qualitative at the edge of the scratch. This test is very simple and rudimentary, but provides a very quick picture of whether there is poor coating adhesion.

Pull-off or shear strength test

Shear strength tests are performed to determine the maximum shear load required to separate the coating from the substrate. The coating is usually applied to an adhesive surface and then loaded with a force to evaluate the adhesive strength.

The quality of the bonding is essential to determine the degree of significance that should be attached to the results of the test. The adhesion of the tension or shear stubs is often not good enough to make a good statement, and is why this test is not very common. Its strength lies in the fact that the component (with good adhesion) can usually still be used after the test. This test can therefore be considered non-destructive.

Scratch test

The scratch test, which is not described in detail here, plays a special role. Here, an indenter is passed over a surface under increasing load. This creates a plastic line impression, the edges of which can be assessed microscopically. A series of critical loads can be derived from this, which are used as a measure of the adhesion and cracking of the coating. However, the formation of these individually defined critical loads depends not only on the coating properties but also on the coating thickness, the notch effect caused by the surface topography, and the shape and wear condition of the indenter.

Nevertheless, this test is very useful for the comparative evaluation of coatings and for the simulation of wear processes [1].

4.13.1 Rockwell Test (DIN 4856)

The Rockwell hardness test is standardized in DIN 4856 (formerly VDI Guideline 3198) and ISO 26443. During the test, a plastic deformation is introduced into the coated substrate by means of a defined hardness indentation (Figure 4.112). This deformation would lead to spalling in the event of poor adhesion of the coating. The results are evaluated by comparing them with a comparative image set in HF steps. In DIN 4856, these comprise six adhesion classes [2]; ISO 26443 suggests four adhesion classes [3].

Figure 4.112
Rockwell hardness tester (Image source: De Martin)

However, this method is only suitable for thinner and brittle coatings. Thicker or more ductile coatings only show a meaningful result in exceptional cases. In these cases, other methods are more likely to be used. The rather subjective nature of the adhesion assessment has repeatedly led to approaches for objective quantification [2, 4].

4.13.2 Thermal Shock Test

The thermal shock test is a test method for evaluating the resistance of materials to extreme temperature fluctuations. The test consists of quickly bringing the material from a high temperature to a low temperature to check whether the material shows cracking or other damage. The thermal shock test is often used to evaluate the quality of coatings, plastics, ceramics, and other materials used in environments with rapid temperature changes. The test is usually performed as follows:

- The material is heated to a high temperature, typically around 150 °C (aluminum) to 250 °C (steel), depending on the material.
- The material is then quickly cooled to a low temperature, typically to around 0 °C to –40 °C. Cooling is achieved by immersing the material in water, a cold bath, or by spraying liquid nitrogen onto it.
- The material is then checked for cracks, deformations, or other damage caused by the rapid temperature changes.

4.13.3 File Test

The file test is used in particular for testing thicker and more ductile coatings. The surface of the coating is processed with a file, and the behavior of the interface between the coating and the base material is observed.

A file with a defined roughness depth profile and grain size is placed on the surface of the coating. The file is pulled with a certain pressure on one edge at a defined angle against the surface normal of the coating. In the event of poor coating adhesion, a gap is created at the interface.

The file test for evaluating coating adhesion is a simple and quick method for assessing the quality of coatings. However, the test is not particularly accurate, and there are other methods that can be used for a more comprehensive assessment of the adhesion of coatings, such as the scratch test or the cross-cut test.

4.13.4 Cross-Section Method

This method is used when the adhesion is not clearly good or bad and the strength of the interface needs to be evaluated. The effort involved is very high and the component or tool is completely destroyed by the necessary preparation. A metallographic section is prepared and a hardness impression is made into the interface between the coating and the substrate, using a small load Vickers tester. The propagation of the resulting crack then shows the quality of the interface bond strength.

References for Section 4.13

[1] Wielage, B.; Lampke, T.; Müller, T.; Nestler, D.: Verifizierung numerischer Verfahren zur Modellierung abrasiver Verschleißprozesse durch Berechnung von Scratchtests, *Materialwissenschaft und Werkstofftechnik*, 2008, 39: 963–966

[2] DIN EN ISO 4856: Carbon-based films and other hard coatings – Rockwell penetration test to evaluate the adhesion, Beuth-Verlag, 2018

[3] ISO 26443:2008: Fine ceramics (advanced ceramics, advanced technical ceramics) – Rockwell indentation test for evaluation of adhesion of ceramic coatings (German version)

[4] Gäbler, J.; et al.: Rockwell-Schichthaftungstest – Der maschinelle Blick, *Journal für Oberflächentechnik*, 2021, 9: 69

5 Application of Functional Coatings

5.1 Hard Coatings

Marko Gehlen

5.1.1 Introduction

Injection molds are subject to constant wear. Due to the cyclical process, in which the melt is injected into the mold at high speed under high pressure and temperature, the surface structure is stressed with every cycle. In addition, due to added abrasive fillers such as glass fibers, this effect is further intensified. It is therefore common nowadays to apply coatings to the mold inserts, which have a protective function due to their physical and chemical properties. In order to improve wear protection, hard coatings can be applied to the molding tools, for example, using the hardness of certain materials to minimize material removal from the tool and extend its service life. Physical vapor deposition (PVD) is a common process for coating. Depending on the application, however, chemical vapor deposition (CVD) can also offer advantages.

5.1.2 Definition and Properties of a Hard Coating

The hard materials considered here are primarily carbides and nitrides, i.e. compounds of metals such as tungsten, titanium, and chromium with carbon or nitrogen. The compounds are characterized by a very high hardness. Today, the established process of PVD in a vacuum is used to coat many machining tools as well as numerous injection mold cavities. For complex molding tool geometries with high aspect ratios or undercuts, CVD, which also takes place in a vacuum, is also suitable as an alternative. Typical coating hardnesses for both processes are around 2000–3000 HV. However, very well-adhering and very smooth coatings can also achieve hardnesses of over 4500 HV [1]. The coating thicknesses are often in the lower single-digit micrometer range. The problem with thicker coatings is that higher residual stresses in the coating system can lead to failure of the coating in the form of spalling.

5.1.3 Areas of Application

The most common hard coating, which is also used in many areas of plastics processing, is a gold-colored coating made of titanium nitride (TiN). In recent decades, however, the range of coating formulations has grown steadily, so that today there are a large number of functional coatings, of which the hard coatings play an important role due to the high stresses in the respective processes.

Tools for machining processes such as drilling, milling, or turning are always exposed to high mechanical loads and are subject to wear on the edges and running surfaces, which can be significantly reduced by such coating systems or hard coatings. When machining soft materials such as aluminium, adhesion (welding or sticking) can occur during machining. The hard coating can prevent this, as it does not form a bond with the soft material.

5.1.4 Requirements and Coating Structure

In general, the application of hard coatings always involves the risk that the applied coatings do not harmonize with the substrate, i.e. that, for example, the adhesion is is not sufficient or that, as described in Section 5.1.2, there is too much residual stress in the bond due to a coating thickness that may be too high. Another reason could be that the substrate is much softer than the applied coating system. In this context, reference is often made to the “eggshell effect” (see also Section 5.1.6). This refers to local overloads that lead to a point stress load on the coating and the substrate, which in turn causes the hard coating material to flake off or crack [2]. The coating system must therefore always be matched to the respective substrate. This suggests that layer developments often take a long time, as the number of possible parameters to be varied is very large.

5.1.5 Process for Applying Hard Coatings

CVD is an essential technology for the deposition of thin hard coatings. The CVD process is shown schematically in Figure 5.1 [3]. The CVD process is therefore the preferred method and the basis for tool coating in plastics processing. CVD has very good properties with regard to deep slits, which appear as ribs on the molded part, as well as undercuts and complex geometries. The reaction gas has a good gap mobility which, depending on the flow conditions, enables aspect ratios much higher than 1:10 to be coated. The result is a good and uniform coating of the surface, and this has also been demonstrated in previous work [4].

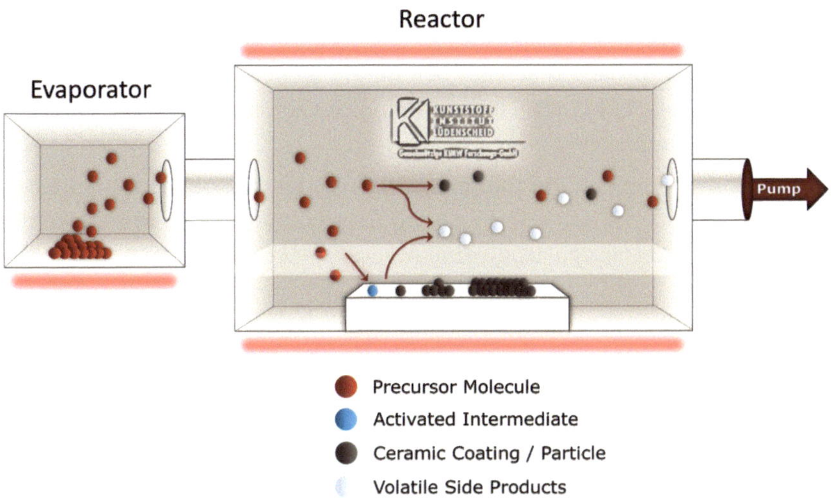

Figure 5.1 Schematic representation of the CVD process

Figure 5.2 shows that areas in the "depth", which are additionally shaded by the helical geometry, are also coated.

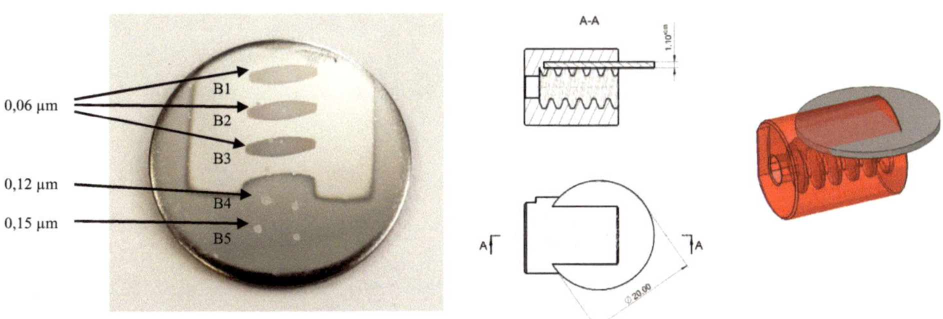

Figure 5.2 Investigation of the 3D capability of the process control

Important precursors and process parameters were researched in the literature as part of the funded AbraCoat project. Important preliminary investigations were carried out at the Texas Materials Institute at the University of Texas in Austin by S. Y. Lee and others as early as 2001 [5]. There, tungsten carbide was deposited on copper at low temperatures as a diffusion barrier. These low temperatures are also of great importance for the tools considered here, as no structural changes to the steel are to be expected in the range up to 350 °C, and therefore the mechanical properties are retained.

5.1.6 Characteristic Values for Evaluating Wear Resistance

There are various characteristic values for determining the wear resistance or the hardness of thin coatings of hard material. The methods for this were described in more detail in the previous chapters. The test methods that are important for wear resistance are briefly summarized below.

From the crater grinding, the volume that has been abraded can be calculated using the determined diameter of the ball grinding point. This volume can be equated to the equivalent unitless abrasion comparison value AV.

Two common methods are often used to assess the adhesion of a thin coating to a substrate. One is the adhesion assessment using the Rockwell hardness test, where the characteristics of the surrounding coatings are used to give a rating between 1 (very good adhesion) and 5 (poor adhesion, "eggshell effect") according to the school grading system (Figure 5.3).

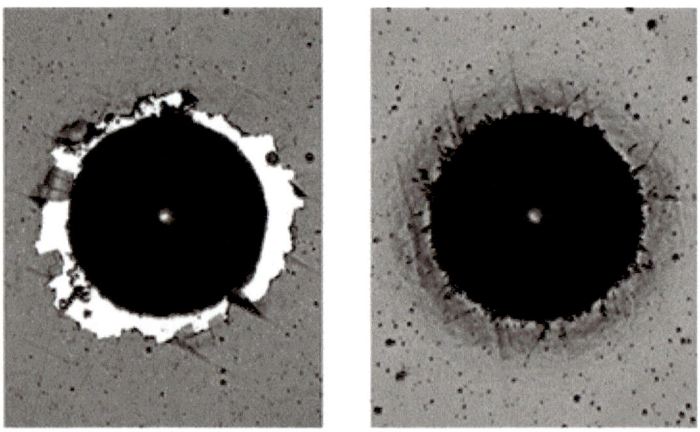

Adhesion class 5 Adhesion class 1

Figure 5.3 Measurement of coating adhesion by means of HRC hardness test

In the other method, a so-called scratch tester is used to measure properties such as the adhesion of the coating, cracking, or delamination effects. The results are the three critical normal forces at which the defects occur:

- LC1 (cracking)
- LC2 (spalling)
- LC3 (penetration of the coating or complete delamination).

When measuring the coating hardness, a hardness indentation test using mN-indentation enables precise hardness measurement on thin coatings. The hardness of the thin coating can be calculated on the basis of the characteristic loading and unloading curve and the determination of the resulting plastic or elastic deformation (Figure 5.4).

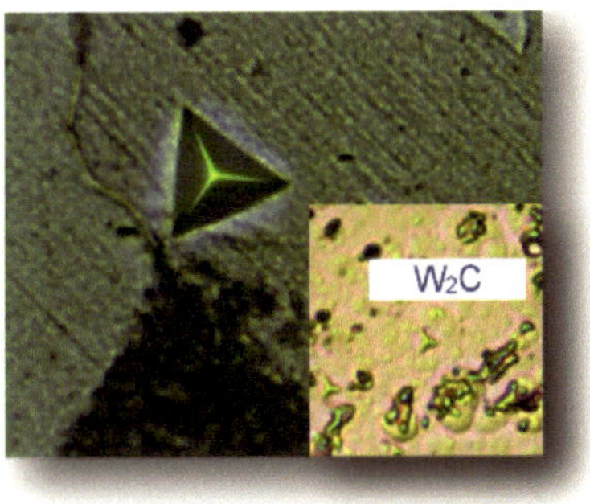

Figure 5.4
Hardness indentation of a "Berkovich" test specimen with a test load of 50 mN

5.1.7 Achieved Abrasion Comparison Values and Hardness

The tungsten carbide coating (W_2C) developed during the AbraCoat project shows a higher degree of adhesion than the PVD thin coatings established on the market, which is a very good result (Figure 5.5). It should be noted here that the CVD coating, whose thickness on the test specimens at AbraCoat was in the lower single-digit µm range, can also be deposited uniformly in deeper areas of the cavity, and does not suffer from shadowing effects like the PVD coatings produced using the directional process. In the case in question, the W_2C coating was deposited as a monolayer.

The hardness of 2000 HV is very high, although available PVD coatings sometimes significantly exceed this value. The coatings values for abrasion and hardness of the PVD coatings (blue) shown in Figure 5.5 correspond to the specifications of the relevant coaters. As the coating thickness is analogous to the coating time, coatings are deposited as thinly as necessary, so that they are usually in the order of 1–3 µm. Depending on the application, it may deviate from this.

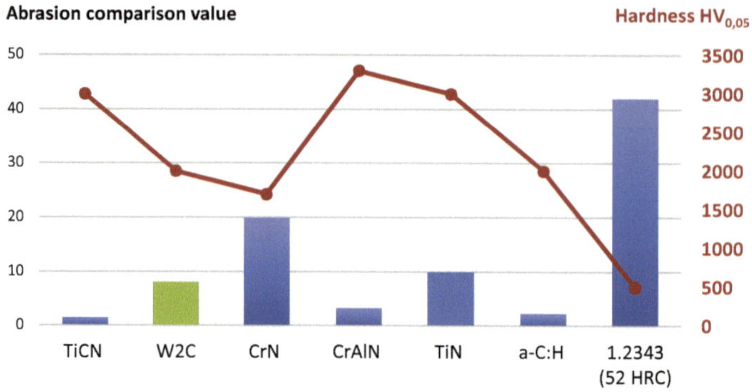

Figure 5.5 Abrasion resistance and hardness of tool coatings

It is also worth noting that hardness and wear are are not always in opposition. For example, the titanium nitride layer (TiN) shows higher abrasion than, for example, the hydrogen-containing amorphous carbon layer (a-C:H), but the hardness determined is higher. Therefore, the principle "the harder, the more wear-resistant" does not always apply.

5.1.8 Summary

The aim of the ZIM project AbraCoat, carried out by the non-profit KIMW Forschungs-GmbH, was to develop and produce wear-resistant coatings for tool inserts using metalorganic chemical vapor deposition (MOCVD).

We have succeeded in depositing tungsten carbide coatings (W_2C) that have relevant properties for protecting molding tool surfaces against wear. With the focus on complex geometries, which is often the case in plastics technology, great progress has been made here. The hardness values of the thin coatings have the potential to be further increased through research and development.

References for Section 5.1

[1] Schade, C.; Käszmann, H.: Hartstoffschichten – Verschleißschutz der höchsten Klasse, *WOTech*, 2014

[2] Weigel, K.: DLC-Beschichtungen für die Umformung [Annual report], Fraunhofer Institute for Surface Engineering and Thin Films IST, 2017, pp. 46–47

[3] Berger, K.-F.; Kiefer, S.: Neuer Verschleißschutz für Spritzguss-Werkzeuge [Jahrbuch 2023], Isgatec, 2023, pp. 170–175

[4] Fornalczyk, G.: 3D CVD [Annual report], Non-profit KIMW Forschungs-GmbH, 2017, p. 12

[5] Lee, S. Y.; et al.: Low temperature chemical vapor deposition of tungsten carbide for copper diffusion barriers, *Thin Solid Films*, 2001, 397: 109–115

5.2 Tribological Coatings and Wear Protection Coatings

Dr. Orlaw Massler

5.2.1 Types of Wear

Wear is a natural process in machine components that occurs due to stress in the movement system of the parts. Wear can lead to unexpected failure of the machine, which in turn can lead to production downtime and expensive repairs. Therefore, it is important to understand the different types of wear in order to take effective measures to prevent or reduce wear [1].

Abrasive wear

Abrasive wear occurs when hard particles, such as sand or dust, hit the surface of the machine part and wear the surface through friction and abrasion. This type of wear is common in machines that are used in dusty or sandy environments, such as construction machinery or agricultural machinery.

Adhesive wear

Adhesive wear occurs when two metal parts adhere firmly to each other due to friction. During movement, the parts separate from each other, causing small particles of the surface to be torn off. This wear often occurs in highly loaded gears or bearings.

Fatigue wear

Fatigue wear occurs when machine parts gradually wear out due to the stress caused by constant movement and loading. This eventually leads to the parts breaking or cracking. Fatigue wear often occurs in parts that are subjected to high loads, such as gears, shafts, or springs.

Tribooxidation

Tribocorrosion occurs when chemical reactions occur between the material of the machine part and the medium in which it is located. In the simplest case, this is oxidation, which leads to the formation of fretting corrosion on the rubbing steel surfaces. The result is often abrasive wear, which then leads to further destruction.

5.2.2 Friction Reduction

Friction between moving surfaces is an important factor that influences the wear of machine components. High friction can lead to increased wear, which in turn can lead to unexpected machine failure. Therefore, the reduction of friction is an important factor that influences the performance and service life of the machine components.

There are various methods to reduce the friction of machine components. One option is to use lubricants such as oils or greases, and another is to use more resistant materials that have a higher sliding capacity.

The reduction of friction also has an impact on the environment and the economic efficiency and therefore the sustainability of systems. By reducing friction, less energy is required to operate the machine, which can lead to a reduction in energy consumption and therefore a reduction in environmental impact. In addition, the reduction in friction can reduce maintenance costs and downtime, which leads to greater economic efficiency of the machine.

Overall, the reduction of friction is of great importance for the performance and service life of machine components. By applying methods to reduce friction, machines can be operated more effectively, resulting in higher productivity, energy efficiency, and cost-effectiveness.

Figure 5.6 shows common coatings and the (dry) coefficient of friction obtained against an uncoated steel surface, plotted against the coating hardness. Depending on the application, there are many ways to reduce friction and wear in the long term. It must be clear that friction as a characteristic value is not a coating property, but a behavior that results from the tribological system.

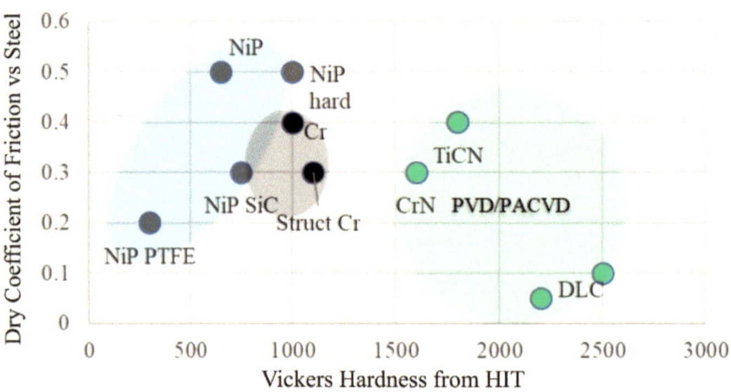

Figure 5.6 Dry friction coefficients versus hardness (vs steel) for different coatings

5.2.3 Galvanic Coatings

Electroplated (galvanic) coatings are used in injection molding tools to improve properties such as wear resistance, corrosion resistance, and hardness. The tribological requirements here are diverse, as are the solutions.

5.2.3.1 Hard Chrome Plating

Hard chrome plating is an electroplating process in which a coating is applied to the surface of the tool to increase its hardness and wear resistance. It forms a dense, hard coating on the molding tool surface that is highly resistant to abrasion and corrosion. In addition, this coating exhibits favorable adhesion behavior against many plastics in direct melt contact. Hard chrome plating is a relatively simple process that is indispensable in many industrial sectors. Figure 5.7 shows an electroplated hard chrome coating.

Figure 5.7 Hard chrome coating (Image source: DeMartin)

Hard chrome can be provided with a tribologically active topography through special process control, and can range from a pronounced crack structure to a fine spherical structure (see Figure 5.8) [2, 3].

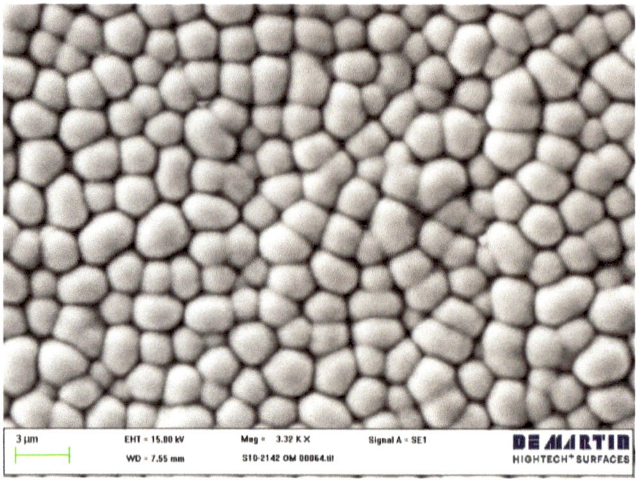

Figure 5.8 Pearl chrome structure (Image source: DeMartin)

The surface produced in this way can also be provided with additional topcoats to further improve the friction and wear behavior. On the other hand, a hard chrome coating can be used as a topcoat for an electroless nickel or dispersion coating sequence (see the following section). The hard chrome process is based on Cr(VI), which is highly regulated due to its health effects on the human body. The end product, on the other hand, is harmless. There have been and continue to be efforts to replace hard chrome with alternative coating systems. Where this has not yet been achieved, the supply chain must be aligned regionally accordingly or a corresponding permit for production must be obtained.

5.2.3.2 Nickel Plating

Nickel plating is an electroplating process whereby metallic nickel can be applied to the surface of the tool to increase corrosion resistance. Nickel plating can be produced in a bright or matt finish and primarily offers good resistance to corrosion. Due to its limited hardness, electroplated nickel coating has rather low wear resistance and is therefore primarily used as corrosion protection, often in combination with other coating processes that provide improved wear resistance.

The choice of electroplating coating depends on the specific application and the requirements for hardness, abrasion resistance, corrosion resistance and other properties. It is important to select the coating carefully and ensure that it meets the requirements of the application in order to achieve the best possible results.

5.2.4 Electroless Nickel and Dispersion Coatings

The deposition of a disperse[1] coating, consisting of a metallic matrix (binding phase) and uniformly distributed solid particles (insoluble in the electrolyte) bound therein, opens up the possibility of synergistically combining the properties of the matrix material and embedded particles [4, 5]. The appropriate choice of coating depends on the specific requirements of the tool surface. All essential requirements for the surface under consideration must be taken into account. Table 5.1 shows an overview of the different types and areas of application of the most important electroless nickel-based coatings.

Table 5.1 Areas of Application for Electroless Nickel-Based Coatings (Source: DeMartin)

Application/coating property	Ni-P Lowphos	Ni-P Midphos	Ni-P Highphos	Ni-P-PTFE Ni-P-hBN	Ni-P-SiC Ni-P-Diamond Ni-P-BC
Wear resistance, moderate attack	Y	Y		Y	Y
Wear resistance, severe attack					Y
Corrosion resistance (ASTM B117)			Y		
Solderability	Y	Y			
Friction coefficient reduction/anti-adhesive				Y	
Friction coefficient increase					Y
Resistance to alkaline media	Y	Y			
Resistance to acidic media		Y	Y		
Optics/gloss		Y			
Machinability Turning			Y		
Non-magnetic			Y		

[1] In this context, we are referring to a composite coating in the form of a solid/solid dispersion, in contrast to a layered composite material.

5.2.4.1 Dispersion Coating

These coatings consist of an electroless nickel coating in which poly(tetrafluoroethylene) (PTFE) particles are embedded. PTFE is a well-known non-stick agent for a wide range of applications. The use of PTFE dispersion coatings on injection molding tools reduces friction, reduces the formation of plastic deposits on the tool surface, and increases the service life of the tool.

5.2.4.2 SiC Dispersion Coatings

These coatings consist of an electroless nickel layer in which silicon carbide (SiC) particles are embedded. SiC is a very hard mineral, which results in high wear resistance. The use of SiC dispersion coatings on injection molds increases the hardness and wear resistance of the mold surface.

5.2.4.3 BC Dispersion Coatings

The incorporation of hard boron carbide particles in an electroless nickel matrix also leads to significantly improved wear resistance. Experience has shown that the round shape of these particles leads to less wear on the mating body in the tribological system.

5.2.4.4 hBN Dispersion Coatings

Hexagonal boron nitride is widely used as a solid lubricant and is also suitable as a functional carrier for reducing friction in a tribological system. Experience has shown that the deposition rate is lower than that of other particle types. On the other hand, the temperature resistance is higher than that of PTFE, which is widely used. It is also possible to combine hBN with other particle types in the same system.

5.2.5 Tribological PVD and PACVD Coatings

PVD and PACVD (plasma-assisted chemical vapor deposition) are of particular importance due to the versatility of the processes and the flexibility of the properties of the coatings produced. In particular, the combination of extremely wear-resistant materials and excellent friction behavior with other coatings is finding more and more applications.

Some important representatives are listed here very briefly:

- **TiN (titanium nitride)** is a widely used coating that has high hardness, high wear resistance, and good corrosion resistance.
- **TiCN (titanium carbonitride)** is a combination of titanium, carbon, and nitrogen and has similar tribological properties to TiN, but is somewhat harder and more resistant to abrasive stress, with lower coefficients of friction.

- **CrN (chromium nitride)** is a coating that has good corrosion resistance and hardness, and is particularly suitable for applications where high thermal stability is required.
- **AlCrN (aluminium chromium nitride)** is a coating that has similar properties to CrN, but also has increased oxidation resistance and thermal stability.
- **DLC (diamond-like-carbon)** coatings are a particularly important group of materials. A DLC coating is a type of diamond-like carbon coating that is usually applied using PACVD or PVD. DLC coatings are characterized by their high hardness, wear, and corrosion resistance, and offer low friction and excellent non-stick properties.

  DLC coatings consist of amorphous carbon and usually have a thickness of 1–5 µm. They are often used in applications where high demands are placed on wear and corrosion resistance, such as in the automotive, aerospace, and medical food industries [6]. These coatings are widely used and have become an integral part of all wear protection and tribological applications. The structure and properties can be specifically influenced within wide limits through process variations.

5.2.6 Hybrid Coatings

Hybrid coatings are used in injection molds to further improve the properties of the mold surface. These are coatings that consist of a combination of two or more materials, and thus combine the positive properties of the individual materials [7]. Some examples of hybrid coatings that can be used in injection molds are presented below.

The choice of hybrid coating depends on the specific requirements of the mold surface. Hybrid coatings can be a cost-effective way to improve the properties of injection molds and extend mold life. Table 5.2 shows an overview of commonly used hybrid coatings.

Table 5.2 Overview of Hybrid Systems

Hybrid system	Structure	Field of application
Hybrid electroplating – electroless nickel	Electroplating base layer (Ni) Topcoat electroless nickel Optional dispersion layers	Combination of wear protection and corrosion protection
Hybrid hard chrome layers	Hard chrome base layer, tribologically active top layer	Combination of wear protection and friction reduction
Electroless nickel Electroless nickel dispersion layers	Electroless nickel base layer(s) Dispersion-type topcoat (SiC, PTFE, BC, diamond)	Combination of wear protection and corrosion protection/friction reduction

Table 5.2 Overview of Hybrid Systems *(continued)*

Hybrid system	Structure	Field of application
Duplex PVD	Plasma nitriding – PVD	Combination of wear protection and friction reduction
Chemical Ni–PVD	Electroless nickel base layer(s) PVD-type topcoat (CrN, TiCN, TiN, AlCrN)	Combination of wear protection and corrosion protection
Chemical Ni–DLC	Electroless nickel base layer(s) DLC-type topcoat	Combination of wear protection and corrosion protection
Chemical Ni–PVD–DLC	PVD DLC-type topcoat	Combination of wear protection and corrosion protection
Hard chrome DLC Pearl chrome DLC	Hard chrome base layer(s) DLC-type topcoat	Combination of wear protection and corrosion protection

The appropriate choice of hybrid coating depends on the specific requirements of the mold surface and the plastics to be processed. Hybrid coatings of electroless nickel and DLC are a cost-effective way of improving the properties of injection molds and extending the service life of the tools. The main objective of a hybrid coating is to combine the corrosion resistance of the thicker base layer, which is easy to produce, with a topcoat that has excellent hardness and wear resistance. This is particularly popular in toolmaking because this coating combination can also be applied with great precision.

The hybrid approach to coating offers many advantages. The functionalization of the surface can be tailored across a wide range by using suitable coating systems. The properties that can be achieved are not attainable with individual systems.

Ni–Cr hybrid

This coating consists of an electroplated or electroless nickel underlayer and a hard chrome top layer. The electroless nickel underlayer serves as an adhesion agent and offers good corrosion resistance, while the hard chrome top layer offers high hardness and wear resistance. Ni–Cr hybrid coatings are often used in applications where high mechanical loads and corrosion occur, such as in the food industry and in the chemical industry.

Plasma nitriding–PVD–DLC

Here, a plasma nitriding process is used to introduce a coating-friendly base material hardening, which gives the thin coating better load-bearing resistance. The disadvantages of this variant are the need for a nitridable steel and the high process temperatures.

Chemical Ni–DLC hybrid coatings

These coatings consist of a base layer of electroless nickel and a DLC topcoat. This system offers high wear resistance, low friction, and corrosion resistance. The Ni–DLC hybrid coating is particularly suitable for applications in plastics processing where high demands are placed on the service life and wear resistance of the injection mold.

Ni–SiC–DLC hybrid coatings

This coating consists of an electroless nickel coating, an electroless nickel-silicon carbide (SiC) dispersion coating, and a DLC topcoat, and offers high hardness, wear resistance, and low friction. The Ni–SiC–DLC hybrid coating is particularly suitable for applications in plastics processing where high demands are placed on the wear resistance and chemical resistance of the injection molding tool.

References for Section 5.2

[1] Rabinowicz, E.: *Friction and Wear of Materials* (2nd edn), John Wiley & Sons, 1995

[2] Podgornik, B.; Massler, O.; Kafexhiu, F.; Sedlaček. M.: Crack density and tribological performance of hard-chrome coatings, *Tribology International*, 2018, 121: 333–340

[3] Massler, O.: Reibungsreduktion durch strukturierte Multilagen-Beschichtungen, *WOMag*, 2022, issue 09

[4] Sörgel, T.; Meyer, J.: Chemische und elektrochemische Dispersionsbeschichtung, *WOMag*, 2013, issue 09

[5] Meyer, J.: Chemische Nickel-Dispersionsschichten in Verschleißschutzanwendungen, *WOMag*, 2015, issue 11

[6] Massler, O.: Diamond-like coatings protection for engine applications, *AutoTechnology*, 2005, 5: 68–71

[7] Massler, O.; Meyer, M.: Neue Wege für funktionelle Oberflächen, *WOMag*, 2018, issue 05

5.3 Corrosion-Protective Coatings

Dr. Anatoliy Batmanov

5.3.1 Definition of Corrosion

Corrosion is the physico-chemical reaction of a (usually) metallic surface with its environment. This results in an undesirable change in properties that negatively affects the function of the corresponding component due to corrosion damage [1].

In the case of chemical corrosion, the metals react directly with the oxygen in their environment and release electrons in the process. The higher the temperature, the faster the reaction takes place. This reaction is also known as dry corrosion [2].

In contrast, in wet corrosion or electrochemical corrosion, several reactions take place spatially separated from each other, whereby an exchange of electrical charges takes place. An electrolyte is required to exchange the electrical charges. This is usually present as an aqueous solution and is more aggressive the more concentrated the ions are in the aqueous solution. The solution tension, i. e. the different solution pressure of the metals, results in a galvanic half-cell. An anodic reaction (Equation 5.1) and a cathodic reaction (Equation 5.2) take place simultaneously at the interface between the metal electrode and the electrolyte in this half-cell [2].

$$\text{Metal atom} \longrightarrow \text{Metal ion} + \text{Electron} \tag{5.1}$$

$$\text{Metal ion} + \text{Electron} \longrightarrow \text{Metal atom} \tag{5.2}$$

In the case of base metals, electrons remain at the electrode, resulting in an excess of electrons. The electrolyte is positively charged by the lack of electrons until an equilibrium potential of the metal–metal ion reaction is formed.

With noble metals, the reaction takes place in the opposite direction, so that the electrode becomes positively charged and the electrolyte negatively charged. Both half cells together form a galvanic element when they are conductively connected to each other. The less noble metal then becomes the anode and the more noble metal becomes the cathode. As the metals are electrically conductively connected and are wetted with the electrolyte, a state of equilibrium cannot be established, so that the corrosion of the anode continues to progress. If the electrolyte is very acidic, e. g. in the case of an acid attack on metals, the excess electrons at the cathode are used to reduce hydrogen. However, the reduction of oxygen from the air, in combination with water from the electrolyte, is the much more common case, as the electrolyte is usually not acidic [2].

A relatively unusual form of corrosion is hot gas corrosion, which is a form of chemical corrosion, and in which a reaction takes place between the hot gas and the metallic surface. The corrosion product is layers on the metallic surface, which can continue to grow through diffusion [1].

In principle, corrosion can occur at various points on a surface and take various forms. The most frequently occurring forms are surface corrosion, pitting corrosion, and crevice corrosion (Figure 5.9).

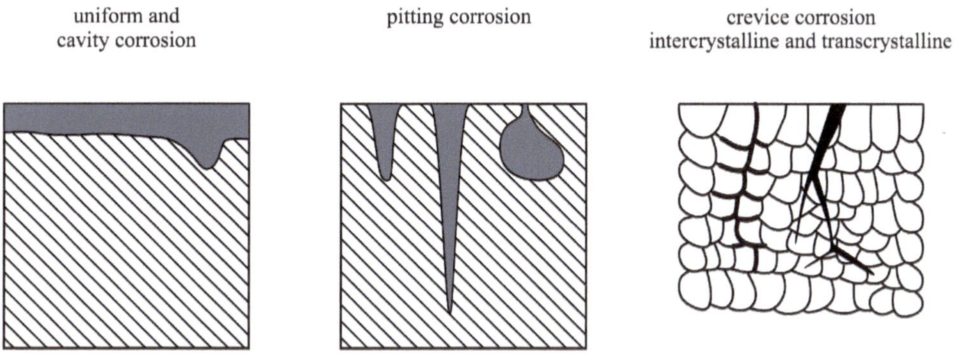

Figure 5.9 Various forms of corrosion (Image based on [2])

In the case of surface corrosion, the surface is corroded evenly, although the formation of depressions is possible if the corrosion progresses unevenly. The removal of the anodic metal is relatively slow so that this form of corrosion can be detected at an early stage and countermeasures can be taken.

In contrast to surface corrosion, in pitting corrosion localized indentations develop, which are needle-shaped and undermine the surface. As few reaction products occur here, pitting corrosion is difficult to detect, and is usually detected when leaks occur.

Like pitting corrosion, crevice corrosion is also difficult to detect. It is the most dangerous form of corrosion, as it leads to stress peaks under mechanical load due to the notch effect. In combination with the reduction in cross-section caused by the crack, the mechanical load can lead to failure of the component. Crevice corrosion can run along grain boundaries (intercrystalline) or through the grains (transcrystalline).

The elements involved in a galvanic cell can be very small. The individual grains in a microstructure can be formed from different components of an alloy, so that different potentials arise and local corrosion can occur at the grain boundaries, as shown in Figure 5.10. This is often caused by the fact that different elements preferentially accumulate at the grain boundaries, so that these are preferentially dissolved in the course of corrosion [2].

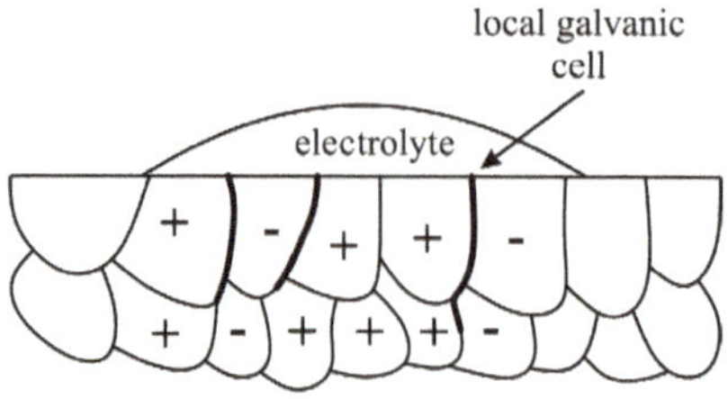

Figure 5.10
Local galvanic cells at the grain boundaries in a heterogeneous crystal structure (Image based on [2])

The appearance of corrosion always depends on the material and the cause of corrosion. An unfavorable choice of material can, for example, lead to different metallic elements being directly connected to each other. The design can also favor the accumulation of electrolytes. In many cases, the occurrence of electrolytes during the product's service life cannot be prevented. Incorrect heat treatment can also lead to unfavorable structural conditions, as can incorrect processing of a material, for example through work-hardening.

5.3.2 Basic Strategies for Preventing Corrosion

Corrosion protection can be taken into account at all points along the value chain and incorporated into the design of components. During product development, materials can be selected to ensure that suitable materials and material combinations are used that are less vulnerable to corrosion. During production, manufacturing processes that increase the tendency to increase corrosion must be avoided. During operation, the choice of coolants and lubricants, regular maintenance, and proper storage have a decisive influence on corrosion [2].

In addition, specific measures can be taken to optimally prevent or delay corrosion. A distinction is made between active and passive corrosion protection. Active corrosion protection directly intervenes in the corrosion reaction by changing the electrolyte, applying sacrificial anodes, or favorably shifting the potential using external voltages. In this way, the oxygen content in a cooling medium can be significantly reduced in order to slow down corrosion in cooling channels. The use of a sacrificial anode the actual anode, i. e. the material to be protected, becomes the cathode [2].

The passive methods of corrosion protection are aimed at separating the surface from the electrolyte. For example, the alloy components nickel and chromium in stainless steels form a dense oxide layer, and this passivation increases the corrosion protection. If the surface is damaged, a new oxide layer forms in the damaged area. However, the oxide layer can be destroyed by halogen-containing compounds [3].

The application of organic coatings, such as wax, grease, or plastic, and the application of ceramic coatings, are also among the passive methods of corrosion protection. The application of metallic coatings is also commonplace. However, attention must be paid to the impermeability of the coating. If a coating made of a less noble material, e. g. zinc, is applied to a steel surface, it acts as an anode when damaged and is slowly removed. The corrosion products are deposited on the surface and slow down the reaction. If a more noble material is used as a coating, e.g. nickel or chromium, the steel becomes the anode, so that pitting corrosion develops when the coating is damaged [2].

5.3.3 Requirements for Corrosion Protection Coatings

The potential for reducing manufacturing costs and times for plastic molded parts and for increasing product quality should not be underestimated. This enables more economical production with higher part quality.

Above all, the materials used for injection molds must be easy to machine, wear-resistant, and corrosion-resistant to the plastic melts used. At present, there is no molding tool steel that meets these requirements. The use of corrosion-resistant steels is therefore limited, as they tend to be brittle and have poor thermal conductivity.

Plastics are generally not chemically aggressive. However, plasticizing and melting can cause the polymer chains to break down or additives to escape from the plastic, and these can contribute to non-aqueous corrosion. For example, during the processing of PVC, hydrogen chloride is released at temperatures above 165 °C. The hydrogen chloride reacts with water from the ambient air to form hydrochloric acid [4]. These chemically aggressive and corrosive waste products impair, above all, the performance and service life of the plasticizing unit and the injection mold.

Various strategies are therefore currently used to protect the cost-intensive injection molds from corrosion. Galvanic coating with nickel is the state of the art, but only high coating thicknesses (>15 µm) are effective. As the injection mold and the cavity surface have to meet a wide range of requirements, such as mechanical stress, precision, and thermal properties, there is hardly any scope with regard to a corrosion-resistant design. In injection molding technology, corrosion protection coatings are sometimes used on the surfaces of the mold cavity. By using these coatings, surface defects on the mold and the molded part resulting from the injection molding process can be concealed or avoided. At the same time, the best possible surface quality of the manufactured part can be achieved.

However, the limitations of conventional coating processes are problematic. For example, uneven coatings are deposited on complex surface geometries using galvanic processes. Complicated electrodes have to be manufactured in order to be able to apply a coating at all. This results in tip effects at the corners and edges of the mold components, which have a negative impact on the deposition [5].

Corrosion protection coatings that are applied without current are dependent on a dense, non-porous coating. They only protect steel parts with a tightly sealed coating. The most important representative of this type of coating is electroless nickel. A coating thickness of >15 µm is required for the nickel coating to provide reliable protection. This comparatively high coating thickness is associated with high costs and leads to technical disadvantages in terms of the precision of the plastic parts. PVD processes have the same disadvantages. In addition, the coating rates of the PVD process are comparatively low, and unavoidable errors in the coating structure result in a poorer protective effect. The ability to coat 3D structures such as holes or undercuts is also limited with PVD process [6].

One promising approach is the use of CVD. Here, the chemical composition of the coatings and the microstructures can be well-controlled during the coating process. Various coating systems based on borides, carbides, nitrides, and oxides can be applied uniformly, almost without defects and with good adhesive strength. The purity of the coating is also very high. However, the high coating temperature (normally over 900 °C) is problematic, because the coating of injection molds or inserts is not possible. Due to phase transformation in the molding, the quality of most tool steels will deteriorate and the mechanical properties will diminish. In addition, suitable precursors must be available with which the desired coating can be deposited.

As the precision of the molded part cavity must not be impaired, the thickness of the coating must be as low as possible. A single-digit micrometer value must be aimed for, which has an influence on the selection of suitable coating systems. Figure 5.11 shows an example of a current density potential measurement with an electroless nickel coating and an oxide ceramic coating. Despite the considerably lower coating thickness, the oxide ceramic coating has a lower current density, i.e. higher corrosion resistance, than the comparatively thick electroless nickel coating [7].

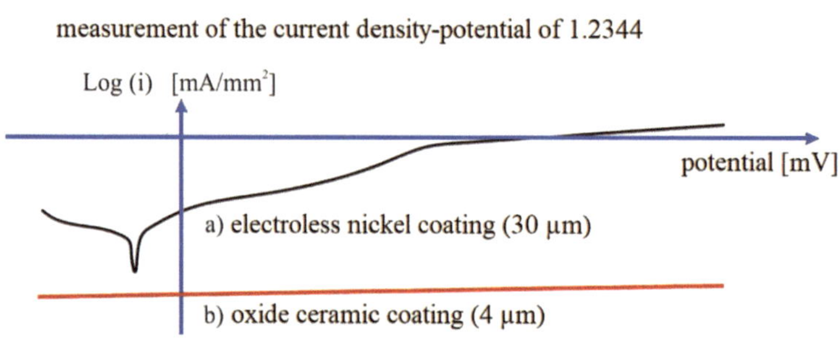

Figure 5.11 Current density potential measurement on 1.2344 steel with (a) a 30 µm electroless nickel coating compared to (b) a 4 µm ZrO_2/SiO_2 layer system. The electrolyte was H_2SO_4 (10%), the resting potential measurement was 3 min, and the potential scan rate was 10 mV/s (Image source: Non-profit KIMW Forschungs-GmbH)

In addition to sufficient corrosion resistance, a suitable corrosion protective coating must also have the ability to coat complex geometries. Good resistance to media is also necessary in order to minimize the coating thickness and at the same time achieve high corrosion resistance. Similar thermal properties as the substrate are required to prevent disintegration and corrosion due to abrasion.

The maximum deposition temperature is approx. 500 °C in order to prevent the structural transformation of the molding tool steel. Metalorganic chemical vapor deposition (MOCVD) is therefore used as the optimum coating process. As CVD generally offers good gap mobility, the coating of complex geometries is possible. Due to the wide range of precursor compounds, a large selection of coating materials is available so that different coating systems can be applied to counter various corrosion situations in plastics processing.

Electrochemical impedance spectroscopy (EIS) is a widely used analytical method that is often used to investigate the kinetics of chemical reactions [8] and to assess the quality of corrosion protection coatings [9, 10]. Basically, EIS provides the frequency-dependent current response of an electrochemical system to a voltage excitation. The interpretation of the data obtained is the real challenge of the method, because the physical or chemical process that has caused the observed spectra must be determined. In order to clarify this question, an equivalent electric circuit is generated whose behavior is simulated during the measurement. A theoretical impedance spectrum is calculated from the equivalent electric circuit and compared with the actual measured data. The parameters of the equivalent electric circuit diagram are adjusted in an iterative process so that the theoretical spectrum is approximated to the measured spectrum, as shown in Figure 5.12.

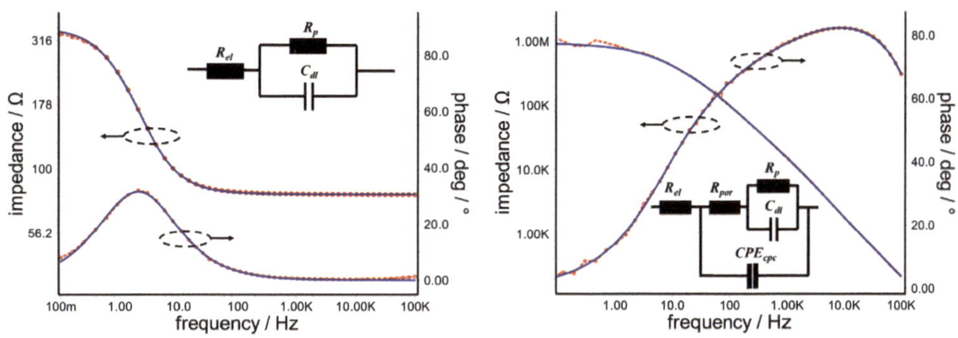

Figure 5.12 Example of a Bode diagram. Left: A bare sample of 1.0330 steel; Right: 1.0330 steel coated with Al_xO_y (red dots: measured; blue line: simulated)

The change in the protective coating in aqueous electrolytes is often investigated, which leads to the exposure of the protective substrate material and thus to corrosion. Changes in the protective coating are usually swelling processes or involve the formation of pores and channels. In some cases, the porosity of the coatings can also change [11]. The measurements can be carried out under real operating conditions so that the impedance data obtained can be directly related to practical processes. This requires a constant comparison of the results from other test methods in order to understand how the behavior of the coatings in an aqueous environment correlates with the resistance under injection molding conditions.

5.3.4 Development of a Corrosion-Protective Coating against Hot Gas Corrosion

To investigate the effect of hot gases on mold corrosion, a PFA plastic was processed in the injection molding process. Processing temperatures of 420 °C were used to sufficiently plasticize the plastic and force the degradation of the polymer chains. Figure 5.13 shows the coated mold insert and the manufactured demonstrator part. The injection mold is made of 1.4112 steel, which has good hardenability and good corrosion resistance. The mold inserts are made of 1.2343 steel, which is frequently used in plastics processing. It has good thermal conductivity and is easy to harden, but its corrosion resistance is poor. As part of the tests, 20 injection molding cycles were run to force corrosion of the affected surfaces. The polymer chain degradation can already be seen on the burnt plastic surface of the finished PFA part.

Figure 5.13
Left: Injection mold with coated mold insert. Right: Manufactured part with burning mark (Image source: Non-profit KIMW Forschungs-GmbH)

The ZrN coating deposited using the PVD process shows a typical decrease in coating thickness along the contour of the demonstrator. In contrast, the multi-layer CVD coating of phosphorus-doped zirconium oxide (P:ZrO) and yttrium-stabilized zirconium oxide (YSZ) has a significantly more uniform coating thickness. Subsequently, the molding tool surfaces were analyzed using a scanning electron microscope to assess their susceptibility to corrosion and the formation of plaques (Figures 5.14 and 5.15).

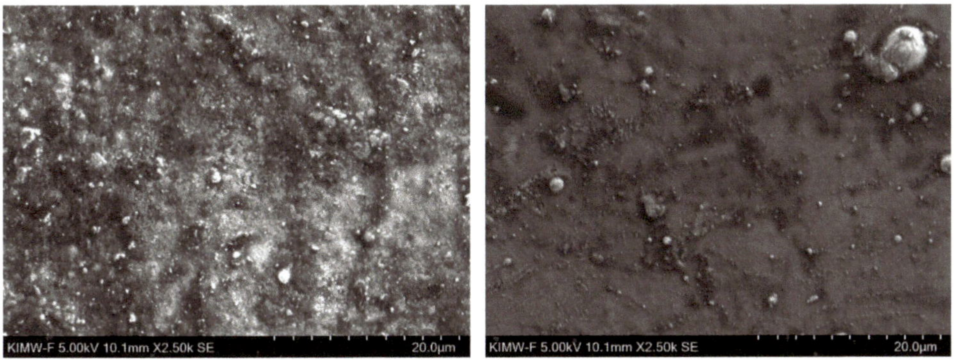

Figure 5.14 SEM examination of the inserts. Left: Uncoated 1.2343 steel; Right: 1.2343 steel + ZrN (Image source: Non-profit KIMW Forschungs-GmbH)

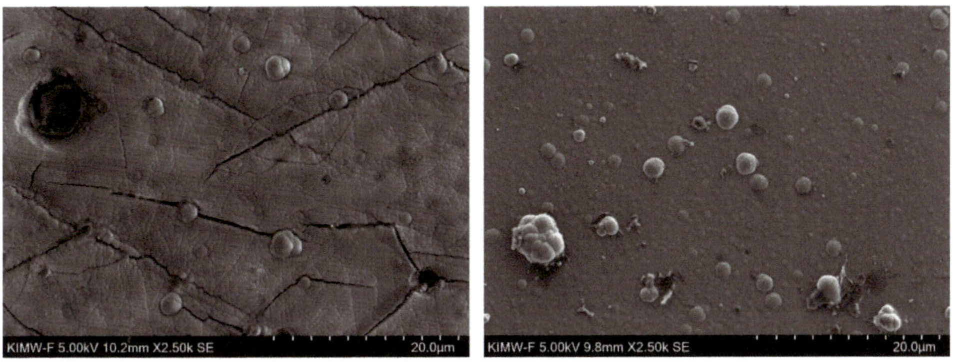

Figure 5.15 SEM examination of the inserts. Left: 1.2343 steel + P:ZrO/YSZ; Right: 1.2343 steel + $AlCrO_x$ (Image source: Non-profit KIMW Forschungs-GmbH)

The uncoated 1.2343 steel shows a massive corrosive attack of the surface due to the released fluorine-containing hot gases. The ZrN layer deposited by PVD protects the steel, but shows some plaques on the surface. The two metal oxide coatings show no changes on the surface, with no plaques visible on the $AlCrO_x$ coating either.

The transfer of the tested coatings was carried out by coating mold inserts for a cap made of PFA, which is used in the medical environment. In the series production process, the melt temperature of the PFA used is 350 °C and the cycle time is 61 s. The currently used chrome coating with a layer thickness of 20 µm serves as a benchmark. This has a service life of 20,000 injection molding cycles. The coated mold inserts each have a coating thickness of 0.6 µm and were operated with a maximum of 10,000 injection molding cycles with the phosphorus- and yttrium-doped zirconium oxide coating (P:ZrO/YSZ), and 1000 injection molding cycles with the $AlCrO_x$ coating. To analyze the coating failure the surfaces were examined using light microscopy and scanning electron microscopy (Figures 5.16 and 5.17).

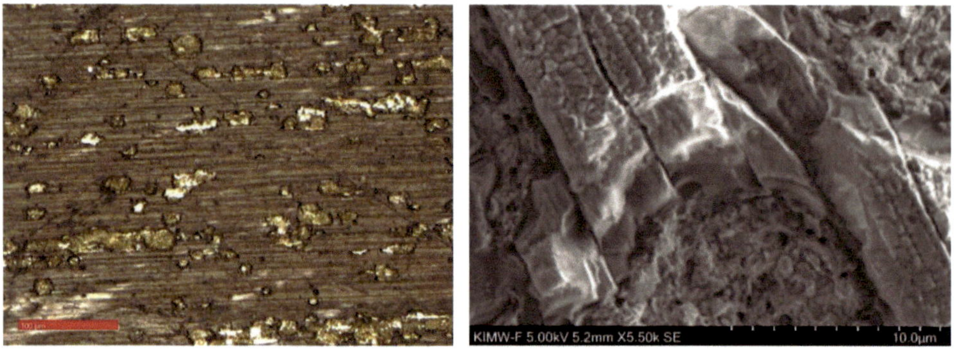

Figure 5.16 Images of P:ZrO/YSZ-coated mold inserts. Left: Light microscopy 1000×; Right: SEM 5000× (Image source: Non-profit KIMW Forschungs-GmbH)

Figure 5.17 Images of $AlCrO_x$-coated mold inserts. Left: Light microscopy 1000×; Right: SEM 5000× (Image source: Non-profit KIMW Forschungs-GmbH)

The reason for the failure of the ZrO/YSZ coating is the cracks integrated in the coating structure (Figure 5.16), which are due to the decomposition behavior of the precursor used. The fluorine released during the injection molding process penetrates through the vertical cracks to the base material and reacts there, forming fluoride and increasing in volume. The affected surface areas break out at certain points and cause the steel surface to become increasingly corroded.

In the case of the $AlCrO_x$ coating, a different mechanism causes the coating to fail. A large number of punctiform artifacts can be seen under the light microscope (Figure 5.17, left). With appropriate magnification, punctiform corrosion can be seen (Figure 5.17, right). This forms a diffusion channel for the released fluorine molecules. The corrosion process is clearly visible in the SEM image. The infiltrated coating is lifted by the corrosion products in the form of bubbles, and then flakes off the surface in a circle. The corrosion process is then intensified. In contrast to the ZrO/YSZ coating, coating growth defects are not the cause; rather, embedded coating particles are responsible for insufficient corrosion protection. These result from the coating process and are related to the process parameters used: temperature, pressure, and the precursor feed rate.

5.3.5 Development of a Corrosion-Protective Coating against Aqueous Corrosion

To investigate the corrosion behavior in an aqueous environment, the cooling system of an injection mold can be used. In the first step, the cooling channel geometry can be simplified to a demonstrator, which has an area through which the cooling medium flows and an area in which the cooling medium stands (Figure 5.18). The temperature of the cooling water used is 40 °C, and the test duration is one week. The test plates are made of 1.2311 steel and were coated by PVD with ZrN coatings of different thicknesses and by CVD with phosphorus-doped $AlCrO_x$. To differentiate and evaluate the corrosion resistance, the areas "contact", "flow-through" and "standing" are defined and evaluated (Figure 5.18, right).

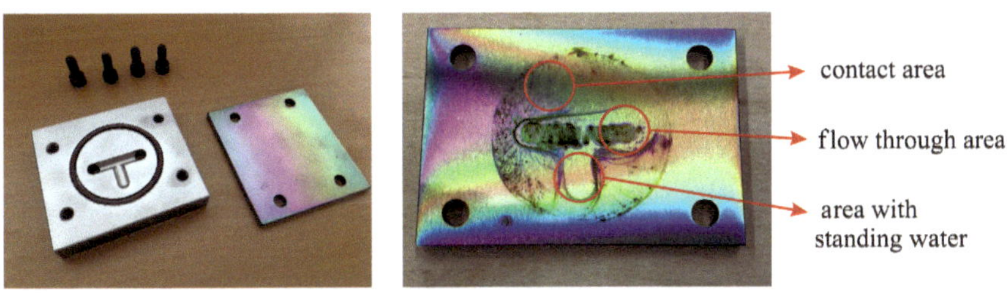

Figure 5.18 Cooling channel demonstrator. Left: Base body made of 1.2083 steel; Right: Coated test plate made of 1.2311 steel with marked corrosion areas (Image source: Non-profit KIMW Forschungs-GmbH)

The 1.2311 steel shows clear corrosion in all three contact areas (Figure 5.19, left). Pitting corrosion is also clearly visible in the flow-through area. The $AlCrO_x$ coating with a MOCVD-Ni adhesion promoter shows very little corrosion. Some corrosion spots are visible in the flow-through area (Figure 5.19, right). In further coating tests, the coating thickness of the $AlCrO_x$ coating could be reduced to 0.6 µm without reducing the corrosion protection effect.

Figure 5.19 Examination of the cooling channel demonstrator after one week of operation. Left: Uncoated 1.2311 steel; Right 1.2311 steel + 1.35 μm $NiAlCrO_x$ (Image source: Non-profit KIMW Forschungs-GmbH)

The ZrN coating deposited using the PVD process with a layer thickness of 1 μm exhibits severe corrosion in the "contact" and "standing" areas. Isolated corrosion spots can be seen in the "flowed through" area (Figure 5.20, left). The ZrN coating with a thickness of 4 μm shows no corrosion in the evaluated areas (Figure 5.20, right).

Figure 5.20 Examination of the cooling channel demonstrator after one week of operation. Left: 1.2311 steel + 1 μm ZrN; Right: 1.2311 steel + 4 μm ZrN (Image source: Non-profit KIMW Forschungs-GmbH)

References for Section 5.3

[1] Weißbach, W.; Dahms, M.; Jaroschek, C.: *Werkstoffkunde Strukturen, Eigenschaften, Prüfung* (19th edn), Springer Vieweg, 2015

[2] Bargel, H.-J.; Schulze, G.: *Werkstoffkunde* (12th edn), Springer Vieweg, 2018

[3] Kann Edelstahl rosten? [Technical information], MirrorINOX GmbH, *https://www.mirrorinox.de/edelstahl/kann-edelstahl-rosten*, accessed on 03 March 2023

[4] Sammt, K. Sammer, J.; Geckle, J.; Liebfahrt, W.: Development trends of corrosion resistant plastic mould steels [Presentation], *6th International Tooling Conference*, 2002, pp. 339–349

[5] Kurowski, A.: Elektrochemische und oberflächenanalytische Untersuchungen zur galvanischen und chemischen Nickel-Phosphor-Schichtbildung [Dissertation], Heinrich Heine University, Düsseldorf, 2003

[6] Cunha, L; Andritschky, M.; Pischow, K.; et al.: Performance of chromium nitride and titanium nitride coatings during plastic injection molding, *Surface and Coatings Technology*, 2002, 153: 160–165

[7] Investigation by Dortmund University of Applied Sciences and Arts as part of the joint project "Innovative Werkzeugoberflächen", 2013–2015

[8] Shao, H. B.; Wang, J. M.; He, W. C.; et al.: EIS analysis on the anodic dissolution kinetics of pure iron in a highly alkaline solution, *Electrochemistry Communications*, 2005, 7: 1429–1433

[9] Deflorian, F.; Fedrizzi, L.; Rossi, S.; Bonora, P. L.: Organic coating capacitance measurement by EIS: Ideal and actual trends, *Electrochimica Acta*, 1999, 44: 4243–4249

[10] Touzain, S.: Some comments on the use of the EIS phase angle to evaluate organic coating degradation, *Electrochimica Acta*, 2010, 55: 6190–6194

[11] Díaz, B.; Härkönen, E.; Maurice, V.; et al.: Failure mechanism of thin Al2O3 coatings grown by atomic layer deposition for corrosion protection of carbon steel, *Electrochimica Acta*, 2011, 56: 9609–9618

5.4 Thermal Barrier Coatings

Vanessa Frettlöh

5.4.1 Understanding Thermal Barrier Coatings

The term thermal barrier coating (TBC) generally refers to a coating that has a significantly lower thermal conductivity compared to the base material and thus insulates it to a certain degree from a heat source. Due to the TBC, the temperature of the heat source penetrates more slowly to the coated base material. Developed in the 1970s, these coatings are also used to protect metallic surfaces in very hot environments (e.g. gas turbines) from the high and prolonged thermal loads that are associated with oxidation and thermally induced fatigue without the protective layer, thus increasing the service life of the components [1].

5.4.2 Influence of Temperature in the Injection Molding Process

By selecting the process parameters, the progress of the injection molding process and thus also the appearance and quality of the injection molded components can be decisively influenced. By changing the temperature of the melt, the mold wall temperature, the injection speed, and holding pressure, the weight, the dimensions of the part, and the quality and properties of the part (gloss, internal stress, crystallinity, and molding of the surface) are influenced.

An increased melt temperature increases the flowability of the plastic, which reduces the injection pressure and allows a longer flow path to be achieved. However, the holding pressure and cooling time are also extended by this step. The increased flowability of the plastic reduces also the visibility of weld line notches [2].

By increasing the mold wall temperature at the same demolding temperature, the achievable flow path length can also be increased. The dimensions of the part, tolerance fluctuations, and the visibility of weld lines are reduced, and the molding of the surface is improved. Residual stresses in the part can be reduced by increasing the mold wall temperature. However, the increased mold temperature also increases the required cooling time. This is approx. 20% per 10 K increase in mold wall temperature [2].

Weld lines in injection molded parts are formed by the meeting of two or more melt fronts, as is the case behind openings (Figure 5.21). The flow fronts of the melt streams must stick together when they meet in order to obtain a homogeneous part surface. If the pressure and temperature are not sufficient to achieve complete bonding of the flow fronts, the corner areas on the flow fronts are incompletely formed and a weld line notch becomes visible on the part surface. With textured surfaces, there is also often a difference in gloss in these areas. In addition, non-homogeneous fusion and mixing leads to a strength weakness in this area.

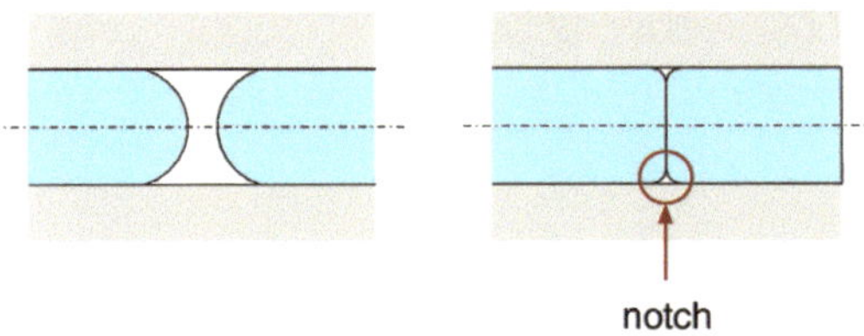

Figure 5.21
Formation of a weld line: flow fronts in contact (left) and stretching of the flow fronts with the formation of a weld line notch (right) (Image source: Non-profit KIMW Forschungs-GmbH)

To effectively avoid visible weld line notches, the cavity temperature must be increased to the softening- or crystallization temperature of the plastic. In most cases, this means increasing the temperature on the mold wall by 30 K or more. However, production at a permanently increased mold wall temperature is not possible, as the part no longer cools sufficiently to be able to eject it with the required strength. With the help of dynamic temperature control technology based on water, oil, ceramic heating elements, steam, or even inductive heating [3], in combination with conformal cooling, it is possible to quickly achieve high changes in mold wall temperatures and thus enable the production of parts with increased surface quality. This can be done over the entire surface as well as partially in the area around the weld line [4], whereby the flow fronts can bond better and the weld line is concealed. However, the surface quality of the tools must also be taken into account. While it is sufficient to heat the area around the weld line for polished surfaces, the entire cavity surface must be heated for structured surfaces in order to avoid differences in gloss [2]. In addition, extensive equipment and knowledge of the molding tool, material, and process technology is essential for the use of dynamic temperature control technology.

The mold wall temperature and the melt temperature of the plastic also have a decisive influence on the gloss level of a plastic part, as differences in gloss are caused by the different molding of the plastic on the cavity wall. A higher gloss impression is caused by light reflection on smooth surfaces. On rough surfaces, the light is scattered more strongly and the surface appears duller. If the plastic melt is too viscous to penetrate the structures on the mold surface because the melt and/or cavity temperature is too low, the plastic surface of the part becomes smoother and the part has a higher gloss level. In order to produce a more matt effect on the plastic surface, the molding of the surface must be improved by increasing the cavity temperature and/or the melt temperature.

A coating that thermally insulates the molding tool surface (e.g. TiAlN, Figure 5.22), also changes the temperature of the mold wall and the plastic melt so that changes in the gloss level on the plastic component can occur. This must be taken into account when applying molding tool coatings, but can also be used specifically to optimize the quality of the plastic parts.

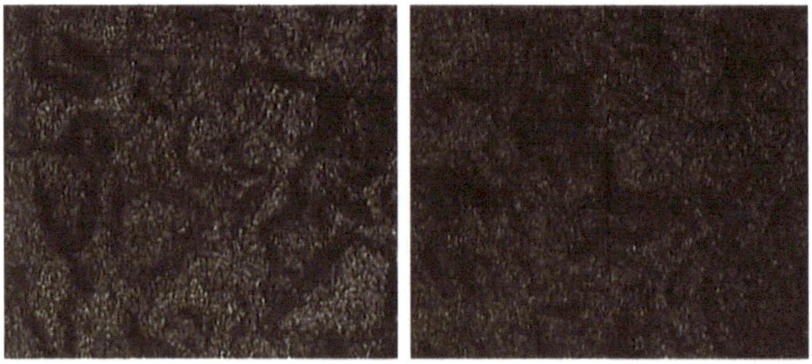

Figure 5.22 Microscopy images of structured plastic parts made of PPO, which were produced with an uncoated tool made of 1.2343 steel (left) and a tool coated with TiAlN (8 µm coating thickness) (right). The coating, which has a thermally insulating effect, improves the molding of the tool surface by the plastic, and the plastic surface becomes more matt, i. e. it appears less shiny (Image source: Non-profit KIMW Forschungs-GmbH)

5.4.3 Application and Properties of Thermal Barrier Coatings

Thermal barrier coatings are primarily known from high-temperature applications such as turbines or diesel engines. When applied to gas turbine blades, TBCs, which were developed in the 1970s to improve the performance of the metal components used (rotor blades, seals, and combustion chambers), have proven their worth [1, 5]. The main requirements for this application are a high melting point, chemical inertness, a similar thermal expansion to the substrate material, good adhesion to the substrate surface and, above all, low thermal conductivity. In addition, the material should not undergo any phase transformation in the temperature range between room and operating temperature, as this is often accompanied by changes in the crystal structure and can lead to delamination of the coating from the base material. The temperatures prevailing in a gas turbine are well above 1000 °C, so the coating applied to the turbine components must be suitable for these conditions.

Metal oxides play a decisive role in the field of thermal barrier coatings. Various metal oxide ceramic coatings can be used as TBCs [6]. Depending on the application, the most common TBCs are yttrium-doped zirconium oxide ($Y{:}ZrO_2$, also mixed with CeO_2), mullite ($3Al_2O_3{\cdot}2SiO_2$), aluminium oxide, silicates, and $La_2Zr_2O_7$. Yttrium-stabilized ZrO_2 with 7–8% Y_2O_3 (7–8 YSZ) is the most widely used TBC in the high-temperature range. The material has a high coefficient of thermal expansion, low thermal conductivity, high chemical inertness [7], and high thermal shock resistance [8], which also makes it an ideal coating material for use in injection molding tools. A disadvantage is the oxygen ion vacancies generated by doping with yttrium oxide, which makes the material oxygen-conductive and can result in corrosion at high temperatures. The formation of thermally grown oxide (TGO) on the bonding layer to the TBC can cause the TBC to peel off. This problem can be overcome by using oxidation-stable bonding layers such as aluminium oxide and mullite, which exhibit very good corrosion resistance because they are not oxygen-conductive. Due to the SiO_2, mullite is a poor conductor of heat, while Al_2O_3 is a good one. Both materials have a much lower coefficient of thermal expansion than steel, which can cause the coating to peel off during temperature changes in the injection molding process. Mullite has a very low density, favorable strength and creep behavior, as well as high stability against chemically aggressive environments [6]. YSZ doped with CeO_2 has the same positive properties as YSZ, but is also corrosion-resistant, as the oxygen ion vacancies are reduced by the addition of cerium oxide. $La_2Zr_2O_7$ is also not oxygen-conducting and is therefore corrosion-resistant and has a low thermal conductivity. However, like silicates, it also has a very low coefficient of thermal expansion, which means that its compatibility with the substrate material must be critically questioned.

5.4.4 Functionality of Thermal Barrier Coatings

Metals have a heat penetration capacity that is 10–20 times higher than thermoplastics [9], which means that the contact temperature (see Equation 4.49) between the plastic melt and the surface of the cavity is slightly higher than the cavity wall temperature. By introducing a thermal barrier coating, which has a heat penetration capacity 4–5 times lower than the mold steel, the contact temperature between the melt and TBC can be raised by approx. 20 K compared to the contact temperature between the plastic melt and the uncoated mold wall, which was determined using simulation (Figure 5.23). The contact temperatures resulting from the contact of the plastic melt with the mold wall/TBC were determined according to Equation 4.49, and the corresponding characteristic values of the materials and the temperature gradient were graphically displayed. In addition, the difference between the heat penetration capacity of the plastic and the TBC is only slight, so that the heat is only retained in the cavity for a very short time during injection. This results in a lower viscosity of the plastic melt, better molding of the surface, better flow behavior, and fewer surface defects.

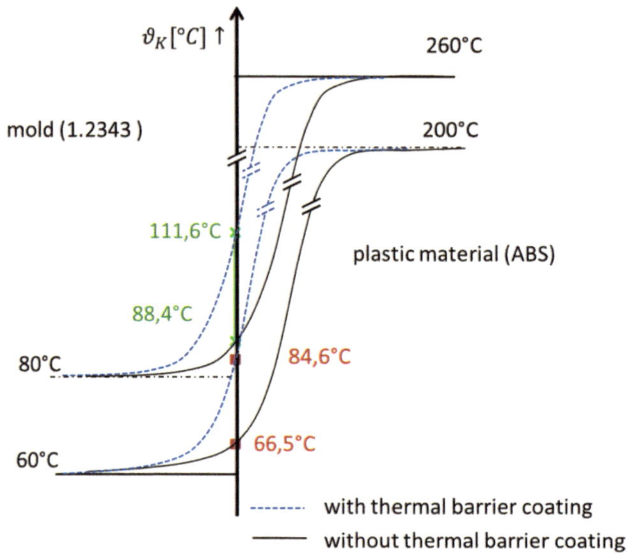

Figure 5.23 Schematic representation of the contact temperature (green/red) between plastic melt and uncoated mold wall (black) and plastic melt and TBC (blue) for two different mold wall and melt temperatures

Depending on the deposition parameters, the applied layers exhibit different thermal insulation properties. On the one hand, a thicker coating can lead to poorer heat transfer, which means that the temperature drop is less significant (Figure 5.24), but on the other hand the coating structure itself can also make heat transfer more difficult.

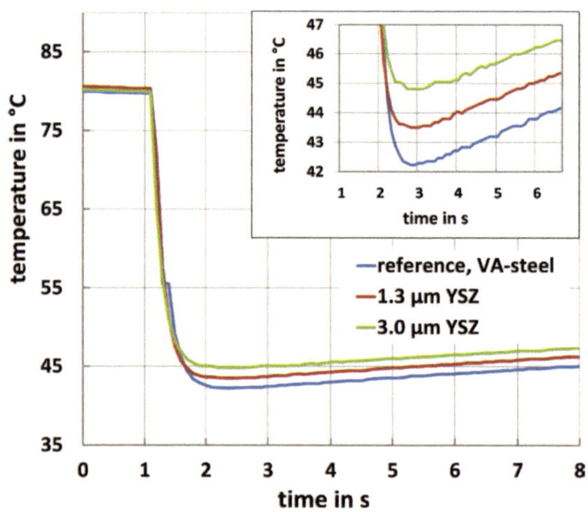

Figure 5.24 Time-dependent temperature measurements of an uncoated sample and two coated samples with different coating thicknesses

The zirconium oxide coating deposited directly using a powdered precursor has a significantly more porous coating structure, and shows a greatly reduced temperature drop at the copper block when measured with the thermal diffusivity measuring device (see Section 4.9 and Figure 5.25).

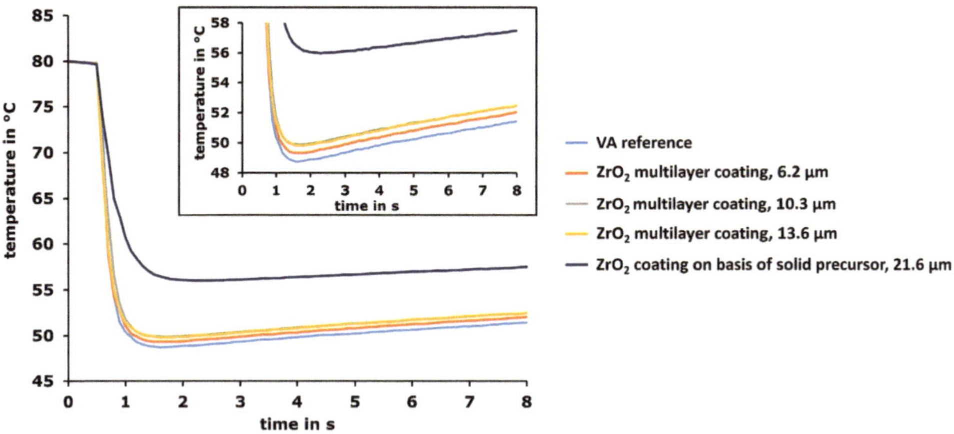

Figure 5.25 Thermal diffusivity measurement on various coatings on VA coins; the multi-layer coatings were deposited with a liquid precursor, and the ZrO_2 coating from a solid precursor

5.4.5 Application of Thermal Barrier Coatings in the Injection Molding Process

If TBCs are to be used in the injection molding process, the coatings must be suitable for the conditions prevailing during the process. The working temperatures are significantly lower than for application on gas turbines, as the melt temperatures of the plastics to be processed lie in a range between 180 °C and 340 °C (up to 400 °C for high-performance plastics). The melting- and glass transition temperatures of the materials are significantly lower, and the coated tools must be tempered to temperatures between 8 °C and 170 °C (up to 210 °C for high-performance plastics such as PEEK).

In addition, injection molding tools have high demands in terms of surface and geometry. The method of application as well as the characteristics of the TBC must be adapted to the requirements of injection molding tool. In particular, the coefficient of thermal expansion of the molding tool steel used as the base material and the coating must be in a similar range so that no delamination of the coating occurs during the thermal shock in the injection molding process. This applies to zirconium oxide, which makes it the ideal candidate for the application of a TBC on injection molds (Table 5.3).

Table 5.3 Thermal Conductivity and Thermal Expansion Coefficients of 1.2343 Steel and Various Ceramic Layers [6, 9–12]

Material	Thermal conductivity (W/(m K))	Thermal expansion coefficient (10^{-6}/K)
Al_2O_3	26–35	8
SiO_2	1.4–1.6	0.54
TiO_2	9–13	7
Mullite ($Al_2O_3 \cdot SiO_2$)	3.2	4
ZrO_2	2–3	7–12
1.2343 steel	26.8–30.3	11.8–12.9

Doping with yttrium oxide prevents the transformation of the high-temperature phase of the ZrO_2 crystal lattice (cubic or tetragonal structure) into the low-temperature phase (monoclinic) in a low-temperature coating process, as a result of which the coating remains stable in the application temperature range.

The deposition of TBC coatings from yttrium-stabilized zirconium oxide can be achieved using various PVD, CVD, and sol-gel coating processes [13–16]. An important advantage of CVD over PVD is the 3D capability of this process, which means that even injection molding cavities with complex geometries can be coated [17]. Aspect ratios of 1:60 and below can be achieved with the process through the targeted selection of process parameters. In addition, the CVD process is relatively easy to handle and does not require complicated equipment. By using organometallic precursor compounds, the coatings can be deposited at significantly reduced process temperatures compared to classic CVD [18–20]. This allows the coating of precisely machined tools at temperatures below 500 °C without changing the dimensions and hardness of the substrate [21, 22]. For these reasons, CVD is the process of choice for applying a TBC to molding tool inserts.

TBCs applied to the mold surface temporarily delay the heat flow from the molten plastic into the mold cavity [23], and can therefore also improve mold filling and the molding of the mold surface. The insulating effect and thus a more homogeneous temperature within the cavity was demonstrated in a simulation [24], and a significant reduction in the temperature drop on a heated copper rod with higher TBC coating thicknesses was demonstrated [25].

Thermal barrier coatings are deposited in the thermal MOCVD process either as pure zirconium oxide, yttrium-doped zirconium oxide, or alternating layers consisting of yttrium- and phosphorus-doped zirconium oxide [21]. The latter results in a multilayer coating structure of alternating porous and smooth, amorphous layers (see Figure 5.26). This structure reduces the internal stresses and makes the coating particularly flexible and resilient [26].

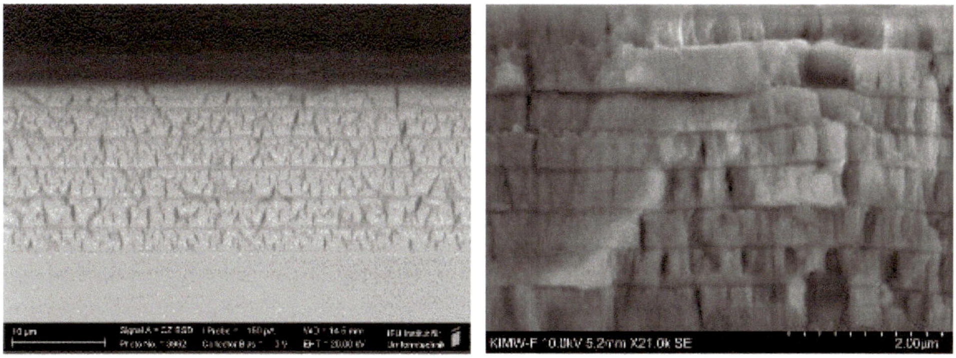

Figure 5.26 Scanning electron micrographs of a multi-layer TBC made of alternating $Y{:}ZrO_x$ (porous, YSZ) and thin, amorphous $P{:}ZrO_x$ layers (PDZ) (Image source: Non-profit KIMW Forschungs-GmbH; the left image was taken with a SEM of the Institute of Forming Technology)

Due to the high coating thickness, the application of a thermal barrier coating on the surface of the injection mold inserts changes both the roughness as well as the gloss level. A surface polished to a high gloss appears significantly duller and has a higher roughness due to the applied TBC. The roughness of structured surfaces are also changed, to which the micro-roughness of the TBC is added. However, as a ceramic coating, the applied zirconium oxide coating can be polished in the doped and undoped state. Due to the internal porosity, however, there is always a slight difference to a metallic high-gloss surface.

As already described, a coating can influence the gloss level of a plastic part by changing the thermal conductivity. When a mold cavity is equipped with a TBC, a change in the surface must always be taken into account. In the case of the grained surface shown in Figure 5.27, the coating slightly reduced the surface roughness from $S_a = 1.10 \pm 0.04$ µm to $S_a = 1.01 \pm 0.03$ µm.

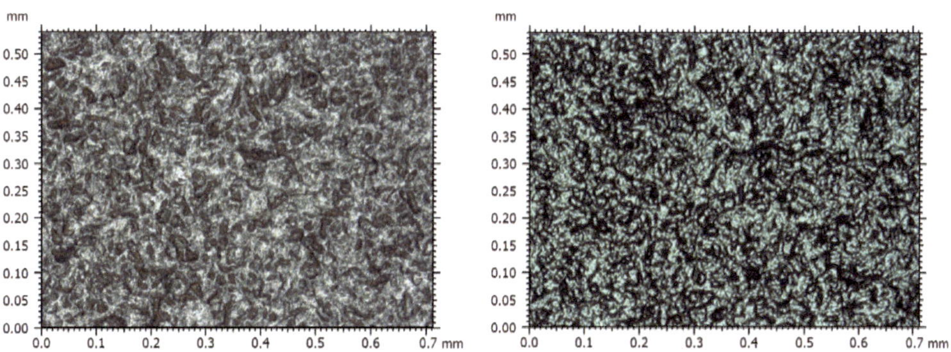

Figure 5.27 Surface images acquired with an optical 3D measuring system of a grained tool surface: uncoated (left), and coated with 12 µm TBC (right)

5.4.6 Use of Thermal Barrier Coatings in Thin-Wall Injection Molding

The CO_2 footprint of plastic products is also increasingly the focus of public interest. Modern packaging helps to preserve food for longer. Furthermore, in view of the global social shift towards small households and the associated changes in many people's lifestyles, e.g. with "convenience food", it is impossible to imagine life without it. The possibilities of saving material in this area and producing thinner-walled packaging has a positive impact on both the environmental impact and economy of packaging, and are therefore attractive from an economic and business point of view. The reduced material consumption lowers product costs and makes the packaging lighter, which is advantageous for the logistics chain and for the volume of waste, regardless of recycling. However, the minimum wall thickness is often not defined by the performance characteristics of the plastic parts, but by the injection molding process, as the injection pressures required to fill thin-walled components on the injection molding machines are limited. Better filling behavior can be achieved by increasing the cavity temperature, but this leads to longer cycle times and therefore has a negative effect on the productivity of the process. The use of TBCs can provide a targeted remedy here.

The use of a TBC reduces the thickness of the frozen surface layer in the mold cavity, which increases the free channel cross-section so that pressure losses are reduced and longer flow path lengths can be achieved. Parts with the same flow path length can therefore be filled with a lower pressure, or the resulting potential can be used to further reduce the wall thickness and save resources. By reducing the wall thickness the necessary cooling time is also reduced (or will not be significantly longer) despite the application of TBC [Wüb74], as would be the case if the mold temperature was increased by dynamic temperature control to achieve the same effect.

Flow path length measurements with test mold inserts for the production of thin injection molded parts (0.3 mm) with a width of 17 mm made of polypropylene, and for injection molded parts made of PC/ABS with a wall thickness of 0.9 mm, show an increase in flow path length of 5–10% with a 10 µm thick TBC under otherwise constant process conditions. Although an effect could be achieved here, the effect is not yet sufficient for some industrial applications.

Nevertheless, the TBC ensures better flow behavior within the cavity, so that narrow gaps can be filled better. This was demonstrated using a corresponding mold insert with cross-rib structures. For this purpose, the mold insert was sampled in an uncoated state and after coating with an approx. 13.5 µm thick TBC. Under process conditions for a filling study, with which the cross-ribs in the uncoated tool can no longer be completely filled (Figure 5.28), the application of the TBC leads to a complete and uniformly molded cross-rib structure. This was examined under a light microscope (Figure 5.29) [27].

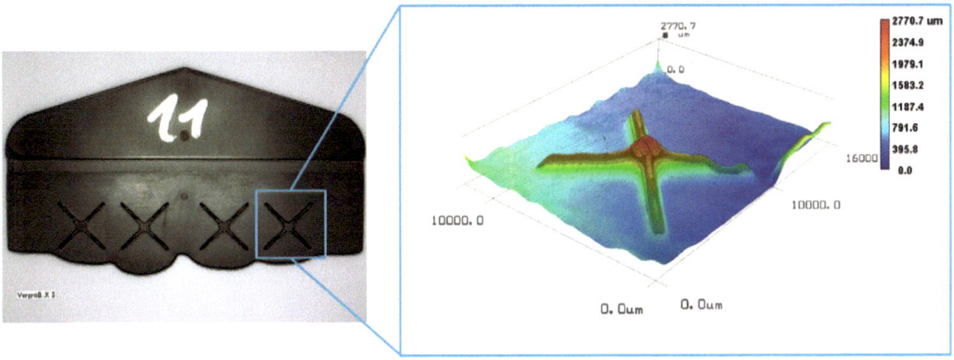

Figure 5.28 Partially filled sample part (left) made of PC/ABS, and 3D profile measurement (right) of a cross-rib produced with an uncoated tool insert. The injection volume flow was 5 cm³/s, and the dosing volume was 15.8 cm³ (Image source: [28])

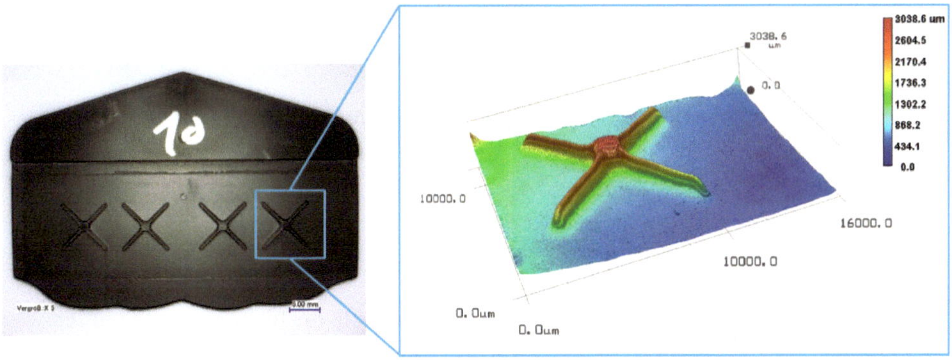

Figure 5.29 Partially filled sample component (left) made of PC/ABS and 3D profile measurement (right) of a cross-rib produced with a TBC-coated tool insert. The injection volume flow was 5 cm³/s, and the dosing volume was 15.8 cm³ (Image source: [28])

Due to the temporarily increased temperature on the mold wall surface and the resulting improved flow behavior, weld line notches on plastic parts can be optically concealed, and the improved fusion of the two flow fronts also makes them mechanically more stable. This was verified using tension rod test specimens that were produced with an injection gate on both sides. A pronounced V-notch can be seen on the test specimen injection molded with the uncoated tool insert. Under the same conditions, the weld line notch on the test specimen equipped with the coated, approx. 20 µm thick TBC is significantly smaller (Figure 5.30). The mechanical properties of the manufactured test specimens were quantified in tensile tests. The test specimens with a weld line were compared with test specimens that were only injection-molded from one side and therefore had no weld line. The tensile strengths determined for the test specimens with and without a weld line are shown in Table 5.4.

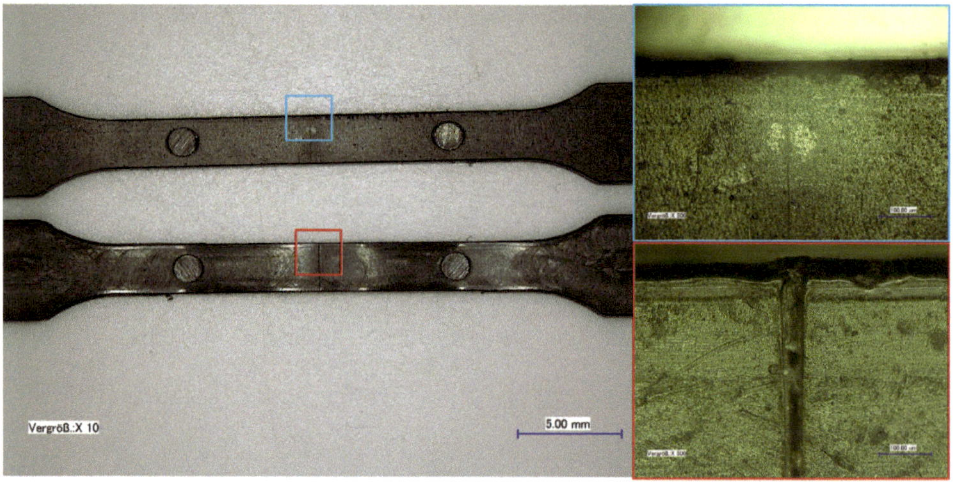

Figure 5.30 Injection-molded tensile test specimens made of PC/ABS (left), and light microscope images of the corresponding weld line notch (right). The top specimen(blue) was produced with a TBC-coated tool insert, the bottom specimen (red) was produced with an uncoated tool insert (Image source: [28])

Table 5.4 Comparison of the Results of the Tensile Tests on Tensile Test Specimens With and Without Weld Lines, Produced With Uncoated and TBC-Coated (approx. 20 μm) Tool Inserts

	Uncoated cavity	Coated cavity
Without weld line	52.3 ± 0.3 MPa	60.5 ± 1.3 MPa
With weld line	25.6 ± 1.7 MPa	49.2 ± 3.3 MPa

The tensile strength of the PC/ABS tensile bars from the uncoated cavity with a weld line is around 50% lower than that of the tensile bars without a weld line. For the tensile bars from the cavity coated with TBC, the tensile strength of the tensile test specimens with a weld line was only 19% lower than the tensile strength of the test specimens without a weld line. This shows that a mechanically significantly higher weld line strength can be achieved with TBC.

References for Section 5.4

[1] Bacos, M. P.; Dorvaux, J. M.; Lavigne, O.; Mévrel, R.; Poulain, M.: Performance and degradation mechanisms of thermal barrier coatings for turbine blades: A review of ONERA activities, AerospaceLab, 2011, pp. 1–11

[2] Guide to defects in moulded thermoplastic parts (12th edn), KIMW NRW GmbH, 2013

[3] Zimmermann, T.; Schinköthe, W.: Kurze Zykluszeiten mit konturnaher Kühlung bei induktiv-variothermer Werkzeugtemperierung – Variotherm, aber trotzdem schnell, *Plastverarbeiter*, 2011

[4] DH System Partielle dynamische Temperierung [Product flyer], Hotset GmbH, 2022

[5] Miller, R. A.: History of thermal barrier coatings for gas turbine engines – Emphasizing NASA's role from 1942 to 1990 (NASA/TM-2009-215459) (prepared for the Thermal Barrier Coatings II conference, 2007), 2009

[6] Cao, X. Q.; Vassen, R.; Stoever, D.: Ceramic materials for thermal barrier coatings, *Journal of the European Ceramic Society*, 2004, 24: 1–10

[7] Minh, N. Q.: Ceramic fuel cells, *Journal of the American Ceramic Society*, 1993, 76: 563–588

[8] Rampon, R.; Marchand, O.; Filiatre, C.; Bertrand,G.: Influence of suspension characteristics on coatings microstructure obtained by suspension plasma spraying, *Surface and Coatings Technology*, 2008, 202: 4337–4342

[9] Optimierte Werkzeugtemperierung [Company publication], Covestro Deutschland AG, 2016, issue 03

[10] Evans, A. G.; Mumm, D. R.; Hutchingson, J. W.; Meier, G. H.; Pettit F. S.: Mechanisms controlling the durability of thermal barrier coatings, *Progress in Materials Science*, 2001, 46: 505–553

[11] Nicholls, J. R.; Lawson, K. J.; Johnstone, A.; Rickerby, D. S.: Methods to reduce the thermal conductivity of EB-PVD TBCs, *Surface and Coatings Technology*, 2002, 383: 151–152

[12] Haynes, W. M.: *CRC Handbook of Chemistry and Physics* (94th edn), CRC Press, 2013

[13] Cubillos, G. I.; Bethencourt, M.; Olaya, J. J.: Corrosion resistance of zirconium oxynitride coatings deposited via DC unbalanced magnetron sputtering and spray pyrolysis-nitriding, *Applied Surface Science*, 2015, 327: 288–295

[14] Wang, D.; Bierwagen, G. P.: Sol-gel coatings on metals for corrosion protection, *Progress in Organic Coatings*, 2009, 64: 327–338

[15] Lei, Z.; Zhang, Q.; Zhu, X.; Ma, D.; Ma, F.; Song, Z.; et al.: Corrosion performance of ZrN/ZrO2 multilayer coatings deposited on 304 stainless steel using multi-arc ion plating, *Applied Surface Science*, 2018, 431: 170–176

[16] Cheng, Z.; Yang, J.; Shao, F.; Zhong, X.; Zhao, H.; Zhuang, Y.; et al.: Thermal stability of YSZ coatings deposited by plasma spray–physical vapor deposition, *Coatings*, 2019, 9: 464

[17] Plummer, J. D.; Deal, M.; Griffin, P. B.: *Silicon VLSI Technology: Fundamentals, Practice and Modeling*, Prentice Hall, 2000

[18] Andrieux, M.; et al.: Residual stress study of nanostructured zirconia films obtained by MOCVD and by sol-gel routes, *Applied Surface Science*, 2013, 276: 138–146

[19] Krumdieck, S. P.; Sbaizero, O.; Bullert, A.; Raj, R.: Solid yttria-stabilized zirconia films by pulsed chemical vapor deposition from metal-organic precursors, *Journal of the American Ceramic Society*, 2002, 85: 2873–2875

[20] Harasek, S.; et al.: Metal-organic chemical vapor deposition and nanoscale characterization of zirconium oxide thin films, *Thin Solid Films*, 2002, 414: 199–204

[21] Fornalczyk, G.; Sommer, M.; Mumme, F.: Yttria-stabilized zirconia thin films via MOCVD for thermal barrier and protective applications in injection molding, *Key Engineering Materials*, 2017, 742: 427–433

[22] Fornalczyk, G.: Verschleiß gebändigt – Neue Lösungsansätze für CVD-Beschichtungen von Werkzeugformeinsätzen, *Magazin für Oberflächentechnik*, 2016, 70: 40–42

[23] Wübken, G.: Einfluss der Verarbeitungsbedingungen auf die innere Struktur thermoplastischer Spritzgussteile unter besonderer Berücksichtigung der Abkühlverhältnisse [Dissertation], RWTH Aachen, 1974

[24] Bobzin, K; Hopmann, CH.; Öte, M.; Knoch, M. A.; Alkhasli, I.; Dornebusch, H.; Schmitz, M.: Tailoring the heat transfer on the injection molding cavity by plasma sprayed ceramic coatings, *IOT Conference Series: Materials Science and Engineering*, 2017, 181: 012013

[25] Atakan, B.; Khlopyanova, V.; Mausberg, S.; Mumme, F.; Kandzia, A.; Pflitsch, C. H.: Chemical vapor deposition and analysis of thermally insulating ZrO_2 layers on injection molds, *Physica Status Solidi C*, 2015, 12: 878–885

[26] Daniel, R.; Meindlhumer, M.; Zalesak, J.; Sartory, B.; Zeilinger, A.; Mitterer, C.; Keckes, J.: Fracture toughness enhancement of brittle nanostructured materials by spatial heterogeneity: A micromechanical proof for CrN/Cr and TiN/SiO_x multilayers, *Materials and Design*, 2016, 104: 227–234

[27] Batmanov, A.; Frettlöh, V.; Ditjo, P.; Kibet, I. K.; Wieser, J.; Frank, M.: Zirkoniumdioxid als thermoisolierende Beschichtung von Werkzeugen im Kunststoffspritzguss, Technomer, 2021

[28] Batmanov, A.; Frettlöh, V.; Ditjo, P.; Wieser, J.; Kibet: I. K.: Höhere Ressourceneffizienz durch erweiterte Dünnwandspritzgießtechnik mittels thermischer Barriere-Schichten (TBC) im Werkzeug [Final report on IGF project no. 20442N], 2022

5.5 Coatings for Plaque Reduction

Mattias Korres

5.5.1 Introduction

Within plastic processing procedures, deposits (plaques) are always formed at different intensities (see Figure 5.31). These can be caused by emissions from the plastics used (see Section 4.11) or by the effects of foreign substances or operating materials such as lubricants, corrosion inhibitors, or mold cleaners. With a technically flawless mold, it can take hundreds of production hours before this becomes visible to the human eye. In the same way, traces can remain on the surface to be molded after just a few cycles. The consequences of this can range from (for example) topographical defects on the surface of the molded part or differences in gloss, all the way up to permanent tool damage due to corrosion or "blue embrittlement". There are also many solutions to this process disruption.

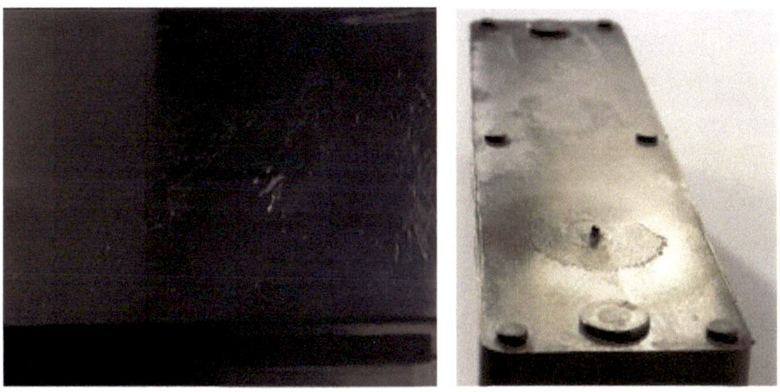

Figure 5.31 Residues of plaques on an injection molded part (left); traces of plaque in the molding cavity near the injection gate (right) [1]

There are various ways to reduce plaque. First of all, it is important to distinguish how and in which areas it forms. In the following sections, the formation and the approach to reduction on different technical surfaces in injection molds will be discussed.

5.5.2 Plaque in the Injection Molding Tool

The formation of plaques in injection molding tools can have various causes. Even the polymers used can support the formation of plaques. These include, for example, POM, PET, ABS, PBT, and PSU. Some material combinations with additives, such as UV absorbers, flame retardants, or lubricants and colorants, can also be sufficient to produce a wide variety of distinct coatings. In addition, the effects on the molded part can vary. The effects range from differences in gloss or streaks to the formation of "diesel effects", which can also cause damage to the mold.

The detection of plaques during ongoing production is often only possible when rejects are produced or through cyclical process controls. In practice, optical sensors can be used to inspect plastic and metal surfaces. With a favorable position and suitable tool geometry, it is also possible to detect plaque formation by means of structure-borne sound measurement. In any case, this must be coordinated for the respective production processes, in order to determine whether it is both economically and technically feasible.

The formation of plaques is frequently observed in colored materials with lubricants. However, the precipitation processes are often not due to the precipitated substance, but to other material additives. When using materials with lubricants, the lubricant itself often leaks out during processing. If a pigment or colorant is released, this is usually due to the incompatibility of the lubricant and colorant combination. The colorant is transported on the lubricant as a "slide rail" and deposited on the tool surface. This process is called "plate-out". The reason for the plate-out effect is insufficient compatibility of a mixture component with the base material. This causes the incompatible component to ooze out during processing and particles to be carried away. Plate-out can lead to surface defects on the tools and thus to waste parts. The presence of stabilizers, just like the presence of lubricants, has a positive effect on the plate-out effect. It is also important that the concentration of colors is less than 0.05% so that complete dissolution in the melt is possible and recrystallization during cooling is prevented [1].

The precipitation process is supported by the shear flow in thermoplastics, in which particles tend to migrate into the low-speed area at the mold wall. Substances that have a tendency to plate-out are, for example, inorganic pigments, stabilizers and fillers. In the case of white pigments (titanium dioxide (TiO_2), which represents approximately 75% of the world production of white pigments), precipitation occurs under the term "chalking", which is due to oxidative degradation of the plastic, accelerated by photoactive titanium dioxide. For materials with colorants, migration occurs in addition to the "plate-out" effect. In the case of the migration of colorants due to incomplete solubility, a distinction is made between solvent bleeding (release of coloring components into an organic liquid), contact bleeding (similar to solvent bleeding, but in the solid/solid system, e. g. film/foil) and blooming (migration of pigment particles out of the plastic/pigment system when stored at room temperature). Migration only occurs over longer periods of time on the finished part (days or months) and is therefore not responsible for the formation of a plaque in the mold. Materials that are modified to be flame-retardant can react chemically under high material stress. The decomposition products released in the process can cause mold plaque and, depending on the structure, also corrosion in the mold [1].

The reduction of plaques can often become a very detailed problem. A quick process engineering solution is often sought without taking the circumstances and process environment into account, because the investigation of all other factors is usually associated with high costs and time expenditure. Simply checking the process window for inaccurate or incorrect settings can lead to success, or at least drive forward the root-cause analysis. However, the following general points should be considered beforehand for an initial assessment:

- The use of lubricants and release agents can lead to a subsequent reaction in the tool.
  - *Which media are used in production and in what quantities (or how often) are they used?*
- Drying the material for too long, or drying it several times, can also lead to material degradation in the case of critical polymers and particularly reactive additives, such as flame retardants, UV absorbers, or blowing agents.
  - *Are the dryer volume and the material throughput coordinated?*
- The residual moisture of the material is too high, and further reactions and outgassing effects are promoted by the evaporation of water.
  - *Is the residual moisture checked by measurement?*
- The material stress caused by melt temperature, dwell time, shear stress, and shear rate during plasticizing and the injection process is too high and results in excessive material degradation and increased emission formation.
  - *Do you already have experience or characteristic values, e.g. from previous productions or material data sheets, which can be compared?*
- The materials used and additives mixed in show incompatibility during processing. These include colorants and batches, blowing agents (for example, during foaming), or demolding aids that are added to the polymer.
  - *Have these combinations been tested by the material manufacturer, and are the mixing ratios correct?*
- There is a lack of mold venting at the end of the flow path and in the area of the transition from the cavity to the parting line. Existing mold venting can also be closed due to excessive closing forces or can become clogged and lose its effect due to heavy plaque build-up.
  - *Has sufficient ventilation been installed (see Section 5.5.4)?*

- The mold geometry, especially in the sprue/gate area or thin sections, is not designed for the achievable flow properties of the plastic and high shear rates occur in these areas.
  - *Are characteristic values or simulation data available that allow a statement to be made?*

Since each injection molding tool is defined by its part geometry and design, and most plastics have a broad processing spectrum, a systematic approach is crucial in order to make a statement about the potential origins of plaques after a general points have been considered. Additional information on plaque formation can be obtained by using the analyses described in Chapter 4. Furthermore, flow and filling simulations can provide information on shear stress and reveal temperature peaks that occur. All this information is helpful for problems that cannot be solved solely by process engineering, in order to pursue a well-founded solution strategy. A permanent reduction in the formation of plaques can then be achieved, for example, by optimizing or adapting the individual tool components, through targeted coatings, or a targeted change in the material or the use of alternative materials [2].

As already mentioned, it is important to follow an efficient approach to process optimization. Systematic adjustment and testing of the corresponding process parameters should always be accompanied by appropriate documentation. A potential procedure is described below [1].

5.5.3 Process Optimization

A simple sequence of queries can be used to reduce any plaques that occur during the process. The resulting melt temperature should be checked first in accordance with the process sequence.

The temperature of the melt is the result of a combination of thermal and mechanical influences in the plasticizing cylinder. A suitable range of cylinder temperatures must be selected first. These should always be within the range specified by the manufacturer. A check of the part geometry with regard to the achievable flow path lengths and any shear zones should also be taken into consideration. In this way, process errors due to excessive stress and the associated material degradation can be reduced if the correct temperature of the melt is selected. The shape of the nozzle tip of the plasticizing cylinder should also be checked to see whether, for example, the heating zone attached to it effectively fills the entire length of this area or whether potential temperature control errors occur (Figure 5.32).

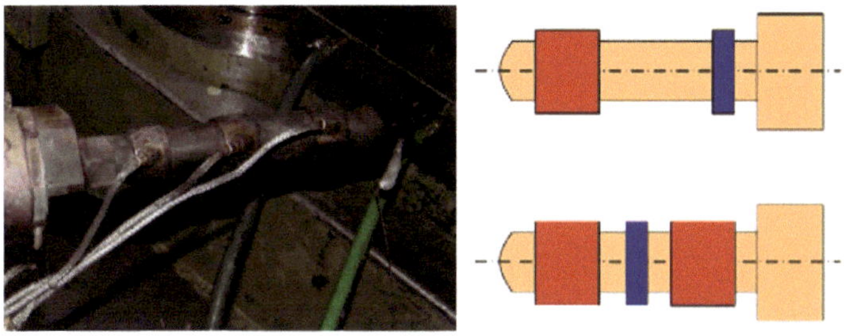

Figure 5.32 Left: Extended machine nozzle. Right: Incorrect design of extended nozzles, which tends to cause freezing effects (top), compared to an optimized design (bottom) [3]

Other measures to vary the temperature of the melt include reducing the screw speed and the back-pressure to prevent excessive shearing during the plasticizing process. The material manufacturer's instructions must also be observed here. When adding color batches, attention must be paid to the formation of color streaks when reducing the back-pressure. The screw speed should be adapted as far as possible to the residual cooling time in order to meter as gently as possible, but also to avoid extending the cycle time. A change in the cylinder temperatures requires a new setting of the switchover point and the associated injection and holding pressure profiles. Reducing the injection speed can have a similar effect to reducing the screw speed and back-pressure if, for example, there is an increased shearing effect in the gate geometries.

If it is not possible to change the temperature of the melt and the associated process parameters, the dwell time should be checked. Here too, the manufacturer's specifications should be checked and adjustments made accordingly. An increased dwell time also occurs when using plasticizing units that are too large. The use of a suitable screw diameter and *L*/*D* ratio should always be aimed for.

If the appropriate plasticizing parameters and the appropriate screw size have been selected, the next step is to check the preparation of the plastic material. In addition to the obligatory measurement of residual moisture, this includes checking the packaging and storage as well as the drying of the material. When transferring the material from other containers, contamination of the material must be excluded. The drying time and the dwell time, both for external dryers and for dryers mounted on the plasticizing unit, must be selected in accordance with the material manufacturer's specifications. An increase in the flange temperature or the temperature at the material feeder can also prevent condensation of water [2].

As a further optimization measure, the thermal stability of this admixture must be checked when admixing colorants or other additives, in addition to checking the compatibility of the carrier material with the named colorants or additives. This also includes the appropriate mixing ratio. This results in defects ranging from color streaks and color deviations to burn marks. A more in-depth assessment of the compatibility between colorants/batches and the material types is usually only possible in conjunction with laboratory tests [3].

The formation of plaques when foaming thermoplastics, due to the mechanisms of the molding process, requires a different approach to optimization. For example, the switchover point must first be checked in order to reduce potential free-foaming before the desired filling level is reached. Especially with flame retardants or other more aggressive additives, the dwell time must also be checked here, as a reaction with the foaming additive may occur. In addition to the switchover point, the injection speed should also be checked, whereby this should be increased during foaming in order to prevent premature gas leakage from the compound. Premature blowing agent expansion in the screw antechamber due to a worn non-return valve or an incorrectly set position control of the screw can also lead to problems [2].

Further measures can be taken by improving the venting of the mold. This includes visual inspection and cleaning of the venting channels, but also reducing the clamping force to optimize the transition to venting. This can either be achieved through "extended clamping force programs" or by a general reduction. In the former case, the clamping force is only increased from the end of the injection phase and the increase in cavity pressure, thus supporting venting. If this is not possible or if flash is already forming when the clamping force is reduced, in most cases only mold-related measures are possible.

5.5.4 Optimization of the Injection Molding Tool

If optimization of the process parameters does not lead to a sufficient reduction of the plaque, the molding tool must be checked with regard to its venting paths. This can lead to time-consuming and costly changes, which must be planned and coordinated accordingly.

First of all, the positions of the required venting paths in the injection mold must be determined. For this purpose, the respective flow path ends of the article geometry must be determined by filling the cavity step by step (a so-called "filling study"). These cavity areas must be connected directly to corresponding venting channels, which also lead from the internal areas of the parting line to the mold edges. The venting channels should be connected to the cavity by grinding the parting line accordingly. The first simple approach is to use a 0.01–0.02 mm cut for semi-crystalline thermoplastics and 0.02–0.04 mm for amorphous thermoplastics. This grinding should be carried out over a sufficient width of the cavity. In the case of complex mold geometries or high requirements due to the viscosity properties of the thermoplastic, it is also possible to build-up a vacuum and vent the entire cavity. This can also prevent any oxidation, i. e. reaction of the emissions with the air [3].

In addition to the venting paths within the parting line, further venting options can be generated by tapering ejectors or venting grooves on slides or jaws. These tapers can be connected to the cavity surface using the same grinding depths that are used in the parting line. Other options can be porous inserts (Figure 5.33) or spring valves, which support the emissions/venting.

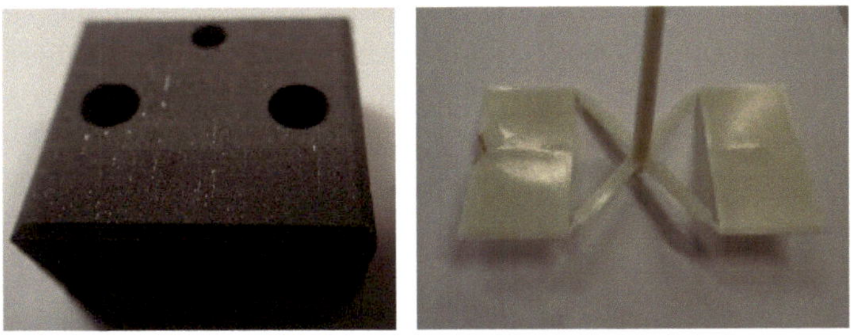

Figure 5.33 Left: Porous mold insert for venting. Right: "Diesel effect" with a dense cavity compared to a cavity with a porous venting module [3]

Although porous venting elements in particular are very effective, they become clogged and blocked by emissions. Accordingly, the inserts must be changed regularly and cleaned using pyrolysis or ultrasonic baths. If these passive venting methods are not sufficient, active venting methods are a further option. In addition to the classic vacuum, venting valves that generate a defined air flow can be sufficient to optimize venting and reduce the formation of streaks, air hooks, or diesel effects. This requires the insertion of seals or pneumatic connections, which should be planned during the design phase. An adequate analysis of the venting points by means of a rheological simulation can help with the planning [1].

5.5.5 Coatings for Plaque Reduction

If a reduction in the molding defects or the plaque caused by the emissions is not possible through process and tooling optimization, coatings can be used to achieve an additional improvement. In this case, direct reduction of the plaque is not always the only concern – protection of the tool against potential signs of wear, such as corrosion and "blue embrittlement", or hot gas corrosion with "diesel effects", may also be considered.

The choice of a suitable coating depends on the plastic materials used, the molding tool steels used, and the surface structures applied. Accordingly, the coating process for existing molding tools should be coordinated in advance with the given molding tool. In the case of new tools, this coordination enables the tool steels to be adapted to the conditions of the coating process. Furthermore, it must be taken into account that a reduction in the adhesion of the plaque does not necessarily result in a reduction in the adhesion of the base polymer. This can have negative consequences for the demolding behavior. A change in the thermal properties of the mold surface and associated changes in the molding properties are also possible, which can lead to differences in the degree of gloss on the molded parts. Coordination with the respective coater and/or surface specialist is recommended, as is consultation with the client [3].

The variation of process properties, such as foaming, the adjustment of process parameters, and/or additives, as well as fillers and reinforcing materials, can lead to a considerable change in the effect of coatings. It is therefore difficult to make a general statement about the effect of coatings. General tendencies resulting from a series of tests to minimize plaque marks at the Lüdenscheid Plastics Institute are shown in Tables 5.5 and 5.6. However, this behavior can change drastically with additives such as flame retardants, flow aids, or demolding aids.

Table 5.5 Overview of Various Polymer Types and Corresponding Coatings for Plaque Reduction [3] (+ Slight improvement, ++ Strong improvement, +++ Very strong improvement)

Plastics		TiN (2–8 µm)	TiCN (2–8 µm)	CrN (2–8 µm)	TiAlN (2–8 µm)	TiAlN (8–10 µm)
Polyolefins	PE, PP, PB	+		+++		+++
Styrene polymers	PS, SB, SAN, ABS, ASA	+++		++		+++
Chlorinated polymers	PVC			++		
Fluorine-containing polymers	PTFE, PVDF			++		
Acetal resins	POM	+		+++		
Polyamides	PA	+++		+++	++	
Linear polyester	PC, PBT(B), PET(P)	+++		+++		+++
Polyaryleneth-enes	PEEK, PPS, PSU, PES, PPE, PPO	+	++	+++	+++	
Polyimides	PI	+++				
Cellulose esters	CA, CP, CAP	+++				
Polyacrylates	PMMA	+++				
Polyurethanes	TPU			+++		

Table 5.6 Overview of Various Elastomer Types and Corresponding Coatings For Plaque Reduction [3] (+ Slight improvement, ++ Strong improvement, +++ Very strong improvement)

Plastics		TiN (2–8 µm)	TiCN (2–8 µm)	CrN (2–8 µm)	TiAlN (2–8 µm)	TiAlN (8–10 µm)
Polyurethanes	PUR			+		
Synthetic rubbers	NBR, EPDM, Si	+		++		
Fluorinated elastomers	Multi-polymer TPE, FPM			++		

It is advisable to check the effect of coatings in simple test tools. In the event of plaque formation on a polished surface, a gloss level measurement can be carried out to detect the formation of a plaque. Furthermore, it should always be kept in mind that this only involves a reduction of the plaque, but never its complete removal. Even when process engineering optimizations, tooling solutions, and the use of coatings to optimize the defect pattern are exhausted, cleaning and adapted maintenance cycles are of crucial importance for trouble-free production [3].

References for Section 5.5

[1] Anwenderhandbuch für Schichttechnologien, KIMW NRW GmbH, 2002
[2] Guide to defects in moulded thermoplastic parts, KIMW NRW GmbH, 2021
[3] Korres, M.: Qualifizierung von Beschichtungen zur Belagsreduzierung, 2020

5.6 Coatings to Reduce Demolding Force

Dr. Ruben Schlutter

5.6.1 Introduction

Demolding describes the ejection of the cooled, molded part from the cavity. Ideally, the molded part would detach itself and fall out of the mold. However, this does not occur due to undercuts and adhesive forces, so the molded part must be pushed out of the mold using a demolding device [1].

On the other hand, robust demolding is essential for a smooth and trouble-free manufacturing process. This challenged by increasingly complex molded parts and molds as well as ever higher demands on the visual and haptic properties of the molded part. This increases the demands on the mold surface, as high-gloss polishes or structures that make demolding more difficult are increasingly required. In addition, the cycle time has to be reduced, so that demolding has to be faster and faster. However, this increases the stress on the molded part and on the mold. General predictions about the demolding forces depending on the mold surface and the plastic used are not yet possible. Therefore, mold–plastic combinations of interest must be investigated empirically with regard to their principle demoldability and the demolding force. Different surface modifications and plastics can be compared with each other [2].

5.6.2 State of the Art

A basic solution for reducing demolding forces is to increase the demolding angles. The conicity on mold cores and internal structures depend mainly on the plastic used. In general, a draft angle between 1.5° and 3° should be provided for amorphous thermoplastics and a draft angle between 0.5° and 3° for semi-crystalline thermoplastics.

For structured surfaces, an additional angle of inclination of 1° per 25 µm structure depth is necessary to ensure demolding. The structure should also run in the direction of demolding [1].

However, it is often not possible to increase the draft angles, so other measures must be taken to reduce the demolding force.

In principle, sleeve-shaped and box-shaped areas of the molded part shrink onto the corresponding mold cores. Cooling of the plastic causes a form-fitting contact surface to be formed between the plastic and the mold cavity. This contact surface is the place where friction occurs during demolding. To demold these areas, a normal force according to Equation 5.3 is required [1].

$$F_E = \mu \cdot p_F \cdot A_F \tag{5.3}$$

F_E = Demolding force (N)

μ = Coefficient of friction (—)

p_F = Surface pressure between molded part and core (MPa)

A_F = Surface area of the core (mm^2)

Various adhesion forces act on the contact surface between the mold cavity and the plastic. Adhesion cannot be described in its entirety, so that various theories have been developed, each of which can describe individual aspects of adhesion. A basic distinction is therefore made between mechanical and specific adhesion [3, 4].

In mechanical adhesion, form-fitting interlocking clamps are formed between the partners involved in the contact surface. A tear-off force is required to break away and overcome these adhesions [5]. In the injection molding process, partial filling or complete filling of the roughness of the cavity surface is possible due to the different process control (Figure 5.34).

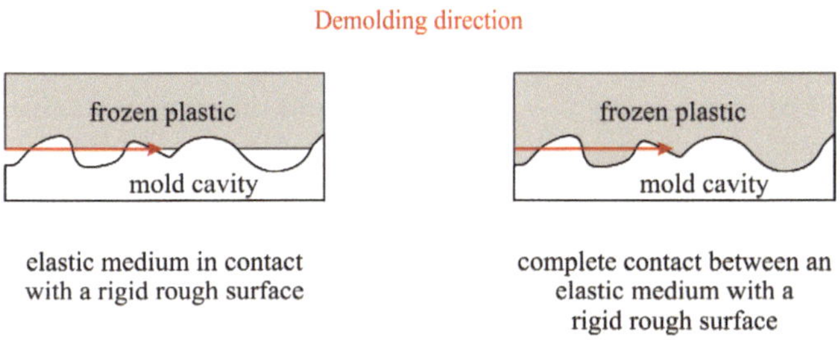

Figure 5.34 Dependence of the demolding force on the surface roughness (Image based on [5])

The most important parameters for influencing mechanical adhesion are the surface energy and the shear modulus of the elastic medium. The energy balance between the elastic deformation of the elastic medium and the surface energy can be estimated according to Equation 5.4, whether adhesion occurs in the contact surface due to gluing or not. Mechanical adhesion can also explain why plastics adhere to highly polished surfaces (small characteristic height and large characteristic wavelength of the surface roughness) [5].

$$\frac{G \cdot h^2}{4 \cdot \gamma \cdot l} > 1 \tag{5.4}$$

G = Shear modulus of the elastic medium (MPa)

h = Characteristic height of the surface roughness (mm)

γ = Surface energy (mJ)

l = Characteristic wavelength of the surface roughness (mm)

The theories of specific adhesion were developed to explain the adhesion of substances to smooth surfaces, since this effect cannot be represented by mechanical adhesion. Specific adhesion is based on chemical, physical, and thermodynamic interactions [3].

The formation of specific adhesion on a chemical basis requires the formation of covalent bonds between the adhesion partners, and would make it impossible to demold a plastic component from an injection mold [6].

Physical adhesion forces are based on the formation of electric fields by the close order of molecules, so that physical bonds can be formed. These can basically be divided into hydrogen bonds, which are based on dipole forces between a hydrogen atom and an oxygen or a nitrogen atom, and van der Waals bonds, which are based on dipole forces between various elements, dispersion forces, and induction forces [7].

In hydrogen bonds, the positive pole is always the hydrogen atom in the interaction of oriented dipoles. Dipoles are caused by an uneven distribution of electric charges. Because of this electric charge, other molecules become oriented so that molecules across the interface can attract each other, as the molecules in the adhesive partner can also align. Dipole forces can therefore only act in polar molecules [7].

In dispersion forces, weak intermolecular forces of attraction occur based on the spontaneous fluctuation of electron density. The electron fluctuation creates an electric dipole moment, which leads to the alignment of a neighboring molecule, as the latter then also forms a dipole. Therefore, dispersion forces also act between completely nonpolar substances. When dispersion forces occur between like substances, they are termed induction forces [7].

The range of the specific adhesion force in each case is decisive for demolding, since these decrease with increasing distance from the bond between the partners involved. Table 5.7 illustrates this relationship. During shrinkage of the plastic component or during demolding, the physical adhesion forces are released.

Table 5.7 Binding Energies of Physical Bonds [4]

Binding type	Binding energy	Average operating range
Dipole forces	20 kJ/mol	0.40–0.50 nm
Induction forces	≤2 kJ/mol	0.35–0.45 nm
Dispersion forces	0.1–40 kJ/mol	0.35–0.45 nm
Hydrogen bonding forces	50 kJ/mol	0.25–0.30 nm

The thermodynamic adhesion theory assumes that surface wetting processes occur during the formation of the contact surface. It is assumed that the plastic has a surface energy that is lower than, or equal to, that of the steel material of the injection mold. However, additional forces often occur, making this theory unsuitable in the context of demolding [3].

The size of the coefficient of friction μ therefore depends mainly on the material pairing between the mold material and the plastic. Influences resulting from the process control of the injection molding process are not considered further at this point.

The demolding force depends, among other things, on the surface roughness of the mold material, as shown in Figure 5.35. Two effects can occur. In the range of very low surface roughness ($R_a < 0.2$ µm [8]), specific adhesion effects mainly occur, which lead to strong adhesion of the molded part. With increasing surface roughness, the influence of the adhesion forces decreases to a minimum.

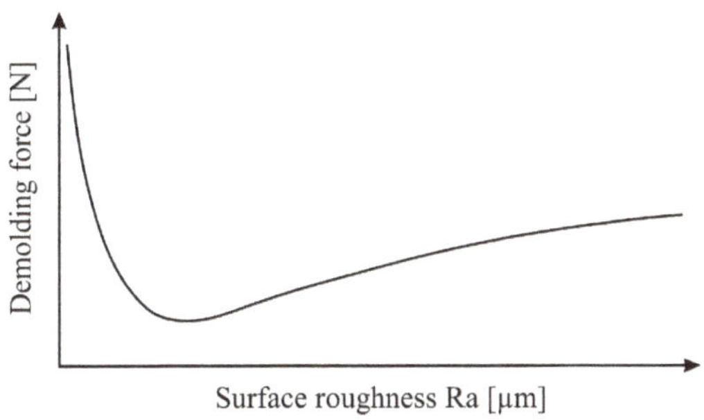

Figure 5.35
Dependence of the demolding force on the surface roughness (Image based on [1, 8])

With increasing surface roughness ($R_a > 1$ µm [1]), the demolding force due to mechanical adhesion increases, since the surface roughness increasingly forms microscopic undercuts into which the plastic melt flows during filling and during application of the holding pressure. These undercuts cause the molded part to clamp onto the mold core, so that the demolding force increases again.

In addition to the adhesion forces, the part is also deformed during demolding. The deformation results from the sliding of the plastic part and the roughness of the mold cavity. The deformation is due to the deformation of the plastic part during sliding. This results in internal damping, so that a frictional force is generated. The conversion of mechanical energy into thermal energy due to thermoplastic deformation is also possible. As the mold cavity and the plastic part slide over each other, several contact surfaces are formed. Due to the viscoelastic behavior of the plastics and the change in load during deformation, the contact areas are not constant. Due to the creep processes in the plastic, the size of the contact areas increases as more and more “roughness mountains” become contact areas. The displacement builds up internal stresses that balance the effect of the external forces [9].

The state of the art for reducing the demolding forces includes the use of lubricants and release agents and the application of PVD or CVD coatings [1, 10]. Both can lead to a reduction of the demolding force.

However, the use of lubricants and release agents interferes with the cleanliness of the process. This leads to contamination of the manufactured parts [11]. Cleaning of the parts for further process steps, such as painting or metallization, is then necessary. The production of medical devices or cleanroom products is not even possible in this way.

In addition to demolding, coatings on the mold surface also influence other properties in process control. For example, PVD and CVD coatings can have a different thermal conductivity than the surrounding mold steel, which improves the molding accuracy of the mold cavity but worsens demolding. Furthermore, defects and pores in the coating can support or even increase the corrosion of the mold cavity [12].

5.6.3 Possible Applications and Potentials

The test setup described in Section 4.10 can be used to investigate various practical issues in the area of demolding of manufactured parts in order to be able to solve problems that arise in a targeted manner, or to be able to make recommendations for modifying the cavity surfaces of an injection mold.

5.6.3.1 Material Selection of the Thermoplastic Material

Figure 5.36 shows the demolding behavior of various unfilled plastics on an unmachined test blank with a high-gloss surface. It can be seen that both the choice of the plastic family and the choice of the specific plastic type have a clear influence on the demolding behavior. In practical terms, this means that the choice of material has a significant influence on the number and positioning of ejectors in an injection mold, and must be adapted to the selected plastic in each case [1].

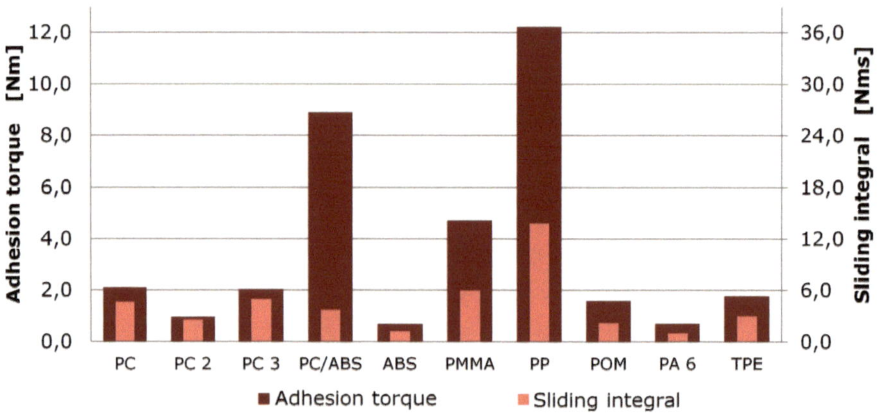

Figure 5.36 Adhesion torque and sliding integral of various unfilled plastics on a high gloss surface (Image source: Non-profit KIMW Forschungs-GmbH)

5.6.3.2 Addition of Additives

The estimation of the level of the demolding force can already be done during the material selection. Figure 5.37 shows the differences in the adhesion torque of an unfilled polycarbonate with different amounts of a demolding additive. The demolding was carried out on a test blank with an unmachined high-gloss surface. It can be seen that the test methodology provides very stable results, as the adhesion torque remains stably constant over several hundred injection molding cycles.

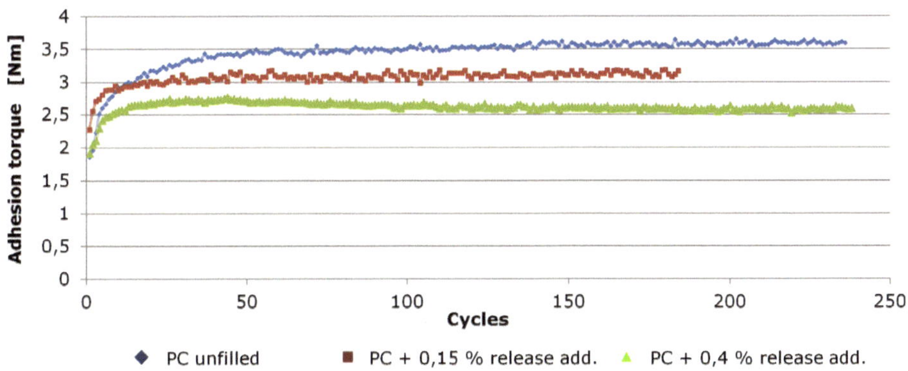

Figure 5.37 Comparison of different additive amounts in PC (Image source: Non-profit KIMW Forschungs-GmbH)

The blue curve shows the adhesion torque of the unfilled polycarbonate. It can be seen that even a small change in the amount of demolding additive has a significant influence on the subsequent demoldability of the part. The addition of 0.15% demolding additive (red curve) leads to faster plaque formation and faster formation of an equilibrium state during demolding after approx. 15 cycles. A disadvantage is the faster contamination of the injection mold. When 0.4% demolding additive is added (green curve), it is first removed as a gas, so that it takes longer for a state of equilibrium to develop during demolding. This is reached after about 30 cycles.

In addition, the variation of the injection molding process has an influence on the measurement results. For example, an increase in mold wall temperature leads to both a higher adhesive torque and a higher dynamic frictional torque. This is mainly due to the molding accuracy, since the molding of the micro-roughnesses of the mold cavity is better at a higher mold wall temperature. This results in the formation of very small undercuts, which have to be demolded by an increased torque. Later demolding and thus a lower demolding temperature reduces the adhesive torque and the dynamic frictional torque.

5.6.4 Modification of the Cavity Surface

Figure 5.38 shows the influence of the mold surface on the demolding process using an unfilled PC as an example. In the course of the tests, the surface of the test rounds was provided with different surface structures.

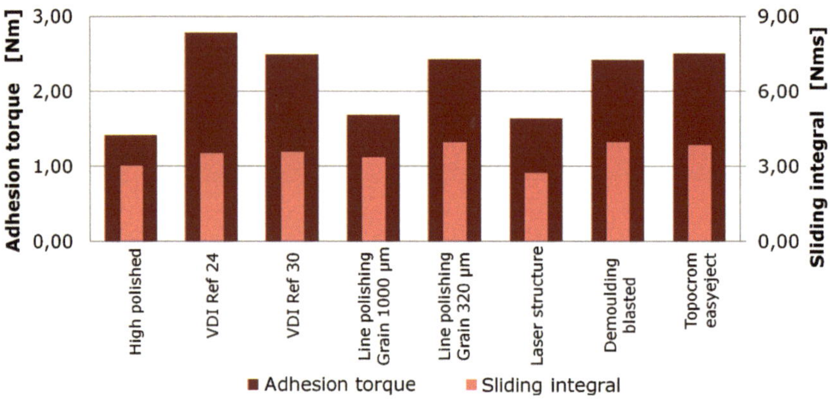

Figure 5.38 Adhesion torque and sliding integral for PC with different tool designs (Image source: Non-profit KIMW Forschungs-GmbH)

This shows that the surface structure has a significant influence on the demoldability. For example, a finer surface finish ("VDI Ref 24" and "Line polishing Grain 320 µm") leads in each case to a higher demolding force than the respective rougher counterpart.

Figure 5.39 shows the influence of different mold coatings on the demolding process. The PC used is not identical to the PC with which the dependence of the tool structures from Figure 5.37 was measured. This can be seen, among other things, in that the adhesion torque and the sliding integral during demolding differ from the unmachined test blank. It can be seen that all the coatings used lead to a reduction in both the adhesion torque and the sliding integral. Since there are significant differences between the individual coatings, the choice of the appropriate coating must be made with care and matched to the plastic in question.

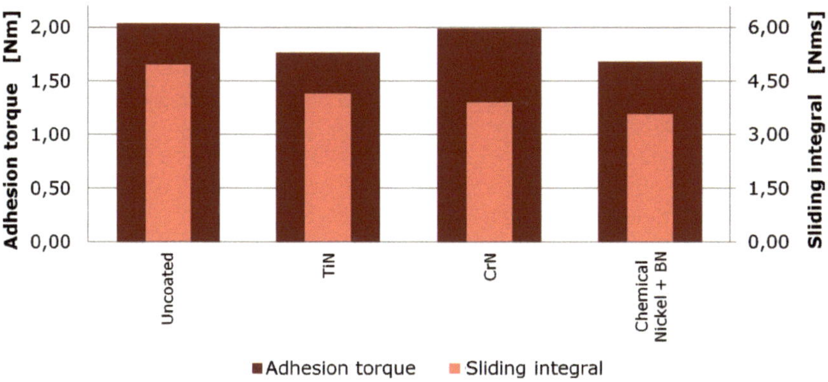

Figure 5.39 Adhesion torque and sliding integral of PC for different mold coating (Image source: Non-profit KIMW Forschungs-GmbH)

5.6.5 Summary

In the scope of the work carried out, it was possible to achieve an accurate and reproducible representation of the friction conditions in the injection mold during demolding under near-series conditions. Thus, the previous imperfections that occur in the representation of the friction conditions during demolding could be overcome. As a result, a simple and robust test was established to model the demolding process. There are no limits to the modification of the mold surface in terms of material selection, the implementation of structures, or the application of coatings through the use of standardized demolding blanks. The plastic material can also be chosen as freely as possible to carry out the tests. The only restrictions are the maximum mold wall temperature, which is 140 °C, and the minimum hardness of the plastic, which must be a minimum of 30–40 Shore A.

References for Section 5.6

[1] Hopmann, C.; Menges, G.; Michaeli, W.; et al.: *Spritzgießwerkzeuge* (7th edn), Hanser, 2018

[2] Korres, M.: Optimierung eines Messwerkzeuges für das Entformungsverhalten im Spritzgießprozess [Bachelor thesis], South Westphalia University of Applied Sciences, 2016

[3] Achereiner, F.: Verbesserung der Adhäsionseigenschaften verschiedener Polymerwerkstoffe durch Gasphasenfluorierung [Dissertation], University of Erlangen-Nuremberg, 2009

[4] Dutschk, V.: Oberflächenkräfte und ihr Beitrag zur Adhäsion und Haftung in glasfaserverstärkten Thermoplasten [Dissertation], Dresden University of Technology, 2000

[5] Popov, V.: *Kontaktmechanik und Reibung* (3rd edn), Springer, 2016

[6] Nikolova, D.: Charakterisierung und Modifizierung der Grenzflächen im Polymer-Metall-Verbund [Dissertation], University of Halle-Wittenberg, 2005

[7] Kaiser, W.: *Kunststoffchemie für Ingenieure* (5th edn), Hanser, 2021

[8] Sasaki, T.; Koga, N.; Shirai, K.; et al.: An experimental study on ejection forces of injection molding, *Precision Engineering*, 2000, 24: 270–273

[9] Burgsteiner, M.: Entformungskräfte beim Spritzgießen in Abhängigkeit von Oberflächenrauigkeit, Stahlwerkstoff, Vorzugsrichtung der Oberflächenstruktur, Beschichtung und Kunststofftyp [Master's thesis], University of Leoben, 2009

[10] Kaynak, B.: Perfluoralkylsilan Beschichtungen für Formen und Werkzeuge in der Kunststofftechnik [Dissertation], University of Leoben, 2017

[11] Michaeli, W.; Pauling, A.; Grönlund, O.; Vissing, K.: Trennmittelfreie Produktion von PUR-Bauteilen, *FAPU*, 2010, 61: 35–37

[12] Fornalczyk, G.: personal communication, 12 September 2018

5.7 Thin-Film Sensor Technology

Dr. Angelo Librizzi

5.7.1 Introduction

The processing of thermoplastics in the injection molding process enables the production of high-quality molded parts under economical conditions. Rising material and manufacturing costs in injection molding companies require efficient, reproducible production with high product quality. This requires transparent processes and precise process control of the quality-relevant parameters, which often cannot be achieved by the injection molding machine alone. By using additional sensors in the injection mold to determine the cavity pressure and the mold wall temperature during the production of the molded part, relevant information can be captured that can be used to analyze, optimize, monitor, and document the process. In complex manufacturing processes, the aim is therefore increasingly to monitor and ensure process quality by continuously recording process information. The primary aim here is to detect quality fluctuations using sensor-based process signals. On the one hand, this enables 100% quality monitoring and, on the other, provides a basis for prompt, quality-dependent interventions in the process. In this context, temperature and pressure sensors are used as standard, which allow point measurements in the injection mold. For this purpose, corresponding holes are usually provided in the mold in the area close to the gate and/or in the area away from the gate, into which the pin-shaped sensors are fitted.

A sensor, transducer or probe is a technical component that measures certain physical or chemical properties qualitatively or quantitatively as a measured variable. These variables are recorded by means of physical or chemical effects and converted into electrical signals that can be further processed. In DIN 1319-1, the term transducer is defined as the part of a measuring device that responds directly to a measured variable. The transducer is therefore the first element of a measuring chain.

The sensors used as standard in injection molds are subject to certain restrictions. For example, to prevent plastic melt from penetrating between the sensor and the bore, it is necessary to fulfil exactly the requirements to the specified bore tolerances and to use additional O-rings for sealing. In this context, the processing of very low-viscosity melts can be problematic. There is a risk of the melt penetrating into the ring gap, which leads to an incorrect measured value, or even prevents measurement completely. The commercially available measuring elements have certain geometric limitations with regard to the radius of curvature of the sensor front. Sensor elements are available that can be mechanically machined and adapted to the tool wall contour, as well as elements that do not allow the front face to be machined. This can make freely selectable positioning, particularly at a quality-relevant part position, more difficult. In the case of temperature sensors that allow mechanical processing, the actual measuring element, usually a wire thermocouple, must have a certain distance from the surface of the fitted pin that comes into contact with the plastic. A deviation between the actual temperature and the temperature interpreted by the sensor must therefore be assumed. Furthermore, the mass of the measuring elements to be heated results in sensor inertia. Furthermore, wear, corrosion, and plaques on tool surfaces and tool elements cause problems during injection molding, which are associated with high costs for reworking.

Sensors for injection molds that are manufactured using surface and coating technologies therefore offer an alternative, in order to circumvent the described limitations and utilize the advantageous properties of thin coatings for the manufacturing process.

5.7.2 State of the Art Sensors

5.7.2.1 Pressure Measurement in the Injection Mold

The cavity pressure curve is a decisive criterion for the analysis, optimization, monitoring, and documentation of the injection molding process [1]. The graphical representation of cavity pressure provides a wealth of information about the interaction of plastic, injection molding machine, injection mold, and environment during the production of molded parts. Processes such as sampling, start-up, and optimization of injection molds can be accelerated, and the quality and consistency of the molded part can be increased.

The manufacturing of molded parts during the injection molding of thermoplastics can be divided into four phases: filling phase, compression phase, holding pressure phase, and residual cooling phase.

The idealized curve of cavity pressure over time, recorded with a sensor positioned close to the gate, is shown qualitatively in Figure 5.40 for amorphous and semi-crystalline thermoplastics.

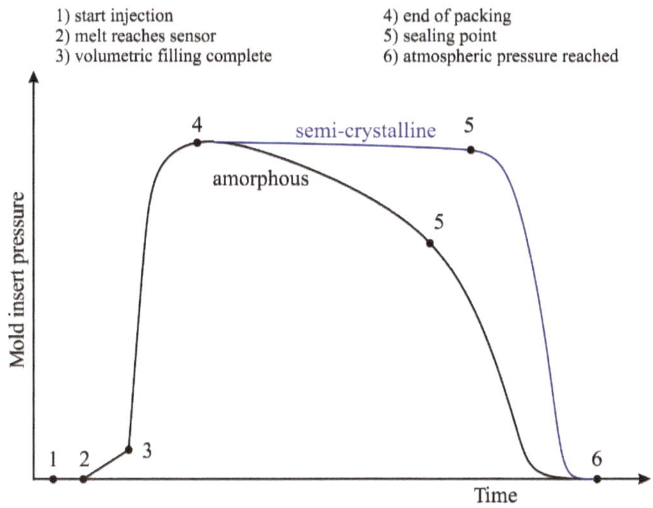

Figure 5.40 Qualitative progression of cavity pressure during molded part production [2]

At point 1, the injection molding machine starts injecting plastic melt into the cavity. At point 2, the melt front reaches the sensor and the pressure increases almost linearly at a constant screw feed speed. At point 3, the volumetric filling phase is complete. This point is determined via a filling study when the injection molding machine is run-in and, ideally, the machine switches from speed-controlled injection to pressure-controlled holding pressure. From point 3 to 4, a steep increase in pressure during the compression phase can be seen, which ends when the maximum pressure is reached at point 4. From point 4 to 5, further material is introduced into the cavity during the holding pressure phase as the plastic cools. As a result of the cooling, the pressure transfer of the holding pressure and thus the melt flow from the screw vestibule into the mold cavity becomes increasingly difficult, as the plastic melt becomes more viscous and a channel constriction occurs, until the sealing point is reached at point 5. At this point, the gate has solidified and no further material can be introduced into the cavity. An extension of the holding pressure phase has no further influence on the quality of the molded part and is uneconomical. The further course of the cavity pressure is then only influenced by the shrinkage behavior of the plastic. At point 6, the pressure in the mold has dropped to atmospheric pressure. The part should only be demolded at this point if the demolding temperature has also been reached.

The measuring chain required for pressure measurement consists of the component's pressure transducer or force transducer, charge amplifier, and measured value converter, with the necessary connector and connection cable. Pressure and force transducers based on the piezoelectric effect are particularly suitable for pressure measurement in injection molds [3–5]. These sensors use artificially grown quartz crystals that emit a proportional electrical charge in response to a mechanical load. This electrical charge is transmitted via a connecting cable to the high-impedance input of a charge amplifier for signal conversion. The pressure-proportional electrical voltage at the output of the charge amplifier is then either transferred directly to the injection molding machine control system or connected to the injection molding machine via the measured value converter. The measured value converter is the link in the measuring chain that is required for process monitoring on a PC for the purpose of recording pressure curves on the screen.

The placement of the pressure sensor at the most metrologically effective point of the cavity must be given the greatest attention in terms of design. As the cavity pressure decreases continuously from the gate to the end of the flow path, it is often recommended to position the sensor in the area close to the gate [4, 6, 7].

5.7.2.2 Temperature Measurement in the Injection Mold

Basically, two different types of temperature measurement are used in the injection molding process. Discontinuous temperature measurement is used after process interruptions or malfunctions, but this has a negative impact on efficiency. Measurements are always taken when a sufficiently high number of faulty molded parts are produced from the injection molding series currently in progress.

The more suitable and economical option is continuous temperature measurement during the ongoing process. It offers the advantage of being able to directly display the temperatures of the sensor in the cavity and thus provide the machine operator with the information immediately. This allows errors, e.g. caused by incorrect setting parameters, to be corrected without incurring high costs due to a standstill of the injection molding machine.

With temperature sensors for continuous temperature measurement, the contact temperature between the mold wall or sensor surface and the plastic melt is recorded over time – the mold wall temperature curve. The cavity temperature has a decisive influence on the surface quality and properties of the molded part as well as the shrinkage behavior of the plastic and thus the part properties. Therefore, in the course of quality monitoring, and especially from the point of view of condition-dependent demolding, its detection is increasingly in demand. The sensor detects the temperature difference between the basic state and the maximum temperature during mold filling. Furthermore, the temperature difference over the flow path can be determined by installing several sensors. The sensor at the end of the flow path provides information about the time of volumetric mold filling. Another possible application of a temperature sensor is the detection of the melt position to control a cascade injection.

The measured temperature curves can be used to check the process reproducibility, as the melt must reach the same place in the cavity again at the same time in the cycle if the viscosity conditions remain constant. In this way, a change in viscosity behavior, e.g. due to batch fluctuations and fluctuating melt or mold wall temperature, can be detected.

The cavity temperature curve (Figure 5.41) forms a sawtooth shape due to the heating during the injection phase and the cooling during the holding pressure and cooling phase. The temperature fluctuations on the cavity wall are physically dependent on the molding tool steel, the molding compound, and its temperatures. The mold wall has the temperature $\theta_{W,min}$ directly before injection. If the molten plastic touches the colder mold wall, a contact temperature is immediately established at the interface that corresponds to the maximum temperature $\theta_{W,max}$. During the cycle, this temperature drops continuously as a result of the mold temperature control. Demolding takes place at the temperature θ_E. If a quasi-steady state is present at the start of the new cycle, $\theta_{W,min}$ is reached again at this point.

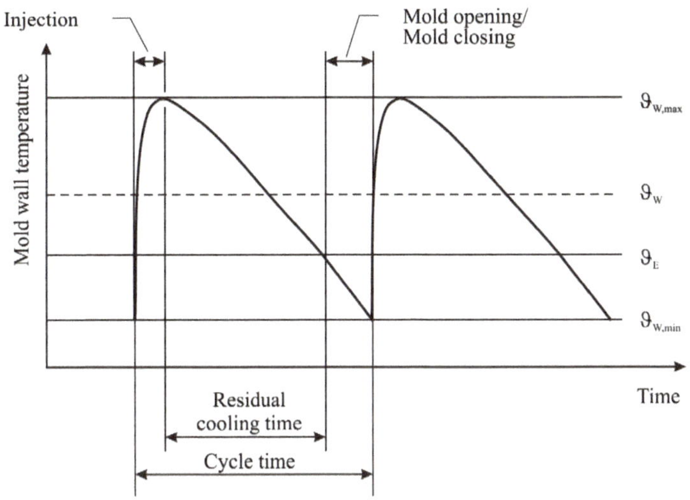

Figure 5.41 Qualitative mold wall temperature curve in quasi-steady state [2]

The coolant inlet temperature of the temperature control unit is kept constant throughout the entire production process. Meanwhile, the actual mold wall temperature rises gradually as production progresses, due to the heat supply of the hot melt, until an equilibrium is reached between the supplied and dissipated heat (Figure 5.42). This means that the course of several successive cavity temperature curves can be used to recognize when a thermal steady state is reached during the start-up of a molding process.

Thermocouples are used for mold temperature sensors. The response behavior of thermocouples is essentially dependent on their mass, i.e. the lower the mass, the faster the thermocouple reacts to a change in temperature. The leads of the thermocouple are individually guided to the sensor surface and welded together there, which ultimately creates the thermocouple. A temperature sensor constructed in this way cannot be machined any further [8].

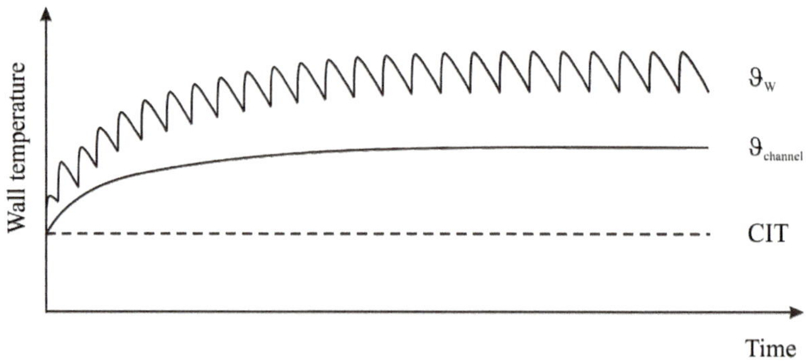

Figure 5.42 Temperature curve of the mold wall (θ_W), the cooling channel ($\theta_{channel}$) and the coolant inlet temperature (CIT) with continuous production time [2]

5.7.3 Measuring Principle for Temperature-Sensitive Thin Films

The thin-film sensors presented below are applied directly to the shaping surface of the mold wall, and therefore enable the temperature to be measured in the direct impact zone of the plastic melt. The measuring principle is based on the Seebeck effect. To utilize this effect, two different electrically conductive materials (thermocouples) are joined together at one end to form a thermocouple and thus the measuring point. As a result of a difference between the temperature of the connection point T_1 and the temperature T_2 at the open end of the conductor materials A and B, a thermoelectric voltage $U_{A/B}$ is generated; see Equation 5.5:

$$U_{A/B} = S_{A/B} \cdot (T_1 - T_2) = S_{A/B} \cdot \Delta T \tag{5.5}$$

The change in thermoelectric voltage with temperature is described using the Seebeck coefficient $S_{A/B}$. The thermoelectric voltage is measured together with the temperature at the open ends and converted into the actual measuring point temperature using a calibration function. The principle is illustrated in Figure 5.43.

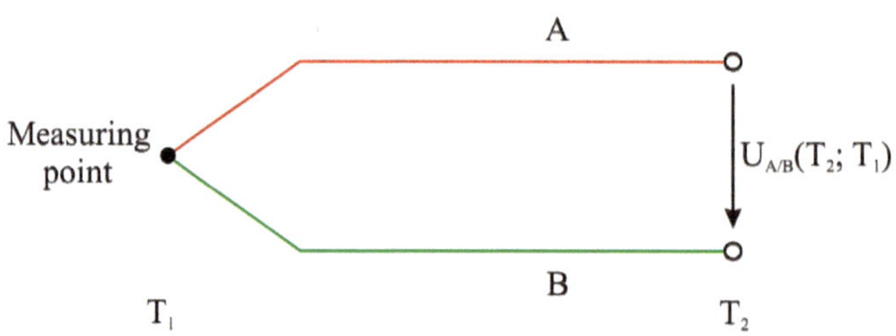

Figure 5.43 Basic element of a thermoelectric circuit (thermocouple) (see [9])

For a correct assignment of the absolute temperature sought, Equation 5.5 is used. It can be seen that the temperature T_2 of the reference point must be known or must be kept constant. A constant temperature at the reference point can be achieved, for example, by placing this area in an ice-water mixture whose temperature is 0 °C. For laboratory applications, this is a common method that offers excellent stability and accuracy. However, this method is completely impractical for industrial measurements. The so-called external reference point correction, on the other hand, is a more elegant method for inferring a thermoelectric voltage with respect to a reference of 0 °C and the absolute temperature of the measuring point. The reference point temperature T_2 is measured by an additional temperature sensor, e.g. a thermistor, and the corresponding thermoelectric voltage, which is stored in tables for standardized thermocouples, is added to the thermoelectric voltage displayed on the measuring device.

In the injection mold, however, there is usually not enough space in the area of the connection point to install the mentioned temperature sensor. In this case, the reference point can be moved to a more accessible location by extending the thermocouple legs (Figure 5.44). This is done using thermocouple cables made of the same material as the actual thermocouple legs. If, on the other hand, thermocouple types are used in which expensive materials, e.g. platinum, are used, the extension can alternatively be carried out using compensating cables. These have the same thermoelectric behavior in a limited temperature range as the material combination used for the thermocouple, but are made of less expensive materials. This means that no new thermoelectric voltage is generated at the connection point, so that the thermocouple can be considered extended to the end of the compensating cable. This new thermoelectric voltage is only generated where the thermoelectric or compensating cables are clamped to normal copper cables or to the connection piece of the measuring device. This is where the additional temperature sensor is attached, which is used for measuring point compensation. Modern measuring devices have such a temperature sensor in the area of the connection points – the so-called internal measuring point correction. This method is the most widely used in practice [10].

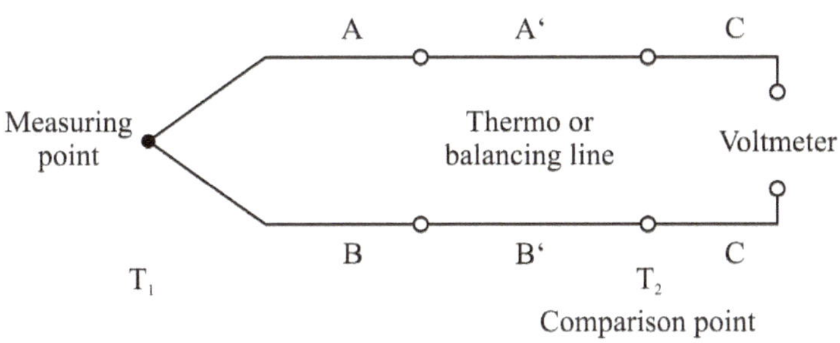

Figure 5.44 Structure of a thermocouple of material pairing A/B with the thermoelectric or compensating leads A′/B′ and the leads C for connecting the thermocouple to a voltage measuring device (see [9])

5.7.4 Coating Structure

For the thin-film sensors described below, the thermocouple combination NiCr/Ni (type K) and Cu/constantan (type T) was used. It should be noted that each additional electrically conductive material introduced into the thermocouple loop generates a new thermocouple, which results in a change in the thermoelectric voltage coefficient. Therefore, the thin-film thermocouple is first electrically insulated from the molding tool steel (1.2343 grade) by means of a fully deposited Al_2O_3 layer. The thermal leg materials are then partially applied to this electrical insulation layer, each with a thickness of approx. 1 µm, using a suitable masking method. To ensure a long service life of the sensor, the thin-film thermocouples can be covered by a wear-resistant passivation layer, also made of Al_2O_3 with a thickness of approx. 2 µm. The passivation layer protects against mechanical wear caused by the forming process and has an electrically insulating effect to prevent short-circuits against conductive materials that come into contact with the sensor. The basic structure is shown in Figure 5.45.

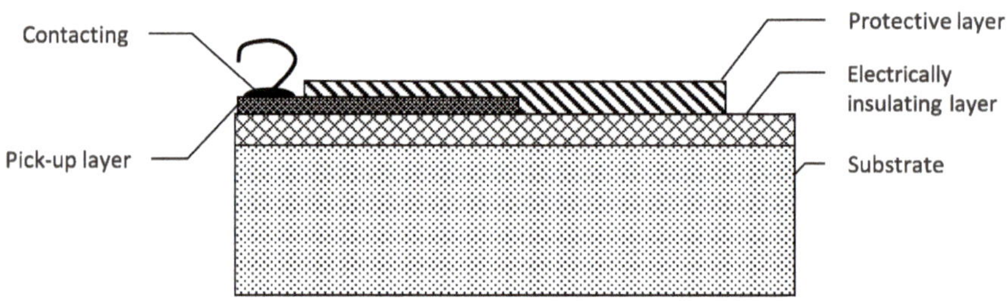

Figure 5.45 Structure of the thin-film temperature sensor [10]

5.7.5 Coating Production

To coat a molding tool insert with the layer stack mentioned, the materials are applied one after the other using magnetron DC sputtering (PVD technology). A high electrical insulation capacity is essential for the first layer of Al_2O_3. Numerous coating tests have shown that layers produced using PVD technology leads to the formation of pinholes (Figure 5.46) [11]. The EDX spectrum in Figure 5.46 shows, based on the peak at Fe, that at the base of such a pinhole the coating substrate is exposed. As a result, the material of the subsequently applied thermal leg is in contact with the substrate and cross-conduction to the substrate occurs. This results in invalidated measurement results or unusable transducer layers.

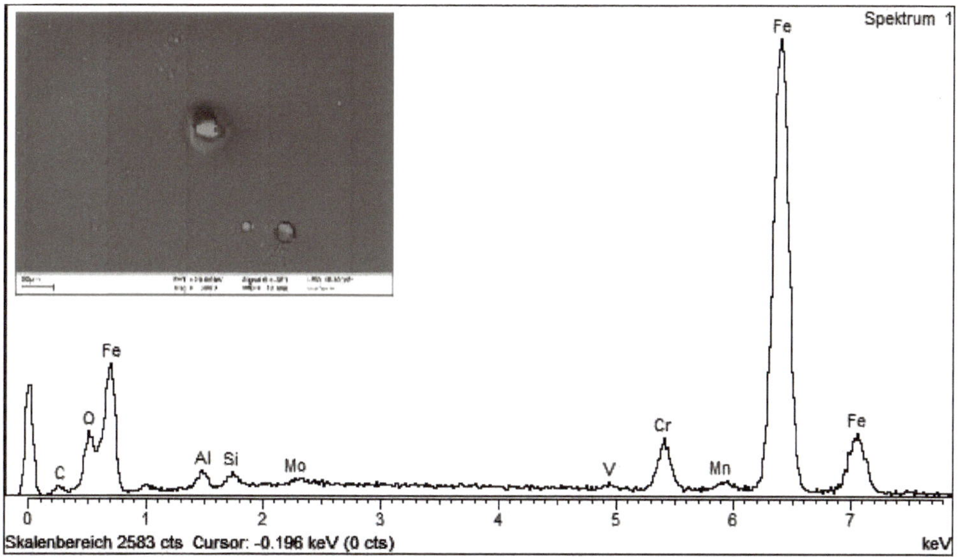

Figure 5.46 SEM image of a pinhole in the electrical insulation layer (left) and EDX spectrum within the pinhole (right) [10]

Successful optimization of the insulation coating can be achieved with a multi-layer coating structure. Although the individual layers have imperfections, these are not directly on top of each other and are each covered by the following layer. This allows insulation resistances to be achieved in the range of 1–2 MΩ compared to the applied thermocouple.

The electrical insulation layer is followed by the actual temperature sensor – the thin-film thermocouple. For the common type K thermocouple the nickel–chromium and nickel are coated one after the other. The actual measuring point is generated in the area where the two materials overlap. Using suitable masking techniques, one or more sensors can be produced simultaneously over a surface in one coating cycle.

The geometric structuring was carried out using shadow masks and alternatively by applying a masking coating with subsequent free lasering. In the first case, a steel mask is applied to the surface to be coated, which has permeable areas for the sensory coating material in the form of the transducer geometry (Figure 5.47).

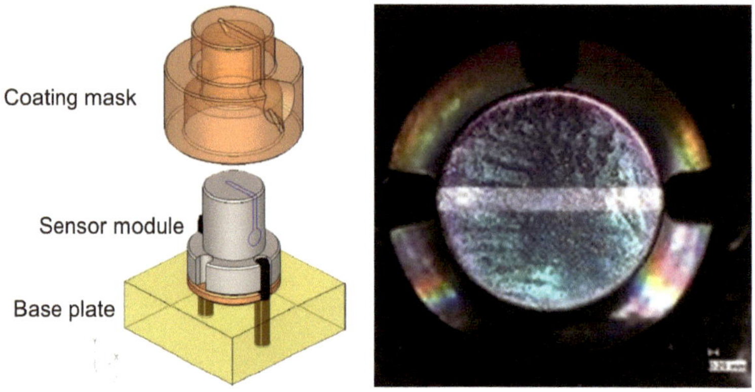

Figure 5.47 Shadow mask for partial coating of the sensor module with the transducer material (left); microscopic image of a thermocouple coating produced in this way (right) [10]

As an alternative concept for producing partial coatings in the form of thermocouple legs, a method has been developed in [12] in which the production and fixing of an additional mask component can be avoided, and masking with a fine geometric resolution can be achieved in a comparatively short time (Figure 5.48). For this purpose, a masking layer is sprayed with a covering lacquer onto a suitable substrate. After the solvent has evaporated from the masking layer, it leaves behind a solid film that is removed from the surface after free laser cutting and partial PVD coating.

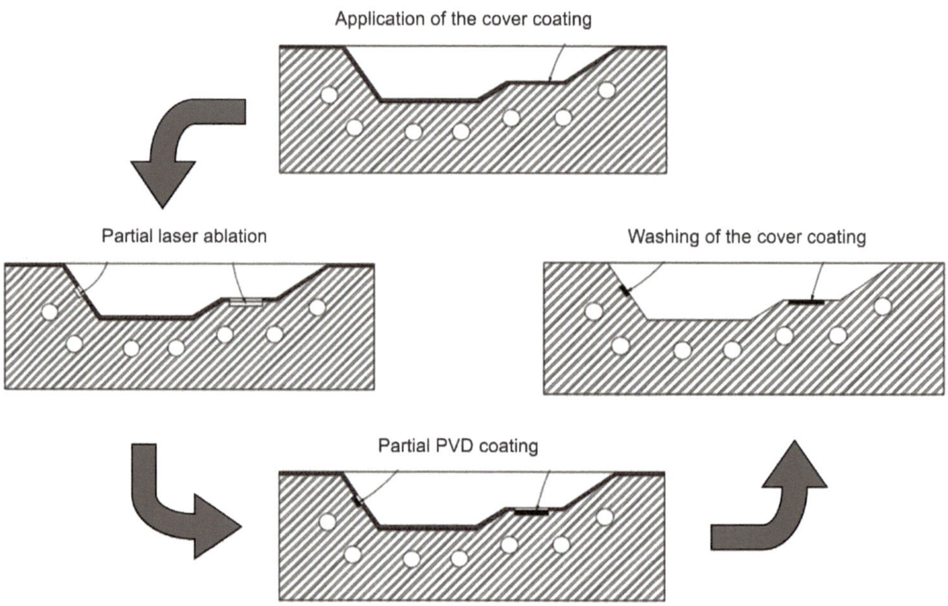

Figure 5.48 Methodology for sensor structuring using a covering lacquer and subsequent free lasering [12]

Figure 5.49 shows four thin-film thermocouples of type T (Cu–constantan) produced in this way.

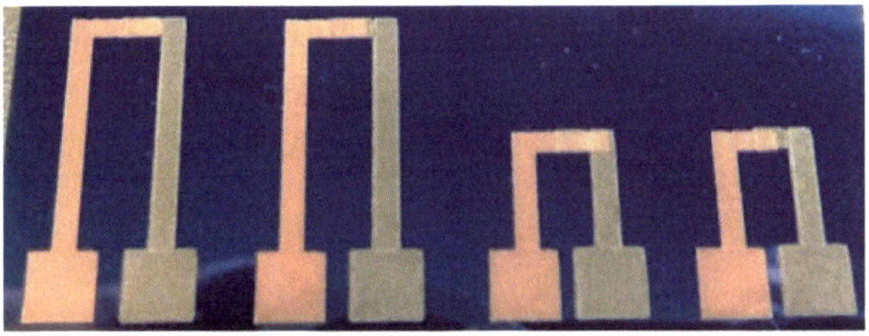

Figure 5.49 Masking (with free lasering) of a covering lacquer for partial coating of type T thermocouples [10]

In order to ensure a long service life of the sensor, as well as electrical passivation, the transducer layers can be protected against external influences by an additional layer. After the thermocouple has been manufactured, the contacting area of the two thermocouple legs is covered in order to apply aluminium oxide with a thickness of approx. 2 µm to the remaining surface as a protective layer using reactive DC magnetron sputtering. Thermal leads are then soldered to the open ends of the thermal legs in order to connect them to the data acquisition system. Figure 5.50 shows an example of two sensor variants.

Figure 5.50 Mold inserts for an injection mold with thin-film thermocouples [10]

5.7.6 Characterization of the Thermoelectric Behavior of the Thin-Film Sensors

The aim of sensor characterization is calibration to determine the thermal-electrical properties with regard to the detection of the prevailing temperature in the injection mold. The characterization can be carried out in a calibration oil bath using the comparison method at quasi-stationary temperature points within an application-specific temperature range. The calibration functions of the temperature values applied to the surface of the mold insert in Figure 5.50 (left) distributed on the surface of the mold insert are shown in Figure 5.51. The thermocouples 1 to 5 each have different widths (b_i) of the thermocouple legs, between 0.5 mm and 2.5 mm. A linear relationship between the thermoelectric voltage and the temperature can be seen. The average sensor sensitivity is in the range of 39.6–40.1 μV/°C, depending on the width of the leg. A t-test for equality of the sensor sensitivities, taking into account the test scatter of the regression functions and their common variance, has shown that this influence is not significant. In addition, in Figure 5.51 the basic values of the type K thermocouple are plotted. For temperatures below 50 °C, there is only a slight deviation from the basic values. As the temperature rises further, an increase in this deviation can be seen.

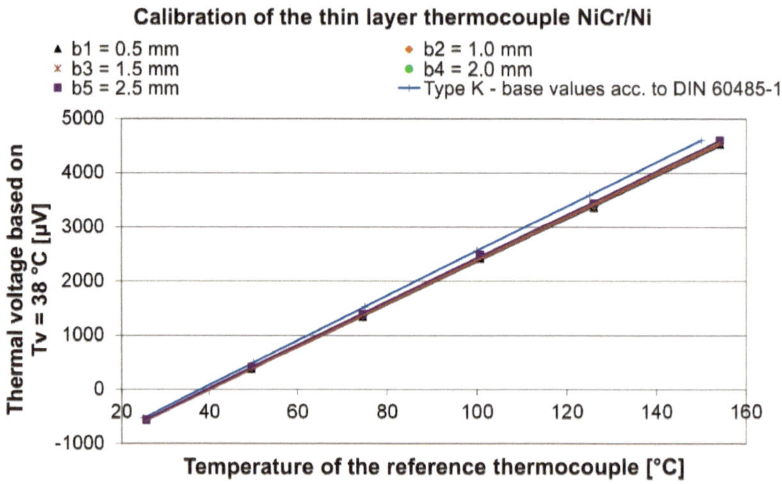

Figure 5.51 Calibration functions of the thin-film thermocouples compared with the basic values of the type K thermocouple [10]

5.7.7 Calculation of the Response Dynamics

The aim of the temperature measurement is to determine the course of the mold wall temperature over time during mold filling with the hot plastic melt. If the actual sensor layer is covered with a layer to protect against mechanical stress, as is the case with the sensors presented here, the following error influences on the measured temperature result:

- The response behavior (step response) of the temperature sensor is delayed by the time it takes for the heat flow induced by the plastic melt to penetrate the protective layer and reach the sensor layer.

- Due to the protective layer, measurements are not taken directly at the contact point between the plastic and the thin-film thermocouple. This means that the measured temperature curve will increasingly deviate from the real temperature curve as the distance between the pick-up layer and the mold surface increases, or as the thickness of the protective layer increases.

The injection molding process is cyclical and is largely dependent on the part geometry to be manufactured. Large temperature changes over time are the result, particularly during the injection phase, when the molten plastic comes into contact with the colder mold wall. The thin-film temperature sensors produced must therefore be able to detect the prevailing temperature with sufficient accuracy and high response dynamics, even with very short injection times and high temperature changes. A high response dynamic is particularly relevant for detecting the melt position in a sensor-dependent holding pressure circuit.

The sensor always follows a sudden rise in temperature with a lag using the so-called step response function. To quantify the measurement dynamics, the nine-tenths time t_{90} (time to reach 90% of the temperature jump) can be calculated using an FEM heat transport simulation with a sudden temperature load as the starting condition. The normalized curve of the step response function for the level of temperature excitation (here T_R = 100 °C) and the calculated nine-tenths times are shown in Figure 5.52 for different protective layer thicknesses.

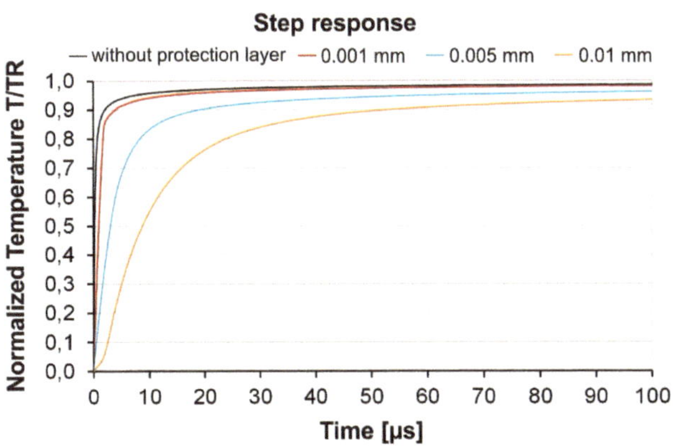

Figure 5.52 Influence of the thickness of the protective layer on the step response of the thin-film thermocouples [10]

Due to the low thermal mass of the thermocouple materials, they initially heat up very quickly. As the heating time progresses, the temperature rise flattens out as a result of heat dissipation in the direction of the substrate depth. Furthermore, an increase in the nine-tenths time with the thickness of the Al_2O_3 protective layer can be seen. Experience has shown that flow front velocities relevant in practice for injection molding of thermoplastics are in the order of 100–200 mm/s. If this value is related to the longest calculated nine-tenths time, it becomes clear that the melt in the sensor variant with a 0.01 mm thick protective layer has only covered a distance of around 0.01 mm in the cavity. This low value reflects the practical suitability of the thin-film sensors, e.g. for detecting the melt position. Furthermore, the calculation results show that a negligible temperature distortion of approx. 0.02 °C occurs during the injection of the plastic due to the protective coating.

5.7.8 Sensor Integration and Application in an Injection Mold

Figure 5.53 shows a mold insert with five measuring points at different positions on its surface. This allows the temperature profile to be captured during mold filling. The contact points are separated from the inflowing plastic melt by a flexible sealing element, which is installed in the die side mold half and dips into the cavity on the closing side when the mold is closed, thereby sealing the contact area.

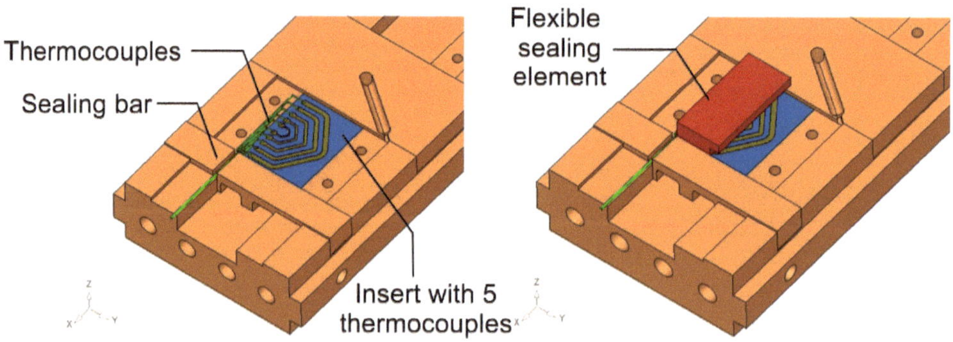

Figure 5.53 Integration of five thin-film thermocouples in the clamping-side mold half of an injection mold (left), and separation of the soldered-on thermoelectric lines from the inflowing plastic melt by means of a flexible sealing element (right) [10]

From the point of view of quick sensor replacement in the event of damage, the use of a modularized variant is recommended. The temperature measuring point is located on the top surface of the pin-shaped module in the area where the thermocouple legs overlap (Figure 5.54, left). These are positioned on the outer surface and contacted there with the thermoelectric cables. The contacted sensor module is then inserted into a sleeve. The lateral recesses on the lower collar of the module serve as a cable duct and allow twist-proof pinning in the sealing element. This assembly is fitted into the mold insert, in the back of which channels have been milled for cable routing (Figure 5.54, center). When installed, the thermal cables are routed via the cable duct in the sealing bar (Figure 5.54, right) and connected to the data acquisition system.

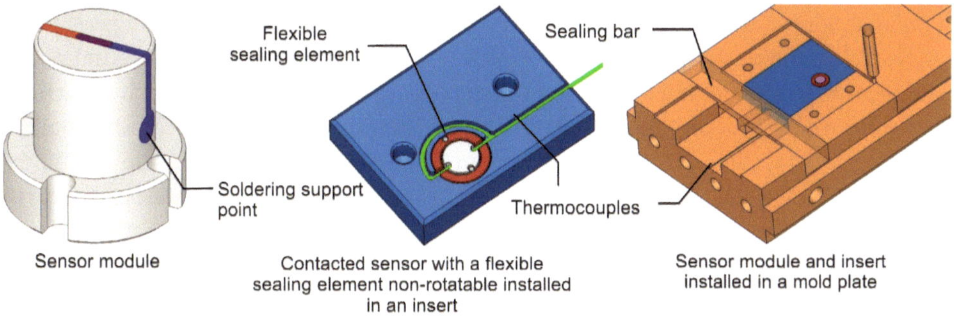

Figure 5.54 Integration of the sensor module into the clamping-side mold half of an injection mold [10]

Figure 5.55 shows the monitoring of the start-up behavior of the injection molding process with the modularized thin-film sensor. After approx. 600 s, at the end of the heating phase of the injection mold, a temperature of 76 °C can be detected. However, this does not correspond to the coolant inlet temperature of 80 °C set on the temperature control unit. This is already an important initial indication provided by the sensor. In the range from 600 s to 1400 s, a slightly varying temperature curve can be seen, which fluctuates around the mean value of 76 °C. The fluctuation is ± 0.6 °C and is due to heat radiation losses on the mold wall and the counter-regulation of the temperature control unit. The first injection molding cycle is started at t = 1400 s. The coolant inlet temperature of the temperature control unit is kept constant throughout the entire process. Meanwhile, the actual mold wall temperature gradually increases as production progresses, due to the heat supply from the hot melt, until equilibrium is reached. This means that the course of several successive cavity temperature curves can be used to recognize when a thermal steady-state is reached during the start-up of a molding process. It can also be seen that a quasi-steady-state temperature profile is only reached after a couple of cycles. Only from this point onwards can consistent and reproducible part quality be expected. The same applies to process interruptions.

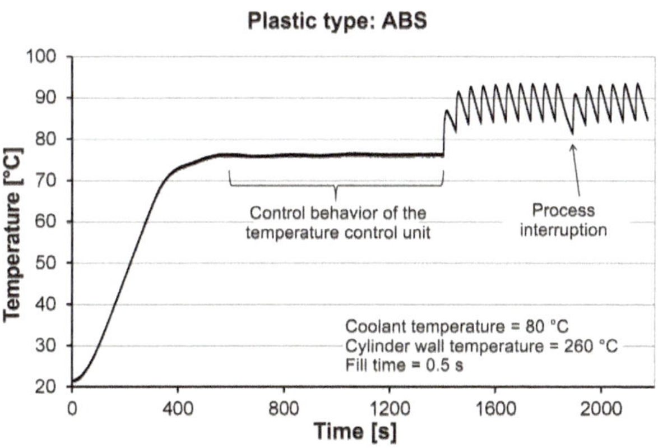

Figure 5.55 Monitoring the start-up behavior of the injection molding process [10]

Example temperature measurements with the sensor module during the production of an ABS molded part are shown in Figure 5.56. The temperature curve is sawtooth-shaped due to the heating during the injection phase and the cooling during the holding pressure and cooling phase. The different curves result from the variation of the cylinder temperature (T_{cyl}) and the coolant inlet temperature (CIT). As soon as the melt touches the sensor surface, the sensor layers are heated and a sudden temperature jump occurs. The temperature curve flattens out slightly until the maximum temperature is reached, as heat is simultaneously removed by the cooling system of the mold. The temperature recorded at this point can be used to check the process reproducibility, as the melt must reach this point in the cavity again in the same time if the viscosity conditions remain constant. In this way, a change in viscosity behavior, e.g. due to fluctuating melt or mold wall temperature, can be detected. As a result of heat conduction, the temperature drops further during the cycle. If a quasi-steady state is present at the start of the new cycle, the initial temperature level is reached again at this point.

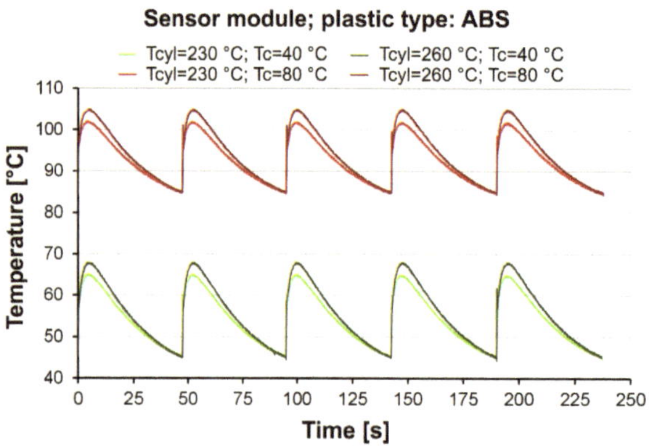

Figure 5.56 Influence of the coolant inlet temperature and the cylinder temperature on the temperature curve on the mold wall [10]

5.7.9 Summary

Temperature-sensitive tool coatings enable 100% on-line monitoring of the injection molding process. The graphical display of the temperature curve significantly speeds up processes such as sampling, start-up, and optimization of injection molds and increases the quality and consistency of the molded part. Temperature sensor technology based on a sensitive mold coating offers numerous advantages and possible applications:

- Monitoring the start-up behavior of the process
- Detection of the melt position for holding pressure switchover
- Detection of viscosity fluctuations and unfilled parts

- Control of a cascade injection and variothermal mold temperature control
- Making conclusions on gloss levels
- Ability to produce miniaturized versions
- Producing multiple sensors in a single coating process
- Fast response behavior.

References for Section 5.7

[1] Stitz, S.; Keller, W.: *Spritzgießtechnik*, Hanser, 2001

[2] Thienel, P.: Fertigungsverfahren Kunststoffe [Lecture notes], Iserlohn, South Westphalia University of Applied Sciences, 2004/05

[3] Bader, C.: Process control: The ABC of mould-sensor systems, *Kunststoffe International*, 2006, 96, 114–117

[4] Hauser, M.: Ein Gebot der Stunde, *Plastverarbeiter*, 2001, 52: issue 01, 60–61

[5] Schnerr, O.: Mehr Transparenz, höhere Qualität, Plastverarbeiter, 2006, 57: issue 05, 64–65

[6] Bader, C.: Einrichten, *Steuern und Regeln von Werkzeugen, in Werkzeugbau für die Kunststoffverarbeitung* (ed. Menning, G.), Hanser, 2008, pp. 634–652

[7] Cordes, D.: Der Werkzeuginnendruck – Der "Fingerabdruck" ihrer Formteilqualität, in "Fachwissen Werkzeugtechnik" (Seminar at the Kunststoff-Institut Lüdenscheid), 2010

[8] Messtechnik und Sensorik beim Spritzgießen [Company publication], Kistler Instrumente GmbH, 2006

[9] Bernhard, F.: *Technische Temperaturmessung*, Springer, 2004

[10] Librizzi, A.: Entwicklung von temperatursensitiven Dünnschichten und Untersuchungen zu deren Einsatz in Spritzgießwerkzeugen der Thermoplastverarbeitung [Dissertation], University of Paderborn, 2015

[11] Librizzi, A.: Temperatursensitive Dünnschichten für den Einsatz in Spritzgießwerkzeugen der Thermoplastverarbeitung, *Proceedings of the 18th GMA/ITG Conference on Sensors and Measuring Systems*, 2016, pp. 656–662

[12] Librizzi, A.: Method for manufacturing electrical functional element e. g. electrode of plastic prototype tool, involves coating exposed surface of tool so as to form metallic layer on tool area through physical vapor deposition coating process, Patent DE 102011002083 A1 (disclosure document), Kunststoff-Institut für die Mittelständische, 2012

5.8 Heating Coatings

Dr. Martin Ciaston

5.8.1 Introduction

When manufacturing plastic parts using the injection molding process, the mold surfaces must be heated and cooled in a targeted manner in order to ensure constant process conditions. Traditionally, temperature-controlled oil or water circuits are used to control the mold temperature. Depending on the process phase, the system switches between hot and cold media in the cavity to regulate the mold temperature. However, there is still room for improvement in terms of cycle time, which should be reduced in order to improve the costs and energy efficiency of the processes.

Another important aspect in the production of plastic parts is the surface finish, especially for black high-gloss surfaces that are considered premium products. However, surface defects (such as weld lines, which can occur during the injection molding process) can be visually noticeable, especially on black, high-gloss plastic parts.

To avoid surface defects, the mold temperature can be specifically adjusted. However, increasing the mold temperature leads to longer cycle times and higher energy costs. Consequently, nowadays, plastics processors use dynamic temperature control methods such as inductive processes or resistance electric heating elements to heat only a part of the mold for a short time and thus avoid the problems mentioned above. However, these methods lead to additional costs for the purchase of equipment or the design of new molding tools.

One way to reduce the energy consumption of common dynamic tempering processes is to keep the heated surface as small as possible. This goal can be achieved by using very thin heating conductors or heat-sealing band coatings. In this section, the possibilities of using heat-sealing bands in the form of thin coatings and thin resistive heat-sealing band elements will be presented. Further information can be obtained from the users/manufacturers.

5.8.2 Basics of Near-Contour Heating Layers

Near-contour heating systems in injection molding are an advanced technology that is becoming increasingly important in the plastics industry. By integrating heating systems directly into the mold cavity of the injection mold, precise and uniform temperature control of the injection molding process is made possible. Near-contour heating improves the quality and efficiency of the injection molding process by improving material distribution and cycle times while minimizing warpage problems. With this in mind, this section explains the characteristics, requirements, and benefits of conformal heating systems in the injection molding process. Near-contour heating systems can be defined as heating systems directly under the mold surface (Figure 5.57, left and center), as well as heating close to the surface on the surface using thin-coating technology (Figure 5.57, right).

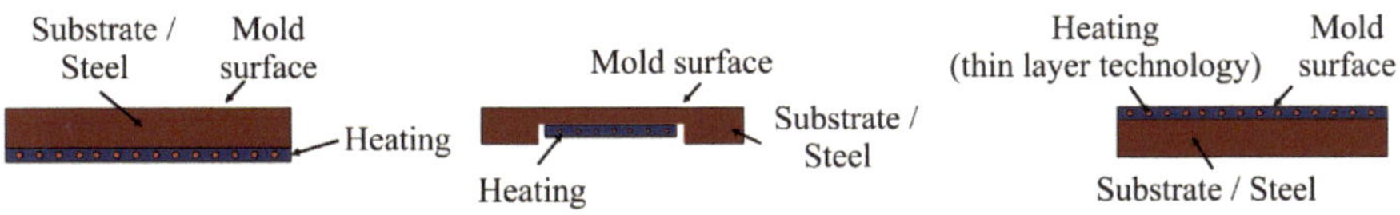

Figure 5.57 Different designs of near-contour heating systems. Left: Flat heating under the mold surface; Middle: Spot heating under the mold surface; Right: Flat heating on the mold surface (Image source: Non-profit KIMW Forschungs-GmbH)

Resistively heated surfaces offer high dynamics for the surface heating of molds, and have a significantly better energy balance than conventional dynamic temperature control processes (Figure 5.58).

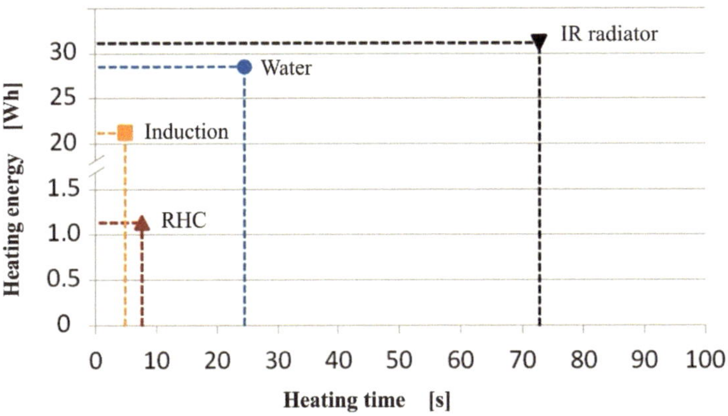

Figure 5.58 Comparison of induction, water, IR emitters, and resistive heating conductors (RHC) as heating methods in injection molding processes (Image source: Non-profit KIMW Forschungs-GmbH)

5.8.3 Requirements for a Coating System for an Application as a Heating Conductor in the Injection Molding Process

For the application of a coating system for the production of electrical heating conductors, several conditions must be met. An electrical-insulating effect against the molding tool must be ensured, which enables electrical separation between the molding tool and the heat conductor coating. In addition, a heat-conductive layer is required to enable effective heat generation. Furthermore, an adhesive bond between the individual layers and to the molding tool must be ensured. Finally, a wear protection layer should be present on the heat conductor coating to ensure increased durability and service life of the coating system.

Certain materials are particularly suitable for electrical insulation due to their excellent properties in terms of resistivity and dielectric strength of several MV/cm. Suitable coatings can be aluminium oxide (Al_2O_3) or silicon oxide (SiO_2), for example. Copper (Cu) is particularly suitable as a heating conductor, as it has high electrical conductivity and low specific resistance. However, heat-conductive coatings can also consist of other materials with high electrical conductivity. Another important aspect is the interaction between the electrically insulating (ceramic) layer and the heat-sealing band layer. If there is no adhesive bond provided, delamination of the layers and destruction of the coating system can occur. Resistant coatings such as hard chrome coatings, PVD coatings, or DLC coatings are suitable for ensuring wear protection of the heat-sealing band. The decisive factor here is the prevention of oxidation of the heat-sealing band coating.

5.8.4 Application of Heating Coatings in Injection Molding Processes

In a project carried out at the non-profit KIMW Forschungs-GmbH, heating conductors in the form of copper foils and an electrically insulating layer consisting of zirconium oxide with a power of 162 W were able to achieve a heating rate of 12 K/s on a mold surface of 16 cm^2. With higher outputs of 430 W, heating rates of 34 K/s could be achieved. In long-term tests, a service life of 6000 cycles was achieved with the system used.

In the 'DynaHeat' project at the same institution, copper heating conductor coatings were deposited on an electrically insulating multi-layer coating consisting of aluminium oxide and chromium oxide, which was also applied using the MOCVD process. In the project, heating rates of 13 K/s were achieved on a surface area of 2 cm^2 with an output of 112 W. Electrical insulation could not be maintained on larger surfaces [1, 2].

In a project at the Kunststoff-Zentrum Leipzig (KUZ) led by Steffen Jacob, a heating conductor coating system was applied to a micro-injection mold. Using a thin-film process, heating elements were deposited directly onto the mold contour of an injection mold, which allowed dynamic variothermal temperature control in injection molding without extending the cycle time. The thin-film elements exhibited high heating dynamics, with heating rates of 100 K in around 0.7 s in the injection molding cycle. Heat outputs of 160 W were achieved on a mold surface of around 72 mm^2 [3]. In another project, cold plasma spraying is currently being used to deposit metallic layers on molded parts, and thus create electrically contactable conductor paths [4].

Previous projects have demonstrated the feasibility of this technology on a small scale. An industrial application is also known as ecoCOAT [5].

5.8.5 Summary and Outlook

Dynamic temperature control techniques are now being used successfully in the processing industry. A major advantage is the optimization of the surface quality of amorphous thermoplastics, whereby the avoidance of weld lines is still a central goal. Although the use of these techniques to increase mechanical properties is obvious, closer examination reveals considerable limitations, which means that this decision is still based on individual cases. The specific choice of dynamic tempering for the application is not initially of decisive importance. In any case, the aim is to keep the cavity wall of the mold at a temperature above the glass transition temperature (for amorphous thermoplastics) during the injection phase, in order to achieve the desired optimization effect on the molded part.

In the second step, the suitable technology is selected based on a series of criteria such as the required surface quality (partial or full-surface), the required temperature (e.g. up to 220 °C for PEI), the cycle time to be taken into account (if necessary), the economic efficiency, and the energy balance. Current advances in dynamic temperature control technology open up a wide range of possibilities for optimizing the quality of injection molded parts. The technical preferences are almost completely covered, ranging from high dynamic cavity temperatures of over 200 °C to cycle-time-neutral or even cycle-time-shortening technologies to improve the economic aspects. There is considerable potential for improving the quality of molded parts and process efficiency through the application of these advanced technologies. Due to the current high energy costs, manufacturers are also striving to increase energy efficiency in injection molding production. The topics presented offer an approach to meeting these ever-increasing demands on energy efficiency.

References for Section 5.8

[1] Ciaston, M.; Batmanov, A.; Mumme, F.: Functionalization of tool surfaces by chemical vapor deposition [Poster], *9th NRW Nano-Conference*, 2021

[2] Annual report 2021, Non-profit KIMW Forschungs-GmbH, 2021

[3] Jacob, S.: Temperieren direkt an der Werkzeugwand, *Mikroproduktion*, 2017, issue 05, 28–32

[4] Dynamisch Temperieren und gezielt Abkühlen im Spritzgießprozess – Lokale Temperierung mit Dünnschichtheizelementen (Funded project of the Kunststoff-Zentrum in Leipzig gGmbH), *https://www.innovationskatalog.de/IK/Redaktion/DE/Solr/Project.html?id=59357378-2bdd-3b44-8cd2-545714c30a4c*, accessed on 10 March 2023

[5] Intelligente Beschichtungen mit Funktion – Ortsselektive Direktbeschichtung ohne Chemie oder Bindemittel, ecoCOAT GmbH, *http://eco-coat.com/*, accessed on 10 March 2023

Index

B

C

D

O

P

U

V

W

X

Z